ESPRIT Basic Research Series

Edited in cooperation with
the Commission of the European Communities, DG XIII

ESPRIT Basic Research Series

J. W. Lloyd (Ed.): **Computational Logic.** Symposium Proceedings, Brussels, November 1990. XI, 211 pages. 1990

E. Klein, F. Veltman (Eds.): **Natural Language and Speech.** Symposium Proceedings, Brussels, November 1991. VIII, 192 pages. 1991

G. Gambosi, M. Scholl, H.-W. Six (Eds.): **Geographic Database Management System.** Workshop Proceedings, Capri, May 1991. VIII, 320 pages. 1992

R. Kassing (Ed.): **Scanning Microscopy.** Symposium Proceedings, Wetzlar, October 1990. X, 207 pages. 1992

G. A. Orban, H.-H. Nagel (Eds.): **Artificial and Biological Vision Systems.** XII, 389 pages. 1992

S. D. Smith, R. F. Neale (Eds.): **Optical Information Technology.** State-of-the-Art Report. XIV, 369 pages. 1993

S. D. Smith   R. F. Neale   (Eds.)

Optical Information Technology

State-of-the-Art Report

Springer-Verlag

Berlin Heidelberg New York
London Paris Tokyo
Hong Kong Barcelona
Budapest

Volume Editors

S. Desmond Smith
Roderick F. Neale
Department of Physics, Heriot-Watt University
Riccarton, Edinburgh EH14 4AS, UK

CR Subject Classification (1991): B.4, J.2, H.1

ISBN-13: 978-3-642-78142-1   e-ISBN-13: 978-3-642-78140-7

DOI: 10.1007/978-3-642-78140-7

Publication No. EUR 14599 EN of the Commission of the European Communities,
Dissemination of Scientific and Technical Knowledge Unit, Directorate-General Tele-
communications, Information Industries and Innovation, Luxembourg
Neither the Commission of the European Communities nor any person acting on behalf
of the Commission is responsible for the use which might be made of the following
information.

Typesetting: Camera ready by authors
45/3140 - 5 4 3 2 1 0 - Printed on acid-free paper

Foreword

Research in optical computing has its roots in the early 1960s and the invention of the laser. Since then, significant progress has been made towards practical optical processing systems, often stimulated by associated technological advances such as the invention of optically bistable devices.

The ESPRIT programme has supported a study group comprising leading European teams working in this area, the Workshop on Optical Information Technology (WOIT), led by Heriot-Watt University in Edinburgh. This book is a compilation of research results presented at a workshop meeting of WOIT and depicts the current state of the art in this field. Numerous significant advances are identified and described both in the development of optical computing concepts and the associated technologies on which they rely.

The unique properties of optics, when suitably coupled to electronic and opto-electronic devices, offer the prospect for improving the performance of processing systems far beyond that possible with any single technology. The considerable progress already made towards a marketable technology inspires optimism that the ongoing research will yield viable alternatives to conventional electronics.

G. Metakides

Preface

This selection of papers, organised by the Workshop on Optical Information Technology, funded under EC Esprit BRA 3199, owes its origin to discoveries by European scientists in the late 1970s. The promise of this earlier work was recognised by the European Community through the European Joint Optical Bistability Project (EJOB) funded under the Stimulation Programme which took place between 1983 and 1986. The rich vein of creative work that this very basic project unearthed led to a 360-page published account [1], as well as more than 200 original papers in the literature. A first answer to a challenge set by the EC Directorate: "Demonstrate a simple optical computational device" was provided in a very primitive way. However, some generic optical computer architectural concepts were devised which have since been used both in Europe and America for more advanced demonstrator projects. Equally important, from a European point of view, the work in the original 8 laboratories attracted as many as 30 research groups that emerged across the continent with an interest in furthering the topic. Unfortunately, the combination of available funds and the state of the research – now beyond the first fundamental basics but well short of proven commercial viability – created conditions such that the investigators have been unable to persuade the Community to fund more than the "information exchange" that the Workshop has provided. It is a measure of the continuing interest in the topic that, despite the lack of funding, the research has continued to be prolific as the 39 papers of this volume illustrate.

It is then perhaps important, since the groups have been attempting to proceed from fundamental scientific discoveries and theories towards practical usable devices, to put this effort in time context. The final aim is to achieve better, faster (or at least different) computational capability than that available in conventional digital time-sequential electronics. The latter technology itself took a long time to reach fruition. Babbage's calculating engine was attempted between 1813 and 1840. Lee de Forest's triode valve was invented in 1908; the first transistor in 1947. The first low cost digital machines began to appear in ca. 1970, following the introduction of integrated circuits.

Optics in the computational context has been researched for less than 20 years and it is only one decade since all-optical digital logic devices were shown to be practicable in 1979. Considerable progress has been made: it has already been shown to be important to attempt to make primitive machines: these have given useful lessons regarding both the requirements for the hardware and the need for new and different architectural approaches to computation. Now, it looks as if the same point made by von Neumann in 1946 will apply to the extension of optics into computation. This point is that the "software", or architecture, or coding, or formulation of the problem will be as important as the "hardware". During the last three years, despite the shortage of funding, we are able to report advances in this area.

Examination of the history of the successful transfer of laboratory discovery to viable product sets our current work in context. One example is the fluorescent tube, an optoelectronic device which took some 79 years to get from an experimental demonstration to a commercial product. In hindsight the reasons can be recognised. A series of crucial sub-technologies was required and fruition could not be obtained until all were in place. The optical processing demonstrator programmes which have been under consideration by members of the Workshop have been designed to expose the necessary sub-technologies to practical testing and thus accelerate the process. Necessarily some dead ends have been encountered but in general there has always been a route forward. To be more specific three crucial areas can be identified:

(i) The input interface – to convey information from time-sequential electronics to 2-D optical format: this requires an electrically addressed "spatial light modulator". As a general comment, progress on this class of device has been disappointingly slow but this has perhaps not acted as a serious delaying factor during this period.

(ii) Closely related technically, but functionally separate, are the optical logic planes in which information is processed with substantial parallelism, with programmable binary logic combinations, and then transmitted with gain to provide fan-out and in such a way that logic restoration is preserved between (say) iterative stages in a processor. From a computational point of view the details of how the light signal is received, processed and transmitted are irrelevant. It is the efficiency of the process that matters, perhaps best characterised by the switching energy for a pixel. Various approaches to advance the available technology were explored but, in contrast to those made in the USA and Japan, they have been greatly handicapped by the lack of funds and indeed also by the consequent lack of coordination of the European capability. From a promising position, 5 years ago, Europe has certainly lost ground in this crucial area. In this volume, several approaches are described some of which are very promising. However, it will be necessary to progress these sufficiently far to make a good judgement. For the time being most demonstrator devices are likely to use American components. Meanwhile, Japanese appreciation of the failure of computational technology has clearly recognised the role of massive parallelism and noted the probable role of optics.

(iii) Given input and logic plane components, the next class of device is the optical interconnect. Here we have a more positive report. The investment required, at least at the early stages, has proved to be within the present ca-

Table of Contents

Nonlinearities in GaAs Systems

Nonlinear optical properties of GaAs/(AlGa)As multiple quantum wells
under quasistationary high laser excitation and transversal electric fields 3
K.-H. Schlaad, Ch. Weber, U. Zimmermann, G. Weimann,
C.v. Hoof, G. Borghs, C. Klingshirn

Electroabsorptive/-refractive effects in asymmetric step quantum wells
and BRAQWET structures.. 13
O. Olin

Electrooptical modulation in vertical multiple quantum well
microresonators.. 20
G. Wingen, S. Zumkley, J. C. Michel, J. L. Oudar, R. Planel, D. Jäger

Criteria for the use of nonlinear GaAs etalons for threshold logic........... 27
B. Acklin, N. Collings, C. Bagnoud

Interconnect Technologies

Synthesis of diffractive optical elements using electromagnetic
theory of gratings 39
E. Noponen, A. Vasara, E. Byckling, J. Turunen,
J. M. Miller, M. R. Taghizadeh

The design of quasi-periodic Fourier plane array generators 47
A. G. Kirk, A. K. Powell, T. J. Hall

Microlenses in PMMA with high relative aperture fabricated by
proton irradiation combined with monomer diffusion 57
K.-H. Brenner, M. Frank, M. Kufner, S. Kufner, M. Testorf

Fabrication of microoptic components by thermal imprinting............... 67
K.-H. Brenner, C. Doubrava, T. M. Merklein

Invariant pattern recognition: Towards neural network classifiers........... 76
G. Lebreton, E. Marom, N. Konforti, D. Mendlovic

Holographic interconnect components for optical processing systems........ 85
A. C. Walker, M. R. Taghizadeh, E. J. Restall, B. Robertson, J. M. Miller

Binary, multilevel, and hybrid holographic optical array illuminators 94
M. R. Taghizadeh, J. Turunen, H. Ichikawa, J. M. Miller,
B. Robertson, P. Blair, N. Ross, A. Vasara, E. Byckling,
T. Jaakkola, E. Noponen, J. Westerholm

demonstration. In the first category the search for usable effects continues, both in materials and structures in III-V and II-VI semiconductor compounds. Interesting new effects have been found, e.g., in doped n-type GaAs and quantum well structures and new devices in the form of integrated detectors-emitters (or pnpn structures) hold considerable promise. In II-VI materials the most significant progress is in p-doping and the realisation of ZnSe pn junction devices for the first time in Europe. This will revolutionise possibilities for these compounds but further work is needed before the emergence of usable devices. As stated earlier, the reports on optical interconnect technology are likely to be seen as seminal in the course of time.

References

1. From Optical Bistability towards Optical Computing – The EJOB Project, Eds. P. Mandel, S.D. Smith and B.S. Wherrett, North-Holland (1987).

S. Desmond Smith
Roderick F. Neale

pabilities of the current participating laboratories and exciting progress has been made both theoretically and experimentally. It is clear that successful fan-out generators can be fabricated using the principle of "Dammann" gratings to provide arrays of equal intensity beams up to at least 10^5. These beams are the power sources for the optical logic planes. The work shows that non-local interconnects can be effected optically and the first 2-D devices to provide this have been included in demonstrators. The more prosaic fan-out requirements (e.g. to nearest neighbours) are proving equally practicable. With further progress reported in bulk optics design and opto-mechanical packaging, we are able to say that the promise of optical interconnection is opening the way to a freedom of computer architecture design previously unavailable via electrical interconnections. This part of the programme is probably the most successful at present. The results within the WOIT reports in this volume will find application wherever optical communication within computational devices is required. This will extend from "optically connected electronics" to the optical interaction of logic planes of the early demonstrator digital optical processors.

The various sub-technologies have come together in a related way and there are at least two digital optical processor sub-systems that are at a state of early testing. This testing is beginning to influence opinion and justify its usefulness although a great deal more needs to be done. Sufficient has been effected, however, to continue to motivate architectural thinking along realisable lines, if only by showing what can and cannot be done optically.

Unfortunately only by cooperating with the Americans, where at least 6 of our community's best young researchers are playing a very significant part in that country's effort, will it be possible to progress to such significant demonstrations as an optical logic plane data rate of 80 GBits/second. The route forward into new computational concepts may well prove to be the most important final outcome but such quantitative steps will probably be necessary to persuade the existing information technology community to invest sufficiently for Europe's technical progress to remain competitive.

As demonstrated in this volume and in the referenced work of European and other researchers, architectures are well under development that can take advantage of the components described, and that are expected to lead to specialised processing machines of performance perhaps 100 times more powerful than electronics can accommodate. The key is recognised to be the use of optics for data communication, exploiting either the high time or high space bandwidths achievable. The question of which processing components the propagating optical signals will be interfaced to is still open – some alternatives are described herein. Optics is moving rapidly into electronic machines (fibre interconnects, disc storage); dedicated fast and massively parallel processing modules within electronic machines, taking advantage of complex optical interconnects, would seem a natural path of investigation.

We have divided this account into sections ranging from studies of basic optical nonlinearities in various semiconductors to the latest state of processor

II–VI-Compound Nonlinearities

Wide bandgap II–IV light emitting devices 103
B. C. Cavenett, K. A. Prior, S. Y. Wang, J. Simpson

Optical nonlinearities of CdS for optical addressing 110
J. Oberlé, B. Kippelen, A. C. Walker, A. Daunois

Optical nonlinearities and switching from excitons
in II–IV semiconductors 118
C. Dörnfeld, C. R. Paton, Z. Xie, J. Erland, J. M. Hvam

Prediction of large optical nonlinearities in quantum well wires 128
S. Benner, H. Haug

Nonlinear optical properties of II–VI semiconductor quantum dots 133
A. Uhrig, A. Wörner, M. Saleh, C. Klingshirn, N. Neuroth,
K. Remitz, B. Speit

Optical Computer Architectures

Optical input and output functions for a cellular automaton
on a silicon chip 143
I. Seyd-Darwish, P. Chavel, J. Taboury, F. Devos,
T. Maurin, R. Reynaud

Demonstration of an optical pipeline adder 153
W. Eckert, C. Passon

Performance and hardware requirements of parallel addition algorithms
for optical implementation using SEEDs 162
D. Rhein

2–D Parallel optoelectronic interconnect using a highly
light sensitive monolithic receiver array 176
K. Zürl

O-CLIP – A demonstrator all-optical processor 184
B. S. Wherrett

Design of a S-SEED cellular logic image processor 194
S. Wakelin, F. A. P. Tooley, G. R. Smith

Thermo-Optic Devices

Thermo-optical logic gate array using SOS waveguide 209
H. Gualous, A. Koster, W. Chi, N. Paraire, S. Laval

Optical switches and oscillators based on thermally induced
optical nonlinearities in II–VI semiconductors 219
J. Grohs, S. Apanasevich, F. Zhou, H. Ißler, A. Schmidt, C. Klingshirn

Thermally induced optical bistability in GaAs/(AlGa)As multiple
quantum wells for application as a temperature sensor 232
U. Zimmermann, K.-H. Schlaad, G. Weimann, C. Klingshirn

Chances for nonlinear optical switching elements 240
H. Bartelt

III–V Bistability and Devices

Subnanosecond switching and recovery in a Fabry-Perot etalon
based on bulk heavily doped n-GaAs 249
D. J. Goodwill, F. V. Karpushko, S. D. Smith, A. C. Walker

Optical bistability in quantum well semiconductor devices 255
R. Kuszelewich, B. G. Sfez, D. Pellat, J. L. Oudar

All optical bistability in a type II heterostructure 265
R. Teissier, R. Planel, F. Mollot

Surface emitting laser diodes and wavelength break
selective photodetectors 272
T. Wipiejewski, K. Panzlaff, K. J. Ebeling

The double heterostructure optical thyristor in optical
information processing applications 280
M. Kuijk, P. Heremans, R. Vounckx, G. Borghs

An architecture for a general purpose optical computer
adapted to PNPN devices 291
N. Langloh, M. Kuijk, J. Cornelis, R. Vounckx

Architectural and Logic Structures

Towards distributed statistical processing – Aquarium: A query and
reflection interaction using magic: Mathematical algorithms
generating interdependent confidences 303
N. Langloh, R. Cottam, R. Vounckx, J. Cornelis

Computer-aided design of digital opto-electronic systems with HADLOP .. 320
D. Fey

Microwave photonics 328
D. Jäger

Si-Based Devices

Avalanche photodiodes for optical bistability 337
A. Koster

On the feasibility of avalanche devices for optical switching in silicon 344
S. Cova, A. Lacaita, M. Ghioni, G. Ripamonti

Sensitivity and switching contrast optimization in an optical
signal processing waveguide structure.................................... 350
N. Paraire, P. Dansas, A. Koster, M. Rousseau, S. Laval

A Passive crystal pixel interchanger 357
C. De Tandt, W. Ranson, P. Schrey, R. Vouckx, R. Cottam

Attendees 367

Part I

Nonlinearities in GaAs Systems

Nonlinear Optical Properties of GaAs/(AlGa)As Multiple Quantum Wells Under Quasistationary High Laser Excitation and Transversal Electric Fields

K.-H. Schlaad,[1] *Ch. Weber,*[1] *U. Zimmermann,*[1] *G. Weimann,*[2]
C.v. Hoof,[3] *G. Borghs,*[3] *and C. Klingshirn*[1]

[1] University of Kaiserslautern, Department of Physics, W–6750 Kaiserslautern, FRG
[2] Walter–Schottky Institut, W–8000 München, FRG
[3] IMEC–MAP, B–3030 Leuven, Belgium

We use the pump and probe beam and the luminescence spectroscopy to study the nonlinear response of *GaAs/(AlGa)As* heterostructures to quasistationary excitation conditions. The carrier induced energetic shift of the 1*hh*–exciton as a function of the quantum well width shows a dimensional dependence of the carrier screening properties. This shift gives a rather good criterion to decide if a system behaves more 2D or 3D like. The high excitation regime is dominated by electron–hole plasma features. Many particle effects lead to a renormalization of the fundamental bandgap. This effect is essential for understanding the physics of *III–V* semiconductor lasers. The carrier density and the reduced bandgap are determined via systematic evaluation of both gain and luminescence spectra. The observed behaviour can be described by a strict 2D theory using effective exciton parameters in order to account for the finite well widths of the structures. The study of the higher sub-bands reveals that both, exciton bleaching and sub-band renormalization are mainly due to direct occupation of the specific sub-band while intersub-band effects are considerably smaller. By coating the two sides of a 50×100Å multiple quantum well with semitransparent Cr–Au electrodes we are able to control the energetic position of the 1*hh*–exciton as a function of the applied electric field *and* of the incoming light power. Several structures to optimize this effect in order to build an electrooptical switch or modulator are discussed.

1 Introduction

This contribution reviews the research activities of our group during the WOIT lifetime concerning the nonlinear optical properties of *GaAs/(AlGa)As* Multiple Quantum Well Structures (MQWS) under the illumination with ns–laser pulses and under the influence of electric fields perpendicular to the MQW layers.

The illumination with ns–laser pulses stands for the realization of quasistationary excitation conditions (i.e. the duration of the light pulses is long in comparison with the intrinsic time constants of the MQWS).

Electric fields perpendicular to the layers lead to an substantial energetic redshift of the excitons without any strong broadening of their shape. This effect is often referred to as the Quantum Confined Stark Effect (QCSE) because of its mathematical equivalence to the problem of a confined hydrogenic system with an electric field perpendicular to the confinement direction [1]. We use this effect to control the energetic position of the exciton by varying the intensity of an incoming light beam.

2 Experimental Setup

For our investigations several undoped MQWS grown by molecular–beam–epitaxy with different well widths L_z between 36 and 190 Å, and a n–type modulation-doped sample with $L_z = 147$ Å and $N_D = 2.8 \times 10^{11}$ cm^{-2} were available. In order to prepare them for transmission experiments, the GaAs substrate was removed by a selective etching technique [2]. The etching solution consists of H_2O_2 and NH_4OH in a volume ratio of about 120/1.

The high excitation experiments were carried out with pulsed excimer laser pumped dye laser systems. The pulse duration was about 10 ns and we reached power densities up to 10 MW/cm^2. The carriers were generated directly in the wells above the fundamental bandgap with photon energies around 1.7 eV. To obtain absorption spectra we used a common pump and probe setup. Lattice temperatures between 5 and 77 K were realized.

To perform the experiments using electric fields perpendicular to the quantum well layers we coated both the cap layer and the etched area of our samples with semitransparent metal electrodes. They consist of thin chromium and gold films with a thickness of about 50 Å each. The optical measurements were made with a tungsten lamp acting simultaneously as pump and probe beam source. In order to reduce the spectral width of photoexcitation to the lowest exciton resonance we used a color edge filter SCHOTT RG830.

The transmission and luminescence signals were detected simultaneously by using an optical multichannel analyser behind a spectrometer.

3 Excitons in Two and Three Dimensions

It was our aim to study the continuous transition from a 2D to a 3D carrier system. We therefore investigated the spectral shift of the first exciton ($1hh$-x) as a function of the optically created carrier density N and of L_z.

To evaluate our spectra we assumed linear relations between the reduction of the excitonic oscillator strength f_{1s}, the electron–hole density N, and the peak value of the excitonic absorption α_{1s} [3,4,5,6]. For the interpretation of the carrier density induced blueshift δE of the $1hh$-x we made use of the relationship [7]

$$\frac{\delta E}{E_{1s}} = C\frac{\delta f}{f_{1s}}.$$

The shift coefficient C is zero for 3D carrier systems, which has been proven experimentally [8,9,10]. This behaviour is due to the compensation of two effects:

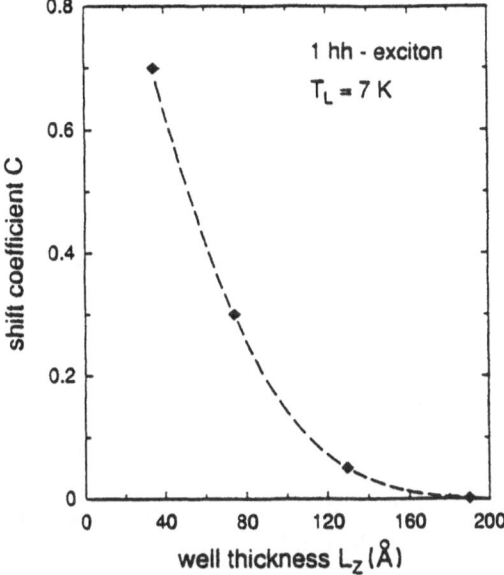

Fig. 1. Dependence of the shift coefficient on the well widths L_z

on the one hand population effects result in an energetic redshift of the bandgap while on the other hand the excitonic binding energy is reduced [11,12]. Theoretical considerations for a 2D exciton gas at a lattice temperature of $T_L = 0$ K [13] give a shift of

$$\frac{\delta E}{E_{1s}} \simeq 3.86 \pi a_{2D}^2 N,$$

with a_{2D} denoting the excitonic Bohr radius.

The numerical value of C is 0.5 in this case. The relation between L_z and C determined for our samples is demonstrated for $T_L = 7$ K in Fig. 1. We give an interpretation of this behaviour as follows: the 2D–limit is characterized by a reduced screening of the Coulomb interaction of the carriers compared to the 3D system. This results in an enhanced influence of the exchange interaction and leads finally to the observed blueshift of the excitonic resonance. With increasing L_z, the screening of carriers becomes more and more important; the 3D–limit is already reached at $L_z \approx 190$ Å.

It should be clearly pointed out that even 100 Å samples, which are usually used for the investigation of 2D systems are – at least with respect to their screening properties – far away from the strict 2D limit.

4 Many Particle Effects; Bandgap Renormalization

By increasing the optical power density the excitonic binding energies decrease because of phase–space–filling, exchange interactions, and screening of the Coulomb interaction. Finally, the excitons vanish completely and a new collective phase – the electron–hole (e–h) plasma – appears [14,15].

Many properties of the e–h plasma are relevant to application. The possibility to create a degenerate plasma and to achieve optical gain leads to the semiconductor laser. In order to reach an optimized spatial plasma confinement, single quantum wells, separate confinement heterostructures, and graded refractive index separate confinement heterostructures ($GRINSCH$) are used. It was our aim to study the many–particle–effects which become increasingly dominant with rising plasma density. They attract attention by shifting the fundamental bandgap to the red and by changing the refractive index, an effect which influences critically e.g. the mode stability. This behaviour can be explained and taken into consideration as follows: the ground state energy of the e–h plasma consists of the kinetic energy E_{kin}, and exchange and correlation terms (E_x, E_{cor}). E_x and E_{cor} both reduce the total energy of the plasma with increasing density. E_x describes the repulsion of electrons and holes with identic spins and depends on the overlap of their wavefunctions. All the other effects which are due to the correlation between e-e, h-h, and e-h are collected in E_{cor}. The contributions to this term are mostly calculated using the random phase approximation method [15,16,17].

In this contribution we show a possibility to determine the plasma density N_p and the renormalized bandgap E'_{gap} by evaluating the gain and luminescence spectra. We assume for simplicity parabolic sub-bands for e and h. The plasma temperature T_p can be obtained from the high energy tail of the luminescence $L(\hbar\omega)$ [18]

$$L(\hbar\omega) \propto \exp(-\frac{\hbar\omega}{k_b T_{eff}}) \quad \text{for} \quad \hbar\omega - \mu \gg k_b T_{eff}$$

with

$$\frac{1}{T_{eff}} = \frac{m_h}{m_e + m_h} \frac{1}{T_e} + \frac{m_h}{m_e + m_h} \frac{1}{T_h}$$

and the denotations $m_{e,h}$ and $T_{e,h}$ for the effective masses and temperatures of e and h respectively; μ names the chemical potential of the created plasma. In this regime the influence of the so-called excitonic enhancement or Fermi–edge singularity becomes negligible [15,17].

We can write $T_{eff} = T_e = T_h$ by assuming that electrons and holes both have enough time to interact and reach the same carrier temperature. To account for well width fluctuations and alloy disorder in the barriers of our samples, we replace the steplike 2D density of states by a more realistic expression which is broadened around the gap of each sub-band [19,20].

With the assumption of momentum conservation for the optical transitions and of a constant matrix element and by neglecting the correlation enhancement due to the high carrier temperatures, the luminescence spectrum $L(\hbar\omega)$ is given in the most simple case [21] by

$$L(\hbar\omega) = \sum_{i,j} \alpha_{ij} \int_{-\infty}^{\infty} D_e^{2D}(E, E_i) D_h^{2D}(E, E_i) f_e(E) f_h(E) \delta_\Gamma(\hbar\omega - E) dE$$

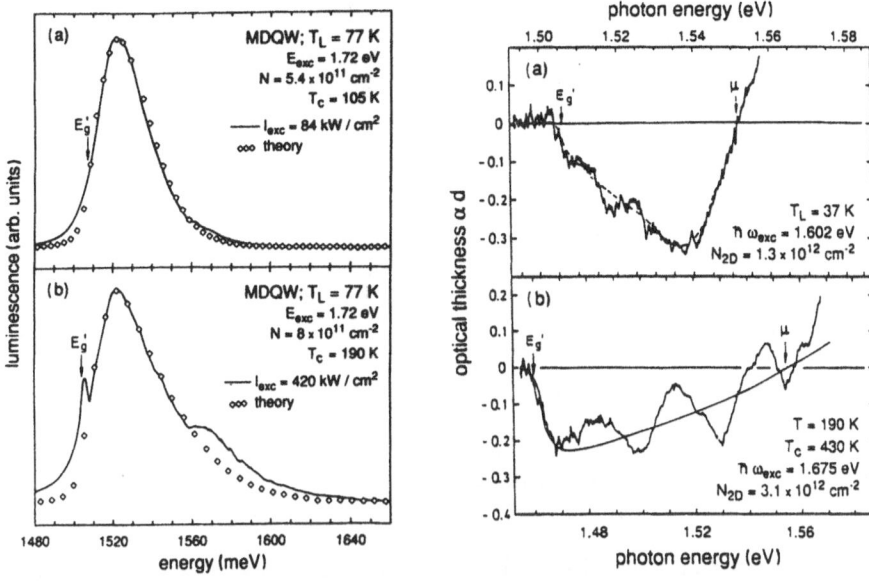

Fig. 2. Line shape analysis of the MDQW luminescence (left) and of gain spectra of a 50 × 100 Å MQWS (right)

Herein D^{2D} denotes the broadened 2D density of states and f the common quasi-Fermi functions for both e and h. The expression $\delta_\Gamma(\hbar\omega - E)$ has a Lorentzian shape and replaces a δ-function which expresses the conservation of energy. δ_Γ takes into account the final-state damping of the photocreated carriers [15,17,22,23,24]; its maximum halfwidth Γ is around the band gap and decreases towards the chemical potential of the plasma [22]. The expression to fit the gain spectra is very similar to the formula given above [20]. Examples for luminescence and gain of a MDQW and of a 50 × 130 Å MQWS are shown in Fig. 2.

The comparison of the described fit procedures is shown in Fig. 3 for the MDQW and expresses an excellent agreement between both experimental techniques. The dashed line represents a result of a dynamical single-plasmon-pole approximation [25].

5 Renormalization of Higher Sub-bands

Concerning the investigation of the renormalization of the higher sub-bands with $n_z \geq 2$ two limiting cases can be distinguished.

1. At low light power densities the excitonic effects dominate the absorption spectra. The higher sub-bands are not occupied by the carriers and therefore are only influenced by intersub-band interactions with carriers in the fundamental sub-bands.

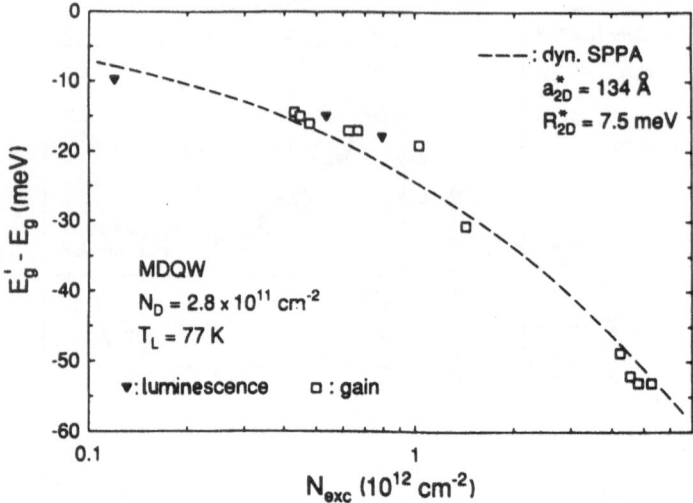

Fig. 3. Comparison of the results for the band gap renormalization from luminescence and gain data with a dynamical single–plasmon–pole approximation

2. Increasing pump power leads to an occupation of the higher sub-bands; for that reason bandfilling effects become important for their renormalization.

To achieve the shifts of the higher sub-bands in the first case, we assumed constant exciton binding energies. This simplification leads to an underrating of the estimated sub-band shifts, but we assume the discrepancy to be small as long as the excitonic resonance of that sub-band is clearly observable. The determination of N was done by evaluating the $1hh$–x oscillator strength for lower densities. Values above the exciton bleaching density are obtained directly from absorption line–shape analysis [26,27].

Figure 4 shows the band–gap shift for the $n_z = 1, 2$ transitions of a 50×130Å MQWS. It is clearly recognizable that in the region of unpopulated higher sub-bands with densities below 8×10^{11} cm^{-2} [28] only a slight shift of the second sub-band can be observed. As soon as the occupation of this band sets in, a drastic decrease of its band edge starts. The shift is in the order of magnitude which one would estimate from the direct occuation density of the $n_z = 2$ sub-band. This behaviour is in good agreement with recent theoretical results [29,30].

6 Excitons in High Transversal Electric Fields

In order to build an all–optical computer the availability of fast *and* low power optical switches or modulators is unavoidable. Several types of low power devices using the QCSE have been developed. Unfortunately they are more or less sophisticated in structure. Furthermore these devices need an external circuit (in the simplest case a series resistor) to operate as an optical switch [31].

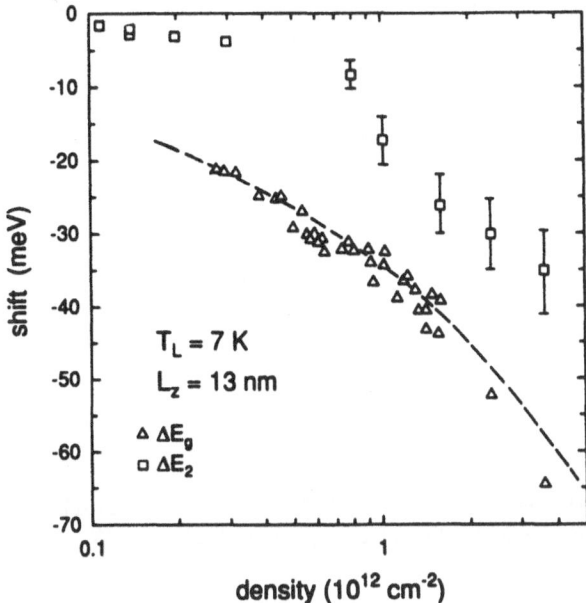

Fig. 4. Shift of the fundamental band gap ΔE_g and of the second sub-band ΔE_2 for a 130 Å MQWS. The curve represents a 2D theory with effective parameters [25].

These facts led us to a new modulator design which is less complicated in structure [32]. The whole structure consists of an intrinsic optically active region which is sandwiched between two metal films, thus forming Schottky contacts at the i–m interfaces. The intrinsic region is a 50×103 Å MQWS in our case.

The operating mechanism of our device is as follows: part of voltage applied between the metal electrodes drops over the MQWS shifting the excitonic resonances to the red due to the QCSE. By illuminating the sample we generate e–h pairs which are separated immediatly via the external field. They are collected on opposite i–m interfaces thus being able to partly screen the applied field. This results in a blueshift of the exiton-resonances as a function of the incoming light power. The effect is demonstrated in Fig. 5; the left side shows the well known QCSE at very low light intensities while the nearly identical spectra on the right are produced by varying the light intensity at constant bias.

The measurements were done on a sample with a transparent area A of about 1 mm². It is our aim to increase A in order to have the possibility to put a whole matrix of light dots onto the sample. It is expected that each dot performs its own nonlinear function and interacts with the dots in the neighbourhood only via lateral carrier diffusion. Therefore, free MQWS areas up to 6 mm² were prepared by the selective etching method described above. Unfortunately, this generation of devices showed until now only thermal bistability under external fields due to electrical heating of the samples. It is not yet clear if this behaviour results from the increased device area or from impurities in the $(AlGa)As$ cap

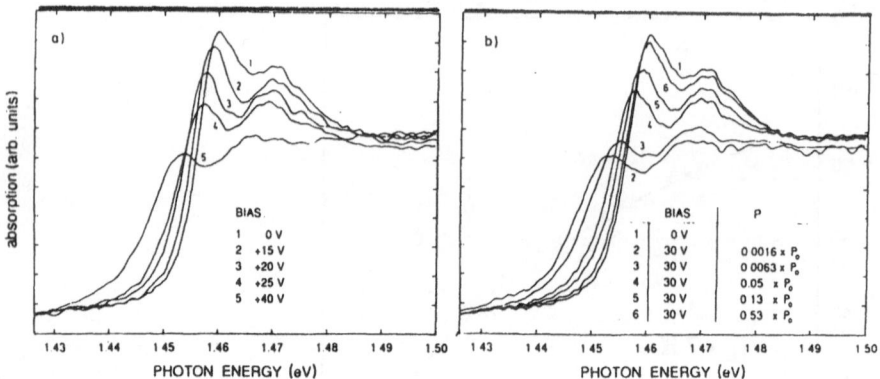

Fig. 5. Absorption spectra of the device at constant light power $P = 0.0016 P_0$ and different bias (left) and constant bias and different P (right), with $P_0 = 1\ \mu$W

layers which prevent the performance of Schottky contacts.

The future plans towards to the optimization of our device are the following:

- One of the semitransparent metal electrodes will be changed into a high reflective one in order to simplify the sample preparation.
- Surface roughness and nonparallel interfaces could lead to large lateral electric field components with undesired carrier drift increasing the crosstalk between neighbouring light spots. This could be prevented by structuring the MQWS via ion implantation or similar techniques.

The dynamical properties of the device are expected to be in the sub–MHz region limited mainly by its capacitance. To proove this experimentally we use a cw–Al$_2$O$_3$:Ti laser which is modulated and stabilized with an extra–cavity electrooptical modulator.

Parts of these research activities concerning thermal bistability and its use to build a temperature sensor are reported elsewhere in this issue.

Acknowledgements

This work has been supported by the Deutsche Forschungsgemeinschaft, by the Materialforschungsschwerpunkt des Landes Rheinland–Pfalz, and by the German Bundesministerium für Forschung und Technologie under the contract No. 0575 7. The responsibility for the contents of this report, however, lies solely with the author. Stimulating discussions with H. Haug, C. Ell, H. Kalt, and W.W. Rühle are acknowledged.

References

1. Schmitt–Rink, S., Chemla, D. S., Miller, D. A. B.: Advances in Physics **38** (1989) 89.
2. LePore, J. J.: J. Appl. Phys. **51** (1980) 6441.
3. Schmitt–Rink, S., Chemla, D. S., Miller, D. A. B.: Phys. Rev. B **32** (1985) 6601.
4. Zimmermann, R.: Phys.Stat. Sol. (b) **146** (1988) 371.
5. Lee, H. C., Kost, A., Kawase, M., Hariz, A., Dapkus, P. D., Garmire, E.: IEEE J. Quant. Elec. **QE–24** (1988) 1581.
6. Park, S. H., Morhage, J. F., Jeffrey, A. D., Morgan, R. A., Chavez–Pirson, A., Gibbs, H. M., Koch, S. W., Peyghambarian, N., Derstine, M., Gossard, A. C., English, J. H., Wiegmann, W.: Appl. Phys. Lett. **52** (1988) 1201.
7. Hulin, D., Mysyrowicz, A., Antonetti, A., Mingus, A., Masselink, W. T., Morkoc, H., Gibbs, H. M., Peyghambarian, N.: Phys. Rev. B **33** (1986) 4389.
8. Fehrenbach, G. W., Schäfer, W., Treusch, J., Ulbrich, R. G.: Phys. Rev. Lett. **49** (1982) 1281.
9. Majumder, F. A., Swoboda, H. E., Kempf, K., Klingshirn, C.: Phys. Rev. B **32** (1985) 2407.
10. Swoboda, H. E., Majumder, F. A., Lyssenko, V. G., Klingshirn, C., Banyai, L.: Z. Phys. B – Condensed Matter **70** (1988) 341.
11. Chang Y. C., Sanders, G. D.: Phys. Rev. B **32** (1985) 5521.
12. Nozieres P., Comte, C.: J. Phys. (Paris) **43** (1982) 1083.
13. Lach, E., Lehr, G., Forchel, A., Ploog, K., Weimann, G.: Surf. Sci. **228** (1990) 168.
14. See e.g. Pokrovskii, Ya.: Phys. Stat. Sol. (a) **11** (1972) 385; Jeffries, C. D.: Science **189** (1975) 955; Rice, T. M.: Solid State Physics, **32**, 1 (Academic Press, 1977); Hensel, J. C., Philips, T. G., Thomas, G. A.: *ibid* p. 88, and the references therein.
15. Haug, H., Schmitt–Rink, S.: Prog. Quant. Electron. **9**, (1984) 3; Haug, H., Koch, S. W.: Quantum Theory of the Optical and Electronic Properties of Semiconductors. World Scientific, Singapore (1990).
16. Brinkmann, W. F., Rice, T. M.: Phys. Rev. B **7** (1973) 1508.
17. Zimmermann, R., Rösler, M.: Phys. Stat. Sol. (b) **75** (1976) 633.
18. Shah, J.: IEEE J. Quantum. Electron. **QE–22** (1986) 1728.
19. Chemla, D. S., Miller, D. A. B., Smith, P. W., Gossard, A. C., Wiegmann, W.: IEEE J. Quantum. Electron. **QE–20** (1984) 265.
20. Schlaad, K.–H., Weber, Ch., Cunningham, J., Hoof, C. V., Borghs, G., Weimann, G., Schlapp, W., Nickel, H., Klingshirn, C., Phys. Rev. B **43** (1991) 4268.
21. Hall, R. N.: Solid State Electron. **6** (1963) 405.
22. Landsberg, P. T.: Phys. Status Solidi **15** (1966) 623.
23. Landsberg, P. T.: Solid State Electron. **28** (1985) 137.
24. Klingshirn, C., Haug, H.: Phys. Rep. **70** (1981) 315.
25. Schmitt–Rink, S., Ell, C.: J. Lumin. **30** (1985) 585.
26. Ell, C., Haug, H.: (unpublished).
27. Weber, Ch.: Ph.D. thesis, Fachbereich Physik der Universität Kaiserslautern, 1989.
28. Knox, W. H., Hirlimann, C., Miller, D. A. B., Shah, J., Chemla, D. S., Shank, C. V.: Phys. Rev. Lett. **56** (1986) 1191.
29. Ell, C., Haug, H.: Phys. Status Solidi (b) **159** (1990) 117.
30. Zimmermann, R.: submitted to Phys. Rev. B.

31. Miller, D. A. B.: In Optical Computing. Wherrett, B. S., Tooley, F. A. P., eds. Proc. of the 34th Scottish Universities Summer Scool in Physics **34** (1988) and references therein.

32. Weber, Ch., Schlaad, K.–H., Klingshirn, C., Hoof, C. v., Borghs, G., Weimann, G., Nickel, H.: Appl. Phys. Lett. **54** (1989) 2432.

Electroabsorptive/-Refractive Effects in Asymmetric Step Quantum Wells and BRAQWET Structures

U. Olin

Institute of Optical Research, S-100 44 Stockholm, Sweden

1 Introduction

The optical properties of semiconductors, for example the absorption coefficient and the refractive index, can be controlled by applying external electric fields and/or currents to the structure. This is technologically important for devices, such as electrically controlled intensity and phase modulators, directional couplers and devices for wavelength multiplexed systems.

Most of the quantum well modulator structures studied so far have been based on the electric-field-induced spectral shift of the fundamental absorption edge, due to the quantum confined Stark effect (QCSE) [1]. Depending on the detuning from the absorption edge, mainly electroabsorptive or mainly electrorefractive effects are obtained.

2 Asymmetric Step Quantum Wells

In symmetric rectangular quantum wells, the shift of the energy levels depends quadratically on the applied electric field strength. Thus, in order to change the effective band gap, and thereby the absorption coefficient and the refractive index, a large electric field strength is required (~ 100 kV/cm). However, if the well is asymmetric, linear contributions to the energy-level shift are obtained. This was experimentally demonstrated by Morita et al. [2], who utilised asymmetric step quantum wells. We have studied how the quantum confined Stark effect for asymmetric step quantum wells (Fig. 1), depends on the specific quantum well parameters. Such parameters are the total well width ($L_1 + L_2$), the width ratio ($L_1/(L_1 + L_2)$) and the step height (ΔE_c and ΔE_v, respectively).

The calculations were performed for wells consisting of one layer of GaAs and one layer of $Al_x Ga_{1-x}As$ and barrier layers of $Al_{0.3}Ga_{0.7}As$. The energy levels and the wavefunctions were calculated using the transfer matrix method (TMM) [3].

Figure 2 shows the field-induced shifts of the effective band gap (transitions from the heavy hole band to the conduction band) for quantum wells of different

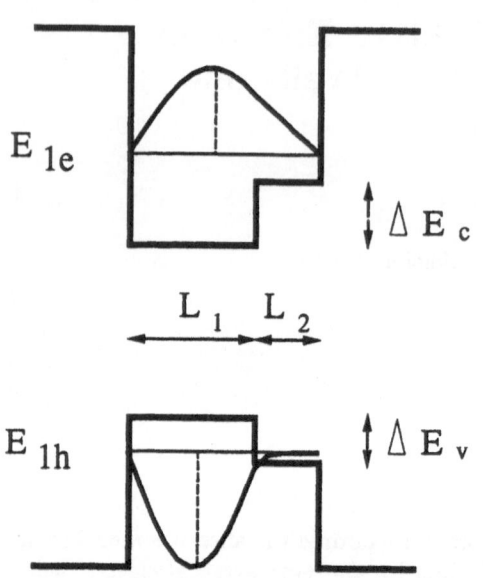

Fig. 1. Asymmetric step quantum well. The thicknesses for the layers in the well and the step heights are indicated

total thicknesses, where the step occupied half of the total well thickness and its aluminium concentration was 6%. The largest shifts were obtained for the widest wells. However, for the widest wells a large separation of the electrons and holes is possible, resulting in a rapid reduction of the overlap integral.

In Fig. 3 results are shown for wells of a total thickness of 75 Å and an aluminium concentration in the potential step of 6%. The different curves represent results for wells having different relative widths of the GaAs layer. The largest shifts are obtained for wells, where 20% of their widths consists of GaAs.

Finally, in Fig. 4 results are presented for quantum wells having a 15 Å thick GaAs layer and a 60 Å wide AlGaAs potential step of different heights. The largest energy level shifts are obtained for aluminium concentrations between 12 and 18%. For these parameters the zero-field separation of the electron and the hole is large, leading to the largest linear contributions to the energy level shift. The shift is nearly twice as large as for a symmetric quantum well.

3 BRAQWET Structures

The barrier reservoir and quantum well electron transfer (BRAQWET) concept was introduced by Wegener et al. in 1989 [4]. In BRAQWETs, electroabsorptive/-refractive modulation due to voltage-controlled carrier density effects is utilized. One period of a BRAQWET structure consists of one or more quantum wells, an n-doped reservoir layer, a p-doped barrier layer and a number of intrinsic

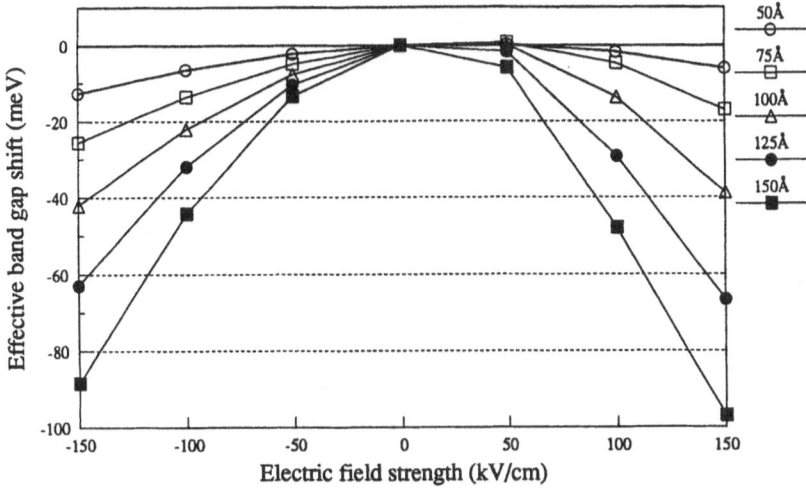

Fig. 2. Field-induced shift of the effective band gap for asymmetric step quantum wells of different total thicknesses

layers. These periods can be stacked to increase the interaction between light and matter.

Figure 5 shows an energy band diagram for one period of a BRAQWET structure with three quantum wells. If a number of such periods are stacked, they can be regarded as a number of equal capacitors. Hence, the voltage across each period will be the same. The n-doped layer provides a reservoir of electrons for the quantum wells. The quantum wells are placed on the potential slope between the n-doped reservoir and the p-doped barrier. By properly choosing the positions and the parameters for the quantum wells, their carrier densities will be low when no external field is applied.

When an external field is applied to the structure, as in Fig. 5b, the energy levels in the quantum wells are pushed downwards relative to the quasi chemical potential and electrons are transferred from the n-doped reservoir layer to the wells.

The electrons thereby occupy the states in the bottom of the conduction band of the quantum wells, resulting in a quenching of the absorption due to the Pauli principle. The changes in the spectrum of the absorption coefficient are, through the Kramer-Kronig's relation, related to changes of the refractive index spectrum. An advantage of the BRAQWET structure compared to structures for current injection modulation is that the modulation time does not depend on the interband recombination time, since mainly electrons are injected into the quantum wells.

The reason for having several quantum wells per period, is the fact that it is mainly the optical properties of the quantum wells that are changed, when an electric field is applied. Thus, it is important that the relative amount of

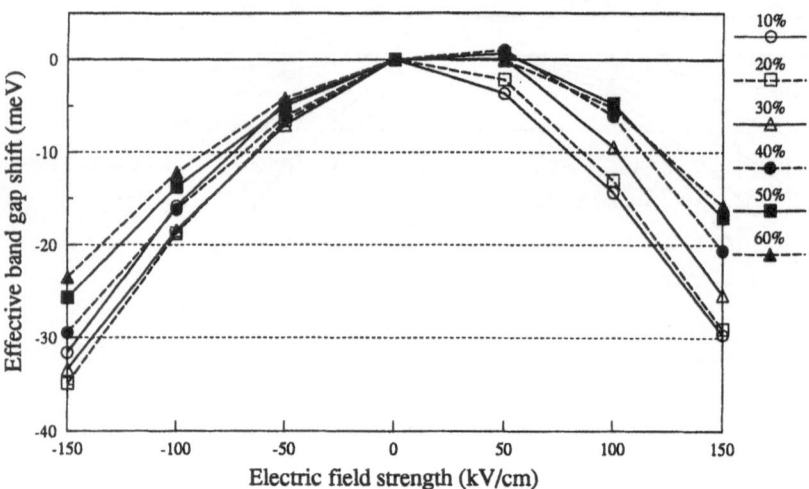

Fig. 3. Shift of the effective band gap energy for 75 Å wide quantum wells having different relative widths of the GaAs layer

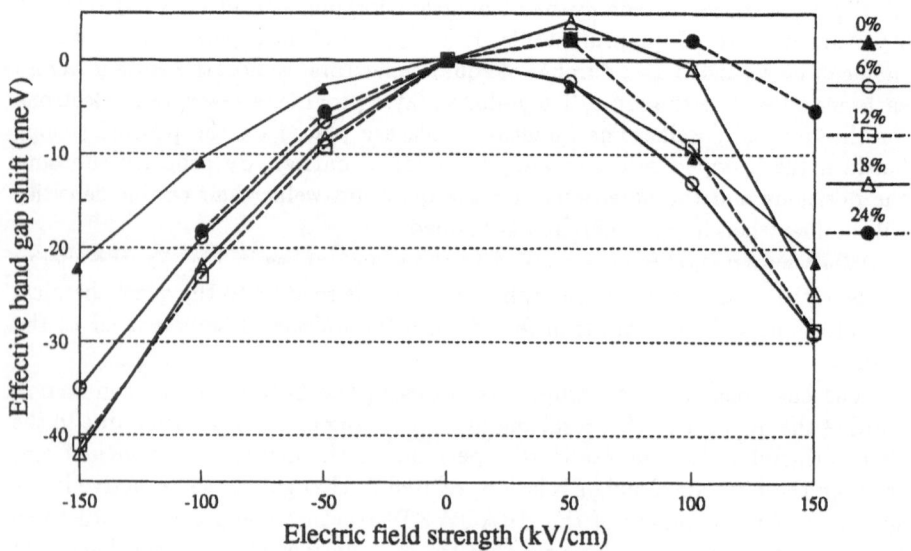

Fig. 4. Effective band gap shift (transitions from level E_{1h} to level E_{1e}) versus applied electric field. The different curves represent different aluminium concentrations in the step layer

quantum well material per period is large. However, having several quantum wells per period causes certain complications. If the well widths are all equal, the electron densities in the wells will be very different. Thus, the modulation originating from the various well layers will be different. In order to fill the wells with electrons more uniformly, the different wells should have unequal widths, as indicated in Fig. 5. However, since the energy levels for the quantum wells depend on the well width, also the compositions of the wells must differ in order to obtain the same effective band gap for all the wells.

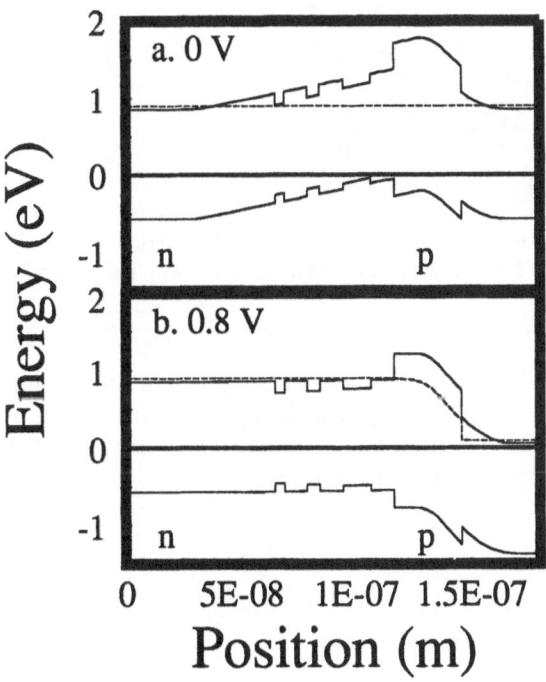

Fig. 5. A BRAQWET structure with three quantum wells for an applied voltage of 0 V and 0.8 V, respectively. The dashed lines indicate the (quasi) chemical potential for the electrons

Figure 6 shows the calculated results for the structure shown in Fig. 5 having reservoir layers of GaAs, barrier layers of AlGaAs and quantum wells of strained InGaAs. The total period length for the structure is 1465 Å. The three quantum wells have different sizes (40, 55, and 120 Å) and different indium concentrations (24.0%, 20.7% and 16.2%).

We have studied the electrorefractive properties for BRAQWET structures in waveguides [5]. An example of a waveguide having five BRAQWET periods in the core is given in Fig. 7. For BRAQWETs consisting of GaAs reservoir

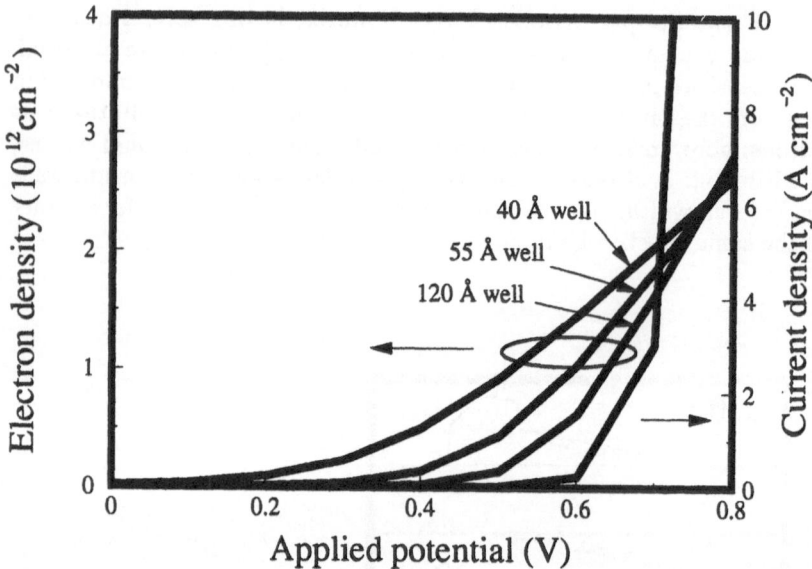

Fig. 6. The electron density inside the three quantum wells and the current density in an InGaAs/GaAs/AlGaAs BRAQWET structure as functions of the applied potential

layers, three strained InGaAs wells and AlGaAs barrier layers, we have found that the product of the voltage to obtain a π phase change times the length of the waveguide, $V_\pi \times L$, could be 0.25 V·mm, which is ten times less than what has been reported for any other semiconductor structure.

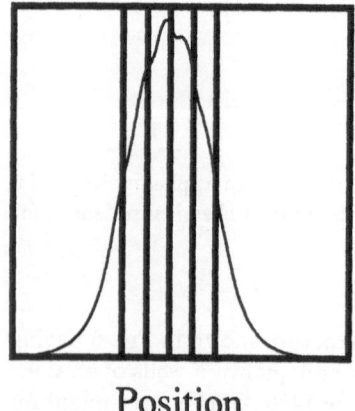

Position

Fig. 7. The lowest optical mode for a waveguide consisting of five BRAQWET periods (the positions of the quantum well layers are indicated by the straight lines)

4 Conclusions

We have shown that band-gap engineering can be used to enhance the electroabsorptive and -refractive properties of quantum well material. Thus, these materials could be of importance for new low-voltage high-speed devices.

References

1. Chemla, D. S., Damen, T. C., Miller, D. A. B., Gossard, A. C., Wiegmann, W.: Electroabsorption by Stark effect on room-temperature excitons in GaAs/AlGaAs multiple quantum well structures. Appl. Phys. Lett. **42** (1983) 864–866.
2. Morita, M., Goto, K., Suzuki, T.: Quantum-confined Stark effect in stepped-potential quantum wells. Jpn. J. Appl. Phys. **29** (1990) L1663–L1665.
3. Jonsson, B., Eng, S. T.: Solving the Schrödinger equation in arbitrary quantum well potential profiles using the transfer matrix method. IEEE J. Quant. Electr. **QE-26** (1990) 2025–2035.
4. Wegener, M., Chang, T. Y., Bar-Joseph, I., Kuo J. M., Chemla, D. S.: Electroabsorption and refraction by electron transfer in asymmetric modulation-doped multiple quantum well structures. Appl. Phys. Lett. **55** (1989) 583–585.
5. Looström, C., Olin, U.: Investigation of modulation-doped tunable-electron-density multiple-quantum-well structures. Submitted to J. Appl. Phys.

Electrooptical Modulation in Vertical Multiple Quantum Well Microresonators

G. Wingen,[1] S. Zumkley,[1] J. C. Michel,[2] J. L. Oudar,[2] R. Planel,[3] and D. Jäger[1]

[1] Fachgebiet Optoelektronik, Universität Duisburg, W-4100 Duisburg, Germany
[2] Centre National d'Études des Télécommunications, F-92220 Bagneux, France
[3] Laboratoire de Microstructures et Microélectronique,
   CNRS, F-92220 Bagneux, France

We demonstrate a novel, efficient and fast vertical **nin** AlGaAs/GaAs multiple quantum well (MQW) microresonator grown by molecular beam epitaxy (MBE). The electrical current of this hybrid device is kept low by using undoped AlGaAs barrier layers cladding the active MQW material. For high-speed operation, we propose a novel traveling-wave modulator with a special coplanar metallic waveguide structure. Electrooptical modulation is studied using a Ti:Sapphire laser pumped by an argon ion laser.

1 Introduction

For future applications in optical signal processing and switching networks vertical photonic components suitable for two dimensional arrays are of most importance. The thickness of the devices, however, is limited to a few micrometers due to the epitaxy process. For sufficient interaction length it is often necessary to built microresonators. Thus, great interest has been paid to optical modulators and switches based on the structure of Fabry-Perot cavities [1,2]. Recently, large progress has been made in the field of epitaxially grown vertical electrooptical microresonator modulators based on the quantum-confined Stark effect (QCSE) [3,4,5]. Usually, the structure is that of an asymmetric Fabry-Perot resonator and the electrical behaviour is that of a *pin*-diode. High contrast ratios with a corresponding low insertion loss and high reflection changes for a low voltage swing have been demonstrated among others by L.A. Coldren et al. [3], J.R. Harris, Jr. et al. [4] and M. Whitehead et al. [5]. In particular, high-speed vertical modulator structures with a cut-off frequency of 6.5 GHz have been reported by L.A. Coldren et al. [6].

In this paper, we report our investigations on **nin** microresonator modulators consisting of a multiple quantum well (MQW) region with two cladding quarter wavelength Bragg reflectors. Complementary to our previous work on Bragg reflector modulators [7] only the MQW region is used as an electrooptical active material.

2 Device structure

The layer structure of the MBE grown sample is sketched in Fig. 1. It consists
of five periods of quarter wavelengths layers of AlAs (72.5 nm) and $Al_{0.2}Ga_{0.8}As$
(62.5 nm) as the top mirror, $20\frac{1}{2}$ periods as the bottom mirror, and an active
region of 24 GaAs quantum wells (10 nm) embedded in $Al_{0.5}Ga_{0.5}As$ barrier
layers (10 nm). The top and bottom mirrors are n-type doped with silicon ($\approx$
2×10^{18} cm^{-3}). The active region between the mirrors is undoped. The current
of this device is kept low (see below) by using undoped AlGaAs barrier layers
cladding the active MQW material. Between the n^+-GaAs substrate and the
bottom mirror a buffer layer consisting of a GaAs layer (500 nm) is provided,
which is also n-type doped with silicon. To prevent oxidation a thin GaAs layer
(5 nm) was grown on top of the epitaxially layers.

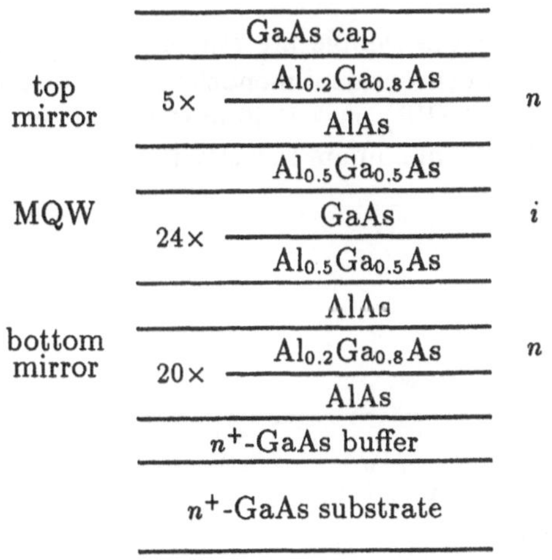

		GaAs cap	
top mirror	5×	$Al_{0.2}Ga_{0.8}As$ / AlAs	n
MQW	24×	$Al_{0.5}Ga_{0.5}As$ / GaAs / $Al_{0.5}Ga_{0.5}As$	i
bottom mirror	20×	AlAs / $Al_{0.2}Ga_{0.8}As$ / AlAs	n
		n^+-GaAs buffer	
		n^+-GaAs substrate	

Fig. 1. Layer structure

Two different electrooptical devices are fabricated using conventional photolithography, lift-off technology and wet chemical etching. First, circular mesa
profiles with a diameter of 1 mm are realized (Fig. 2(a)). The top contact has
a diameter of 500 μm and the bottom contact located on the backside of the
substrate covers about 1 cm^2. Second, for high-speed operation a traveling-wave
modulator shown in Fig. 2(b) has been realized. As can be seen, a broad center
contact (width 100 μm) of a special coplanar metallic waveguide structure is
evaporated on top of a 200 μm mesa ridge. The ground contacts are located on
the substrate on both sides of the ridge. The length of the traveling-wave modulator is about 4 mm. The position of illumination is on top of the microresonator
near the center contact.

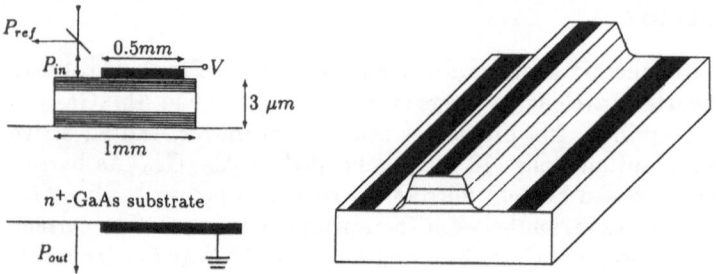

Fig. 2. Sketch of the dc-modulator (a) and traveling-wave modulator (b)

3 Experimental results

The optical properties of the sample are studied at wavelengths between 700 nm and 1000 nm using a conventional monochromator setup. As can be seen in Fig. 3, a clear Fabry-Perot (FP) peak (859 nm) and an exciton (Ex) resonance (843 nm) are detected. The maximum reflectivity of the sample is about 98% at a wavelength of 880 nm.

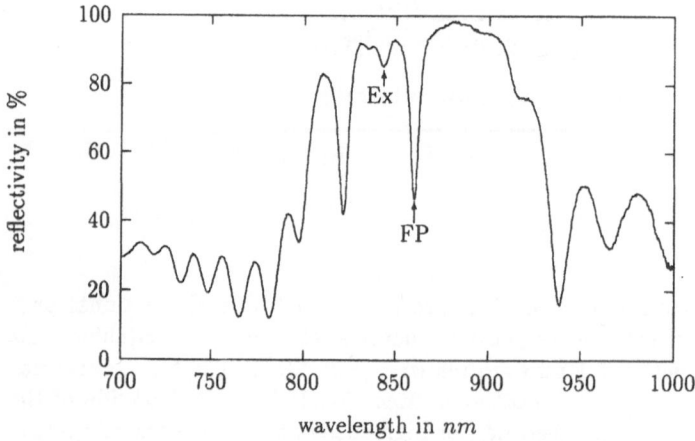

Fig. 3. Measured reflection spectrum of the microresonator

Figure 4 shows the I-V-characteristic of the device. Clearly, a typical **nin**-diode characteristic can be recognized. For example, the electrical field in the i-region exhibits a value of 130 kV/cm at a voltage of $V = -10$ V with the

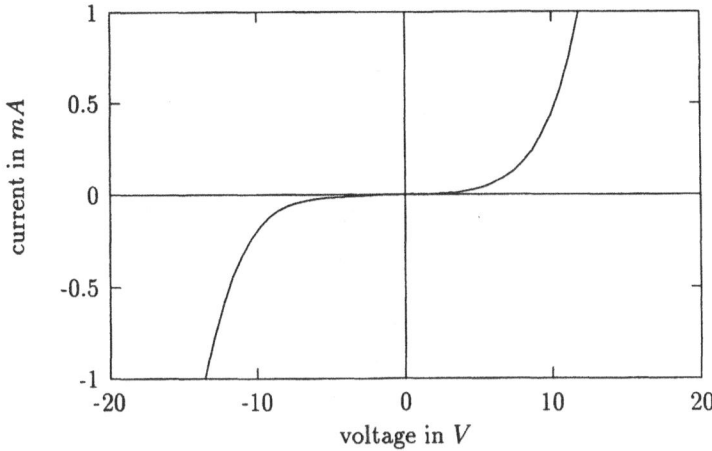

Fig. 4. I-V-characteristic of the **nin**-microresonator of Fig. 2 (a)

assumption of a voltage drop in the i-region only. The applied field leads to a current of less than $I = 170~\mu A$ driven through the device. A current flow of 1 mA is reached for an applied voltage of -14 V.

In Fig. 5 the reflection spectra for different applied voltages of the microresonator are shown. A clear Fabry-Perot and exciton resonance can be recognized at zero voltage. In this case the wavelength separation of the two resonance points in the spectra is about 9.5 nm. With increasing voltage from 0 to 12 V a red shift of the exciton resonance of about 10 nm can be detected. As can be seen, at a bias of 12 V the exciton resonance is shifted through the Fabry-Perot resonance. Additionally, a shift of the Fabry-Perot peak (about 2 nm) and an increasing of the full width half maximum from 4 to 9 nm is observed.

In Fig. 6 the reflectivity is plotted as a function of the applied bias voltage for three different wavelengths. A maximum reflectivity change of more than 42% is achieved at a voltage swing from 0 to 10 V at a wavelength of 854 nm. On the short wavelength side of the Fabry-Perot peak (849 nm) a comparable reflectivity change of $\approx 40\%$ is reached for a voltage swing from 0 to 12 V. The minimum reflectivity of the device is obtained at 852 nm for an applied bias voltage of 8 V.

The highest contrast ratio of 11 dB is measured at a bias voltage of 8 V (Fig. 7). The corresponding insertion loss is with 8 dB likewise high. With regard to low losses a more interesting result can be seen at a wavelength of 853 nm, where a contrast ratio of 6 dB is obtained for a voltage swing from 0 to 10 V, with an associated insertion loss of only 2.5 dB at zero bias.

In a first and preliminary high-speed experiment an electrical microwave signal is applied across the device of Fig. 2(b) and the reflected optical signal

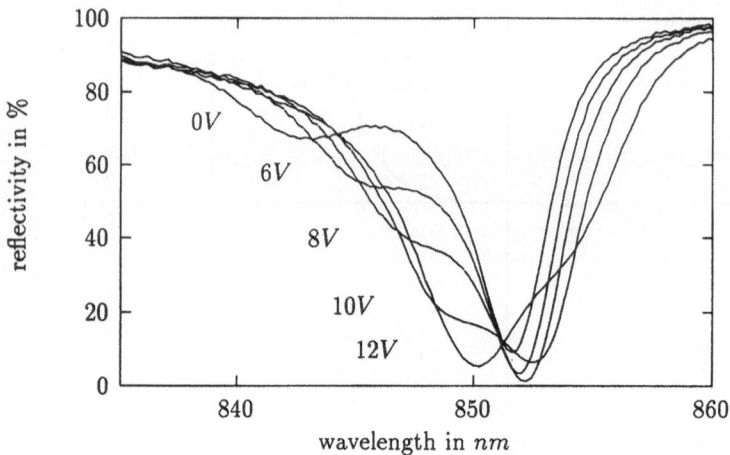

Fig. 5. Reflection spectrum, parameter is the bias voltage

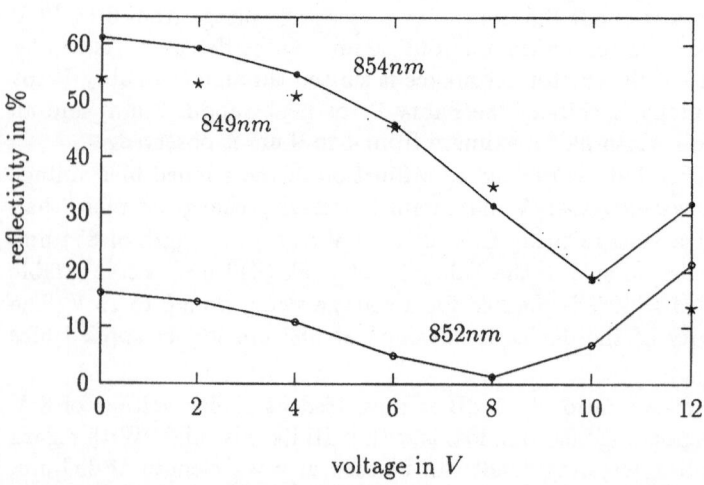

Fig. 6. Measured reflectivity as a function of applied bias voltage at different wavelengths

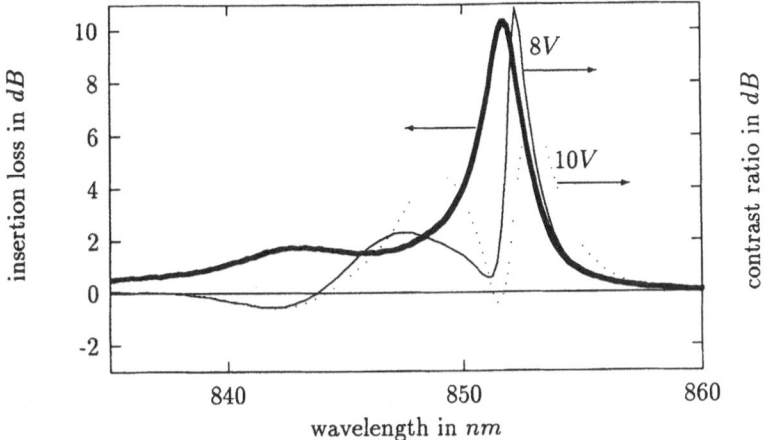

Fig. 7. Contrast ratio and insertion loss of the microresonator

is recorded by a spectrum analyzer. The modulation of the reflected optical signal versus the position of illumination is shown in Fig. 8 for frequencies of 100 MHz and 1.1 GHz. As can be seen, a clear change of the modulation by moving the position of the incident beam is measured. This change is traced back to a combined effect of standing waves and the spatial inhomogeneity of the reflectivity of the microresonator. It is noteworthy, that by moving the position of the incident beam no attenuation in the modulation is observed.

4 Conclusion

In summary, we demonstrate an efficient and fast **nin** electrooptical modulator. A maximum reflectivity change of 42% has been achieved. At a wavelength of 853 nm a reflectivity change of 40% and a contrast ratio of 6 dB for a voltage swing from 0 to 10 V with a corresponding insertion loss of only 2.5 dB at zero bias has been found. Additionally, high-speed electooptical modulation in the GHz-region has been demonstrated. Hence the results are comparable to those of the usual *pin* devices but **nin** structures provide some technological advantages. On the one hand the possibility of integration with common GaAs ICs and MMICs (monolithic microwave integrated circuits), which are based on FETs (field effect transistors) using only n–doping is facilitated. Furthermore, p–type doping is often not available and not recommended, because p–doping would decrease the quality of electronic devices.

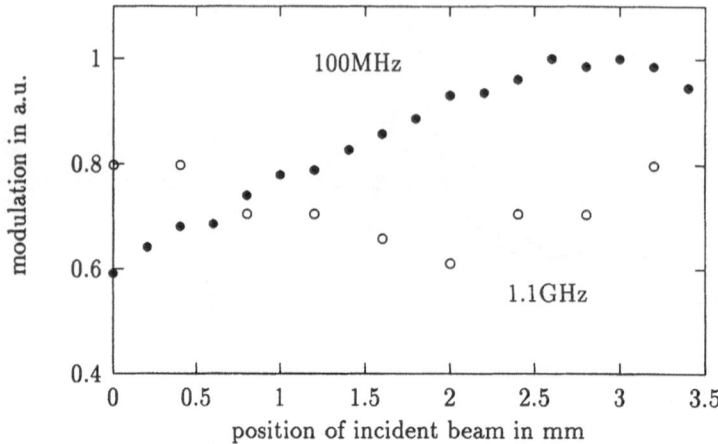

Fig. 8. Modulation vs. position of illumination of the traveling wave modulator

References

1. Sfez, B.G., Oudar, J.L., Michel, J.C., Kuszelewicz, R., Azoulay, R.: External beam switching in monolithic bistable GaAs quantum well etalons. Appl. Phys. Lett. **57** (1990) 1849–1851.
2. Zumkley, S., Wingen, G., Oudar, J.L., Michel, J.C., Planel, R., and Jäger, D.: Electrooptical modulation in nin AlGaAs/GaAs MQW microresonator. IEEE Lasers and Electro-Optics Society 1991 Annual Meeting November 4-7, San Jose, CA, USA, paper No. OE8.5 (1991).
3. Yan, R.-H., Simes, R.J., and Coldren, L.A.: Surface-normal electroabsorption reflection modulators using asymmetric Fabry-Perot structures. IEEE J. Quantum Electron. **QE-27**, (1991) 1922–1931.
4. Pezeshki, B., Thomas, D., and Harris, J.S., Jr.: Optimization of modulation ratio and insertion loss in reflective electroabsorption modulators. Appl. Phys. Lett. **57** (1990) 1491–1492.
5. Zouganeli, P., Whitehead, M., Stevens, P.J., Rivers, A.W., Parry, G., and Roberts, J.S.: High tolerances for a low-voltage, high-contrast, low-insertion-loss asymmetric Fabry-Perot modulator. IEEE Photon. Technol. Lett. **3** (1991) 733–735.
6. Simes, R.J., Yan, R.H., Barron, C.C., Derrickson, D., Lishan, D.R., Karin, J., Coldren, L.A., Rodwell, M., Elliot, S., and Hughes, B.: High-frequency electrooptic Fabry-Perot modulators. IEEE Photon. Technol. Lett. **3** (1991) 513–515.
7. Zumkley, S., Wingen, G., Scheffer, F., Prost, W., and Jäger, D.: Electrooptical modulation in AlGaAs/GaAs DFB structures. Proc SPIE **1280** (1990) 202–208.

Criteria for the Use of Nonlinear GaAs Etalons for Threshold Logic

B. Acklin, N. Collings, and C. Bagnoud

Institute of Microtechnology, University of Neuchâtel, CH-2000 Neuchâtel

1 Introduction

In optical computing it is often difficult to place a device concept in a system context. The hardware of digital computers is based on logic and memory devices. It is within this context that work on GaAs etalons has been historically pursued [1] and the demonstration of transistor action and bistability [2] have been important events. The systems aspect of optical computing has meantime broken new ground and we need figures of merit for such devices that reveal new avenues for exploiting them. The three figures of merit which we discuss are the gain-bandwidth product (GBWP), the time-bandwidth product (TB), and the space-bandwidth product (SBWP). A section is devoted to each.

The latter two figures of merit differ slightly in meaning from their more common usage, in the context of input spatial light modulators, where they signify the number of channels which can be generated. Here they are applied to nonlinear Fabry-Perot devices used as integrating/thresholding etalons, and the TB and the SBWP quantify the number of signal channels which can be fanned-in to the device temporally and spatially, respectively. We restrict the consideration to incoherent fan-in because we judge that coherent fan-in is not practical. In order to provide gain, the NLFP has to be used in a three-port addressing scheme (Fig. 1), where a separate hold beam supplies the output power.

In the first section, we present the current status of our practical research on all-optical switching devices based on GaAs. In the following three sections, we discuss each figure of merit in the context of the experimental measurements which have been made on GaAs etalons. A conclusion section highlights new directions of research.

2 Experimental results on bulk GaAs etalons

Our nonlinear optical devices are based on the large dispersive nonlinearity below the bandgap energy of intrinsic GaAs (1.425 eV). The nonlinear refractive

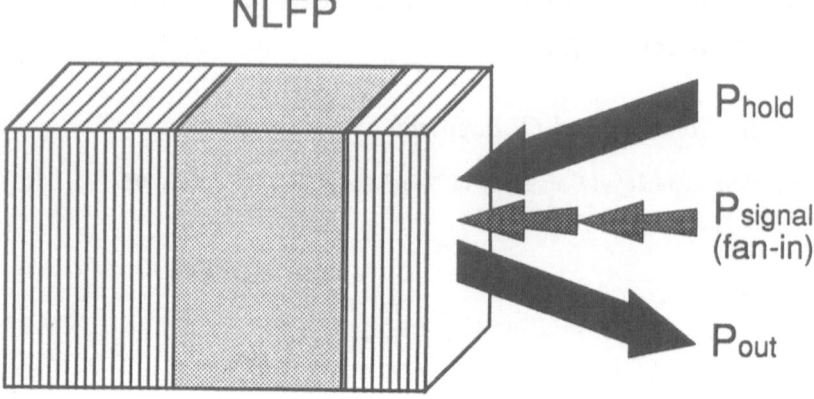

Fig. 1. Sketch of a NLFP device operated as a three-port logic gate: a hold beam biasses the NLFP close to threshold, and provides the output power. The output state is controlled by the signal beams which are fanned-in to the NLFP as temporal or spatial modes

index change is enhanced in resonant optical AlGaAs structures, which are fabricated in a single epitaxial growth process. Two types of reflective devices have been investigated: nonlinear Fabry-Perot devices (NLFP), where the nonlinear material is restricted to the spacer layer between two impedance-matched dielectric mirrors, and nonlinear Bragg reflectors (NLBR), where the nonlinearity is distributed throughout the device, in the high index layer of a dielectric mirror.

We have reported bistable operation of a NLFP at a threshold power of only 1 mW for a spotsize of 6 mm diameter [3]. The structure consists of a 2 mm thick GaAs spacer between mirrors with estimated reflectivities of 91% and > 99%, giving a cavity finesse of more than 30. The device shows a contrast ratio of about 10:1 with an attenuation of less than 1 dB in the high reflectivity OFF-state. Due to the low threshold power, this NLFP can be operated in a thermally stable manner, without any heat sinking other than through the 0.5 mm thick substrate. Figure 2b illustrates that the device remains latched for more than 0.4 s, until the bias pulse is removed, after a short signal pulse switched it from high to low reflectivity.

Recently we have demonstrated bistable operation in a NLBR structure. The nonlinear reflector with 30 nonlinear dielectric layer pairs has been designed for an operating wavelength of 885 nm. Although simulations predict a lower threshold for a NLBR with the same nonlinear length, the performance of the first structure grown is not yet comparable to the above NLFP: a threshold of 4 mW and a contrast ratio of 3:1 have been measured. The device remains in the ON-state for only about 1 μs, due to thermal effects.

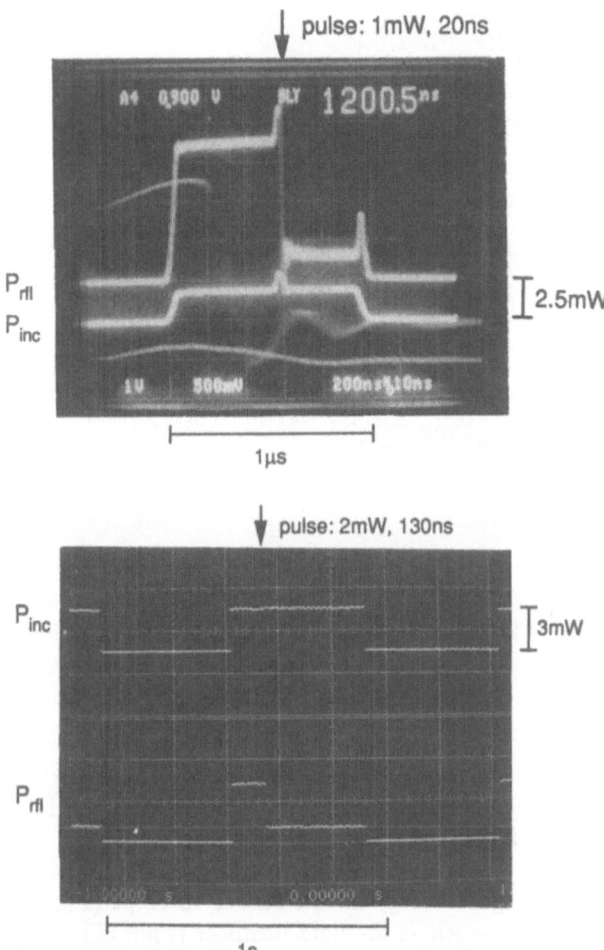

Fig. 2. Thermally stable latched operation of a NLFP: incident (P_{inc}) and reflected (P_{rfl}) power vs time. (a) The incident power is composed of a bias pulse (2.5 mW) of 1 μs duration and a short signal pulse (1 mW, 20 ns) which switches the reflected power from high to low. (b) The incident power is composed of a bias pulse (3 mW) of 0.5 s duration and a short signal pulse (2 mW, 130 ns, not visible on this time scale) which switches the output

3 Gain-bandwidth product (GBWP)

The GBWP is the product of the gain and the frequency of operation. The gain is the ratio of the hold beam intensity to the intensity increment for switching. The GBWP is a better figure of merit in these devices, because the switching energy remains constant under given operating conditions.

For pulses which are shorter than the carrier recombination time $\tau \approx 10$ ns, the switching energy is determined by the need to generate a critical carrier den-

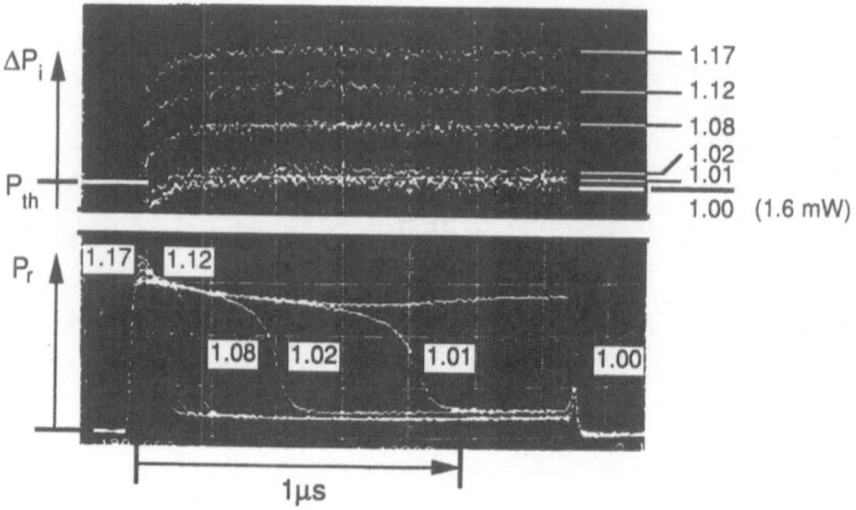

Fig. 3. Critical slowing down at switch-ON of a NLFP: response of the reflected signal (below) to an incident pulse (above) of 1.01, 1.02, 1.08, 1.12 and 1.17 times threshold (1.6 mW)

sity within the nonlinear spacer. For longer pulses this energy increases because of the losses due to carrier recombination. However, if the device is biassed close to the threshold, the above energy requirement remains valid for the switching increment (signal energy). This is a consequence of "critical slowing down" and allows the use of very small signal power at the expense of long transition times (up to μs in Fig. 3). The incident signal energy in this case is 10-15 pJ (Fig. 3). Since the signal is at the same wavelength as the hold beam, it is not efficiently coupled into the cavity. The signal energy actually absorbed is about $E_{sw} \approx 4$ pJ [4].

Figure 4 sketches the I/O characteristic of a NLFP used with the three-port (off-axis) addressing scheme shown in Fig. 1. A critically biassed hold beam keeps the NLFP close to threshold, and small switching signals are fanned-in on-resonance for maximum absorption. The gain-bandwidth product is

$$\text{GBWP} \approx \frac{P_{\text{hold}}(R_H - R_L)}{E_{\text{sw}}} . \tag{1}$$

In view of the high contrast and the low attenuation of the etalon, $(R_H - R_L)$ is about one. The threshold of 1 mW and the above switching energy requirement give a GBWP of approximately 100 MHz. The GBWP illustrates the trade-off between gain and cycle speed for the single device. According to this it appears possible to increase the gain of the device, by reducing the switching increment required and operating in the regime of critical slowing down. The minimal switching increment will then be limited by either the "forbidden region" of the device or fluctuations in the power and signal beams. An example of the forbidden region is the width of the bistable hysteresis loop which occurs when the device is detuned beyond the critical detuning. It is preferable to operate without bistability if high gain is required, for example close to the critical detuning, where the contrast and the threshold are optimal. The intensity fluctuations of the beams will become important when the width of the forbidden region is less than a few per cent of the total switch power.

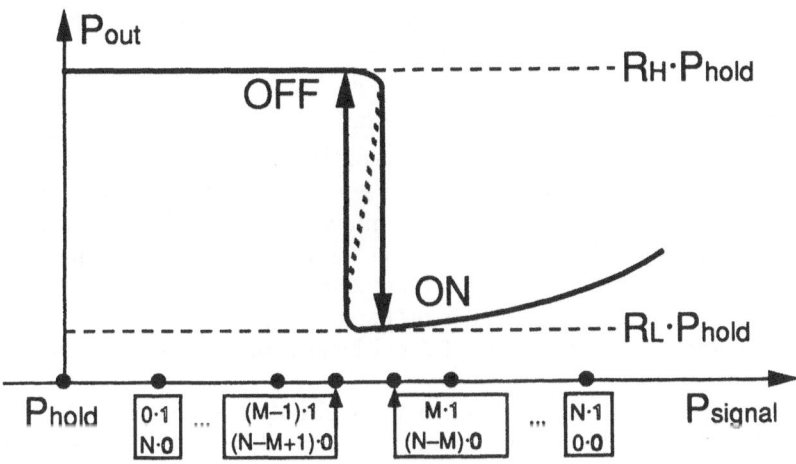

Fig. 4. I/O characteristic of a thresholding element based on a three-port addressed NLFP device

Once we can achieve gain in the device, we can consider fanning-out the output beam and switching similar devices with the fanned-out beams (cascadability). More powerful optical computing architectures, where the number N of beams fanned-in to the device equals the number of fan-out beams, become accessible. In the context of thresholding logic, the device is required to distinguish between $[M - 1] \cdot 1$, $[N - M + 1] \cdot 0$ and $M \cdot 1$, $[N - M] \cdot 0$, where M can be any number between 1 and N. In such circumstances, the gain must be reduced in order to accommodate the fan-in. Unfortunately, no practical estimation has yet been made on how much gain is required for a given fan-in in this type of system (because all-optical devices in array format with gain do not yet exist). However, if we require a fan-in of 3^2, we demand a gain of at least 9 in a system where fan-in equals fan-out. Therefore, we can operate at about 10 MHz, according to the GBWP, provided that neither the width of the forbidden region nor the fluctuation of the light source exceed 10%.

4 Time-bandwidth product (TB)

The common drawback of nonlinear devices, like the NLFP and the NLBR, which rely on a resonant optical structure is their wavelength sensitivity. In the context of threshold devices with fan-in, this sensitivity limits the number of signal beams which can be fanned-in to the device, and it also determines the wavelength tolerance for the hold beam which has to be detuned from the cavity resonance (critical biasing).

The TB is the product of the bandwidth and the time period over which the device integrates the signal. It quantifies the number of longitudinal modes (temporally incoherent signals) which can be fanned-in to the device. The bandwidth is determined by the width of the resonance $\Delta\eta_{cav}$, and the nonlinear material averages over variations which occur within the relaxation time τ:

$$\mathrm{TB} = \tau \Delta\nu_{cav} = \frac{\tau c}{F 2nL} , \tag{2}$$

where F is the finesse and nL the optical length of the Fabry-Perot resonator. For our NLFP sample with a 2 μm spacer, the finesse is 30 and the length $L \approx 2.8$ μm includes the dispersion in the dielectric mirrors. The resulting bandwidth $\Delta\nu_{cav} = 500$ GHz (FWHM ≈ 1 nm), together with a recombination time $\tau \approx 4$ ns imply a maximum time bandwidth TB ≈ 2000. It is unrealistic to operate several incoherent sources within this narrow bandwidth, but it might be feasible to use different (acousto-optically) modulated beams from the same stabilised laser.

The second consequence of the wavelength sensitivity is the need for a critically biased hold beam. The critical detuning of the hold beam (Fig. 5) is about one resonance width, which is 1 nm for the present NLFP. For larger detunings the threshold increases strongly. A threshold increase of about 1 mW/nm, measured close to the minimal threshold, indicates that wavelength stability of the hold beam to within 0.1 nm (40 GHz) is necessary to ensure threshold fluctuations of less than 10%.

Clearly, the marked wavelength dependence of the threshold is a severe limitation to the use of these devices with low-cost diode laser systems. In order to reduce the wavelength dependence, one would need to reduce the finesse, which would compromise the device performance in other areas, such as switching threshold. In fact, the trend is towards making shorter, higher finesse devices, i.e. steeper wavelength dependence. In this case, the spectral width $\Delta\nu_{cav}$ will decrease further, due to the constant "thickness" contribution from the mirrors. If we confine our attention to stabilised laser systems, we could envisage utilising the good wavelength stability to achieve channel separation in the wavelength domain. Temporal fan-in within the resonance linewidth, as described above, then becomes feasible. A possible application could be to integrate in the spectral domain the transmission of a hole burning memory material [5]. The actual fan-in would probably be constrained by source intensity fluctuations (as described in the previous section). In addition, the beam intensity used to read such materials non-destructively falls short of the power required to switch the etalons, at present.

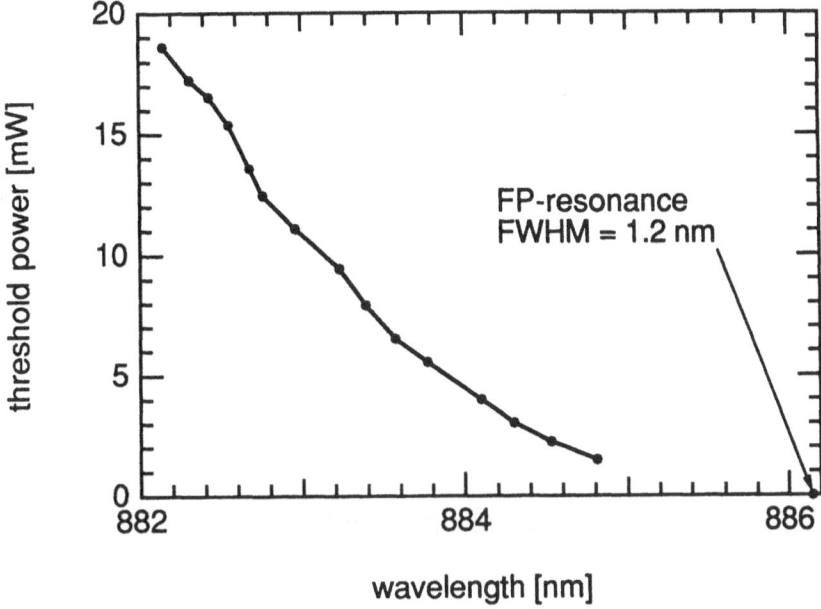

FP-resonance
FWHM = 1.2 nm

Fig. 5. Increase of the threshold power with detuning of the wavelength of the hold beam from resonance

5 Space-bandwidth product (SBWP)

The SBWP is a product of two components, the number of spatial modes (distinct spots or angles) which can be fanned-in to each device, and the number of devices which can be accommodated in an array. Both of these figures are restricted by the finesse of the optical cavity, and the desire to limit the power dissipation.

Devices with small transverse dimensions are desired to reduce the threshold power. However, the threshold of a NLFP does not scale proportionally to the spot area. Due to carrier diffusion, the threshold begins to saturate as the beam waist approaches the diffusion length of the spacer material $L_D = \sqrt{D\tau} \approx$ 2.5 μm. Due to diffraction, the threshold begins to increase when the beam waist is reduced below $w_0^{\min} = \sqrt{FL\lambda/n}/\pi$. This is because the effective finesse F is reduced and, therefore, the detuning of the hold beam must be increased to maintain critical biassing. The detuning of the hold beam increases the threshold in accord with Fig. 5. The two effects lead to a minimum threshold of about 1 mW at a spot size of 3 μm radius (Fig. 6).

The carrier diffusion length L_D defines a device area within which carriers excited by incident fields will integrate. On the other hand, $\pi \left(w_0^{\min}\right)^2$ gives the minimal spot area of the signal beams. Therefore, the space bandwidth product

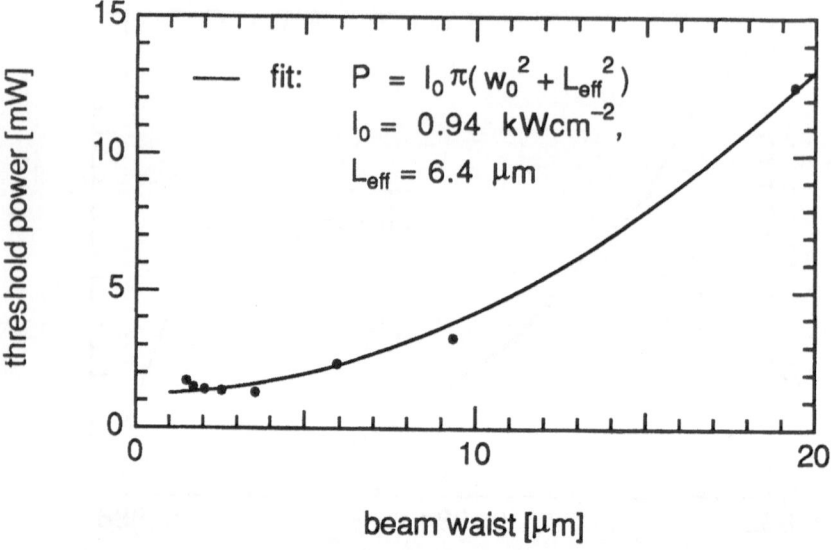

Fig. 6. Measured decrease of the minimum threshold of a NLFP as a function of the beam waist. The simple fit indicates a phenomenological limit radius for small spots of $L_{eff} \approx 6$ μm, which accounts for diffusion and diffraction effects, and a threshold intensity of 1 kW/cm^2 in the large spot limit

is the quotient of the diffusion spot area and the diffraction spot area.

$$\text{SBWP}_{\text{device}} = \frac{L_D^2}{\left(w_0^{\min}\right)^2} = \frac{n\pi^2 L_D^2}{F L \lambda} \ . \tag{3}$$

For our NLFP device the SBWP$_{\text{device}}$ is 2, and it decreases further with increasing finesse F, when the phase dispersion in the spacer becomes comparable with that in the mirrors ($F \geq 100$). Therefore, even if we account for the two possible orthogonal polarisations of each beam, the spatial fan-in of these resonant devices is very limited and hardly practicable. This trend towards spatial monomode devices is particularly true for pixellated devices, which promise reduced power requirements by the use of waveguiding to eliminate diffraction losses.

Alternatively, if one is interested in increasing the spatial fan-in to a single device, this entails larger integrating device volumes, higher switching energy, and lower speed. Two possible approaches are either to work in the regime of critical slowing down, or to develop material with longer diffusion length. In an array, the switch power, and therefore the thermal cooling requirement, can be kept constant if the density of devices is reduced accordingly.

The number of devices in an array is constrained by the power dissipation capability of the heat sink, the minimum separation of devices, and the growth tolerances which can be achieved. The energy absorbed to switch our device is about 4 pJ. The cycle time, on the arguments presented in the time-bandwidth section, can be variable. However, the power dissipated per device, which is determined by the hold beam power in the high gain regime, remains constant at 1 mW. The minimum separation of devices is about three times the diffusion length for a 1D array [6], and, presumably, about the same for a 2D array. This minimal figure has to be increased if the device is to be used in the critical slowing down regime. It is in accord with experiment [7], where less than 10% reduction of the switching threshold was observed for two spots separated by 15 μm.

On our wafer, we have measured a change of the resonance wavelength of less than 2 nm in a 10 mm diameter disk. Assuming, that threshold fluctuations of 10%, and therefore a wavelength deviation $\Delta\lambda_{tol} \approx 0.1$ nm can be tolerated, this would allow the use of 0.25 mm^2 of the wafer which yields a space bandwidth product of

$$\text{SBWP}_{\text{array}} = \frac{\text{tolerance area}}{\text{device separation}^2} = \left(\frac{\Delta\lambda_{tol}}{15\,\mu m}\right)^2 \left(\frac{\partial\lambda}{\partial x}\right)^{-2} \qquad (4)$$

for the array. Such an array, of ≈ 1000 elements, would be limited by power dissipation (1 W/mm^2).

6 Conclusions

We have fabricated nonlinear Fabry-Perot (NLFP) devices, which operate at a threshold power of 1 mW with a contrast ratio of about 10:1 and less than 1 dB attenuation. The absorbed switching energy is about 4 pJ in the NLFP. Due to critical slowing down this figure is independent of the transition time, which can vary from a few ns to μs, depending on the magnitude of the switching signal used. The threshold power, together with the dissipated switching energy, corresponds to an estimated GBWP of 100 MHz.

We have investigated theoretically the suitability of these GaAs etalons as integrating/threshold units. Due to the wavelength selectivity of the resonant structure, the device should be used with stabilised laser systems, until adequate frequency stability and tunability is achieved with semiconductor lasers. These devices have minimal spatial fan-in capacity. Nonetheless, there is a possibility for exploiting temporal fan-in in conjunction with very high density optical memory materials based on spectral hole burning.

The trend in NLFP devices is towards shorter, higher finesse cavities, to reduce the switching intensity. However, because of the limited optical confinement provided by the integrated dielectric mirrors, diffraction losses become increasingly important for shorter cavities. Calculations for devices based on AlGaAs material indicate that the switching energy tends to saturate at an asymptotic value of ≈ 0.1 pJ for finesse values larger than about $F_c \approx 100$. To eliminate diffraction losses and to improve the switching energy, it is necessary to use waveguiding. Such devices will accept a single spatial mode, and their temporal fan-in will decrease proportionally with finesse, beyond F_c.

At present, arrays containing about 1000 devices on an area of less than 1 mm^2 are conceivable from our tolerance studies.

We would like to thank Prof. R. Dändliker for his helpful comments, and acknowledge the support of F. Morier-Genoud and D. Martin who grew the crystals.

References

1. Gibbs, H. M.: Optical bistability: controlling light with light. Academic Press, Orlando 1985.
2. McCall, S. L., Gibbs, H. M., Venkatesan, T. N. C.: Optical transistor and bistability. J. Opt. Soc. Am. **65** (1975) 1184.
3. Acklin, B., Bagnoud, C., Dupertuis, M. A., Martin, D., Morier-Genoud F.: Low threshold optical bistability in bulk GaAs etalons. OSA Proceedings on Photonic Switching, H. Scott Hinton and Joseph W. Goodman, eds. (Optical Society of America, Washington, DC 1991), Vol. 8, 231–234.
4. Acklin, B., Bagnoud, C., Dupertuis, M. A., Martin, D., Morier-Genoud, F.: Thermally stable latched switching in AlGaAs-etalon. Appl. Phys. Lett. **60** (1992) 3099–3101.
5. Wild, U. P., Renn, A.: Spectral hole burning and holographic image storage. Mol. Cryst. Liq. Cryst. **183** (1990) 119–129.
6. Firth, W.J., Galbraith, I.: Diffusive transverse coupling of bistable elements-switching waves and crosstalk. IEEE J. Quant. Elect. **21** (1985) 1399–1403.
7. Jewell, J. L., Lee, Y. H., Duffy, J. F., Gossard, A. C., Wiegmann, W.: Parallel operation and crosstalk measurements in GaAs etalon optical logic devices. Appl. Phys. Lett. **48** (1986) 1342–1344.

Part II

Interconnect Technologies

Synthesis of Diffractive Optical Elements Using Electromagnetic Theory of Gratings

E. Noponen,[1] A. Vasara,[1] E. Byckling,[1]
J. Turunen,[2] J. M. Miller,[2] and M. R. Taghizadeh[2]

[1] Department of Technical Physics, Helsinki University of Technology,
SF-02150 Espoo, Finland
[2] Department of Physics, Heriot-Watt University,
Riccarton, Edinburgh EH14 4AS, UK

1 Introduction

Synthetic diffractive optics has proved to be an efficient method of realizing the optical interconnections needed in modern digital optical computers [1]. The design and operation of periodic holographic diffractive elements has been widely studied on the basis of scalar (Fraunhofer and Fresnel) diffraction theory [2]. These theories are applicable if the smallest transverse features of the grating structure at least an order of magnitude larger than the wavelength of light. However, the trend towards miniaturization and integration of optical circuits calls for more compact interconnects, which no longer obey the approximate scalar theories. In particular, Fourier optics has been shown to fail when the characteristic feature sizes are comparable to the wavelength of light [3,4]. The analysis and design of gratings operating in this region, which we call the resonance domain, must be based on the electromagnetic vector diffraction theory. Several numerical methods have been proposed and applied to solve the rigorous diffraction problem [5,6,7].

In the resonance domain, rigorous design methods can utilize the strong scattering and polarization effects absent in the scalar treatment, and diffractive component structures can be found with enhanced performance compared to their conventional counterparts [8,9]. On the other hand, new types of components can be synthesized, the operation of which depends on e.g. the state of polarization or the angle of incidence. In this paper we present designs for high-efficiency fan-out elements connecting one source to N detectors and star couplers connecting N sources to N detectors by using both transmission and reflection type gratings. We also introduce a polarization-sensitive switch which either passes through or exchanges two beams incident at different angles.

2 Electromagnetic Theory of Resonance Domain Holograms

We consider here periodic lamellar structures (Fig. 1) with a binary refractive-index modulated region, where $n(x, -H<z<0) = n_1$ or $n(x, -H<z<0) = n_2$.

Let the grating be illuminated by a linearly polarized unit-amplitude plane wave $U_I(x, z) = \exp[ikn_0(x \sin\theta + z \cos\theta)]$, where the symbol U refers to the y-component of the electric field E (TE polarization) or the magnetic field H (TM polarization). The reflected field $U_R(x, z \leq -H)$ and the transmitted field $U_T(x, z \geq 0)$ can be expressed as Rayleigh expansions [6], which are infinite linear combinations of plane waves corresponding to the diffraction orders,

$$U_R(x, z) = \sum_{m=-\infty}^{\infty} R_m \exp[i(\gamma_m x - r_m z)] \tag{1}$$

and

$$U_T(x, z) = \sum_{m=-\infty}^{\infty} T_m \exp[i(\gamma_m x + t_m z)] . \tag{2}$$

Here R_m and T_m are the amplitudes of the reflected and transmitted diffraction orders, respectively, $\gamma_m = kn_0 \sin\theta + 2\pi m/d$, $r_m = [(kn_0)^2 - \gamma_m^2]^{1/2}$, and $t_m = [(kn_3)^2 - \gamma_m^2]^{1/2}$.

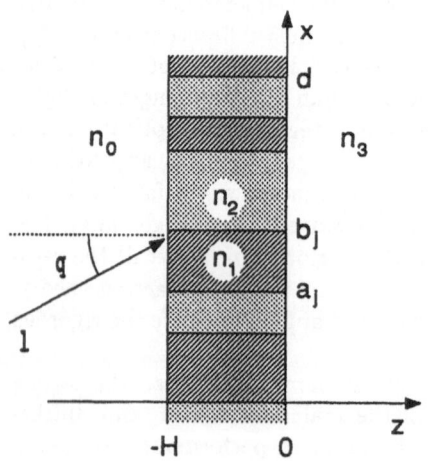

Fig. 1. Diffraction of a plane wave by a periodic lamellar diffractive element

We are interested primarily in the diffraction orders contributing to the far field, i.e. the propagating orders $\mathcal{R} = \{m \mid r_m \text{ real}\}$ and $\mathcal{T} = \{m \mid t_m \text{ real}\}$. Their relative intensities (defined perpendicular to the xy-plane) are given by $P_{R,m} = (r_m/r_0)|R_m|^2$ and $P_{T,m} = C(t_m/r_0)|T_m|^2$, where $C = 1$ for TE polarization and $C = (n_0/n_3)^2$ for TM polarization. Although the waves with imaginary values of r_m and t_m, known as evanescent waves, do not carry energy in the z-direction, they contribute significantly to the field inside the grating and thus modify the amplitudes of the propagating orders.

Within the grating, the analysis is applied directly to the real physical surface profile, not e.g. to the refractive index modulation or the phase function as is the case in Fourier optics. In order to find the field in the modulated region, it is represented as a suitable expansion, which, when inserted in the wave equations, yields an eigenvalue problem [4,8]. Applying the continuity conditions at the boundaries between the modulated and the unmodulated regions then gives a system of linear equations, which can finally be solved for the amplitudes R_m and T_m. The method and its numerical implementation are described in detail for transmission gratings in Ref. [4,8], and for reflection gratings in Ref. [10].

The method for binary gratings can be generalized [4,8] for multi-level gratings, or kinoforms [11]. Gratings with a sufficient number of levels can be used to approximate continuous surface relief profiles with a good accuracy.

3 Design Method

We concentrate on binary intensity signals, i.e. the desired intensity distribution is described by a goal power spectrum $\mathcal{P} = \{\hat{P}_m \mid \hat{P}_m \in \{0, 1\}, m \in \mathcal{W}\}$, where $\mathcal{W}$ is the image window containing the specified diffraction pattern ($\mathcal{W} \subseteq \mathcal{R} \cup \mathcal{T}$). The number of equally intense signal orders within the image window is denoted by $\mathcal{N} = \sum_{m \in \mathcal{W}} \hat{P}_m$.

We define the diffraction efficiency

$$\eta = \sum_{m \in \mathcal{W}} \hat{P}_m P_m \tag{3}$$

as the total power diffracted into the signal orders relative to the power of the incident beam. Another quantity characterizing the fidelity of the element is the reconstruction error

$$E = \max_{m \in \mathcal{W}} \hat{P}_m |1 - \mathcal{N} P_m / \eta| \ , \tag{4}$$

which gives the maximum relative deviation from the goal $\mathcal{P}$. In the design procedure we wish to find a grating structure that maximizes the diffraction efficiency η, and simultaneously minimizes the reconstruction error E. Instead of applying these quantities directly, it is advantageous to minimize the quadratic merit function

$$Mf = \sum_{m \in \mathcal{W}} (P_m - \hat{P}_m \hat{\eta} / \mathcal{N})^2 \ , \tag{5}$$

where $\hat{\eta}$ denotes the goal efficiency. If the component should be designed for several different inputs waves, the total merit function can be derived by summing up the merit functions for each individual input.

We use parametric optimization to minimize the merit function. The free parameters available for optimization include the relief depth H, the grating period d, and the transition points $\{a_j, b_j\}$. Due to the computational complexity of the rigorous diffraction analysis, sophisticated optimization algorithms like simulated annealing [12,13] exceedingly time-consuming. We have chosen a conventional gradient algorithm, where each grating parameter is perturbed slightly to evaluate the gradient of the merit function in the parameter space. Next the parameter set is translated in the direction opposite to the gradient, until no further decrease of the merit function is achieved. Then a new cycle is started by evaluating a new gradient. Since the optimization problem is highly nonlinear by nature, a large number of local minima generally exists in the parameter space, in which the gradient method may stagnate. Consequently, to find an acceptable solution, the optimization must be repeated a number of times with randomly selected initial values for the parameters.

4 Optimization Results

In the resonance domain, we usually aim to control the diffracted field in a full half-space. In other words, unlike in the conventional scalar designs, we restrict the period length such that all propagating transmitted or reflected orders are considered. This is actually the ultimate level of miniaturization achievable, since gratings with smaller period lengths do not generate a sufficient number of propagating diffraction orders, if the wavelength is kept constant.

In dielectric transmission gratings we choose $W = T$. For the examples presented in this section we use the refractive indices $n_0 = n_1 = 1.5$ and $n_2 = n_3 = 1.0$. For metallic reflection gratings we choose $W = R$. In this case the refractive indices are $n_0 = n_1 = 1.0$ and $n_2 = n_3 \rightarrow i\infty$. The latter values correspond to a perfectly conducting medium, whose reflectivity is 100%. Thus, for an ideal perfectly conducting grating, a 100% diffraction efficiency is achieved automatically, since no power is lost in unwanted diffraction orders. Real metals have a finite conductivity, and a portion of the incident power is absorbed. The diffraction efficiency is nevertheless equal to the reflectivity of the medium. For a comparison of perfect conductors and real metals in the present situation, see Ref. [8].

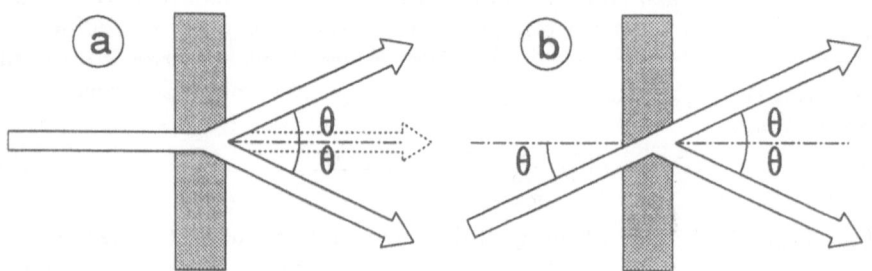

Fig. 2. Geometries for gratings with fan-out of two. (a) Normal incidence and (b) Bragg incidence

We consider first the design of a beamsplitter generating two equal-intensity orders. Two geometries depicted in Fig. 2 can be used. For normal incidence (Fig. 2a), we want the three central orders to propagate, of which the zero order should vanish, i.e. we use the goal spectrum $\mathcal{P} = \{1, 0, 1\}$ in the image window

$\mathcal{W} = \{-1, 0, 1\}$. The correct number of propagating orders results provided a period length in the range $\lambda/n_3 < d < 2\lambda/n_3$ is chosen. Assuming a simple lamellar grating structure, we found the solution

$$\{d/\lambda, H/\lambda, c/\lambda\} = \{1.769, 1.396, 0.322\} \,, \tag{6}$$

where $c = b_1 - a_1$ is the width of the groove. The transmitted spectrum is then $\{47.3\%, 0.2\%, 47.3\%\}$, which implies that 5.2% of the incident power is reflected.

In the second case illustrated in Fig. 2b, the light is incident at the first Bragg angle $\sin\theta = \lambda/2n_0 d$. We now have two propagating orders, $\mathcal{W} = \{-1, 0\}$, with the goal $\mathcal{P} = \{1, 1\}$, and require that $\lambda/2 < d < 3\lambda/2$. The solution

$$\{d/\lambda, H/\lambda, c/\lambda\} = \{1.000, 0.784, 0.468\} \tag{7}$$

gives the efficiencies $\{48.8\%, 48.5\%\}$. The reflected power is 2.7%, which is actually less than the specular reflection by a boundary between indices $n = 1.5$ and $n = 1$.

The design principles for fan-out of two can easily be generalized for higher fan-out. Odd-numbered fan-out to $N = 2M + 1$ equal-intensity beams is obtained by choosing $\hat{P}_m = 1$ for $\mathcal{W} = \{-M, ..., M\}$ and $M\lambda < d < (M+1)\lambda$. Using Bragg incidence, we can get even-numbered fan-out if we choose $\hat{P}_m = 1$ for $\mathcal{W} = \{-M, ..., M-1\}$ and $(M - 1/2)\lambda < d < (M + 1/2)\lambda$. Period values near the upper end of the range are preferred in order to minimize the highest fan-out angles. In general, more than one groove is needed.

For odd fan-out we use a symmetric grating profile, which is required to establish a symmetrical diffraction pattern ($P_m = P_{-m}$) in the rigorous treatment. For example, a dielectric transmission grating with period $d = 4.86\lambda$, depth $H = 1.26\lambda$, and three grooves defined by the boundaries

$$\{(a_j/\lambda, b_j/\lambda)\} = \{(0.00, 0.69), (1.41, 1.87), (2.59, 3.28)\} \,, \tag{8}$$

generates nine beams with $\eta = 93.3\%$ and $E = 3.0\%$. The reconstruction error E can be made arbitrarily small by increasing the optimization accuracy. A perfectly conducting reflection grating with $d = 4.47\lambda$, $H = 0.30\lambda$, and the three grooves

$$\{(a_j/\lambda, b_j/\lambda)\} = \{(0.00, 0.55), (1.07, 1.63), (2.41, 3.69)\} \,, \tag{9}$$

generates eight beams with $\eta = 1$ and $E = 3.5\%$ at an angle of incidence $\theta = 6.42°$. Here, for oblique incidence, no special rigorously valid symmetry exists.

As an example of the synthesis involving several input waves, we consider the so called star coupler [14]. This is a diffractive element that connects each of the N mutually uncorrelated sources A, B, C, etc. to all of the N receivers A', B', C', etc. Here again, we use a symmetric lamellar profile for both odd and even fan-out star couplers (Fig. 3). This simplifies the optimization with the implication that the diffraction patterns at angles of incidence θ and $-\theta$ are rigorously mirror images, and thereby equivalent for our purpose.

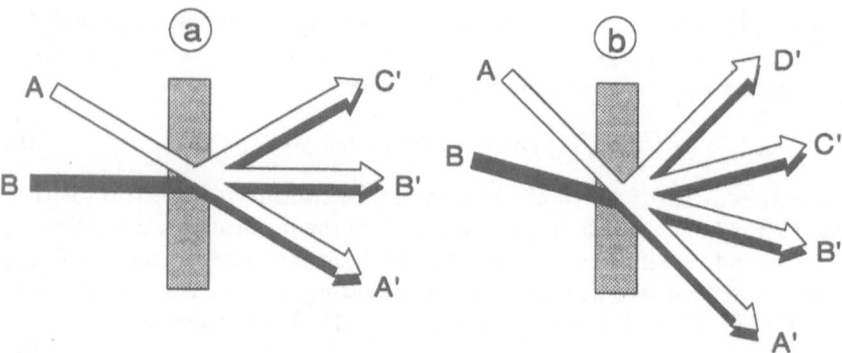

Fig. 3. The operation of a star coupler connecting N sources to N detectors. (a) Odd fan-out. (b) Even fan-out at Bragg incidence (the omitted inputs are mirror images of the illustrated inputs)

The design of star couplers is analogous with the single-input case, with the extension that the response must be evaluated separately at each non-negative angle of incidence. For dielectric and perfectly conducting 3-to-3 and 4-to-4 star couplers, we found the solutions presented in Table 1.

Table 1. Designs for dielectric and perfectly conducting 3-to-3 and 4-to-4 star couplers

	Dielectric		Perfect conductor	
	3-to-3	4-to-4	3-to-3	4-to-4
$a_1/\lambda, b_1/\lambda$	0.104, 0.397	0.556, 0.676	0.000, 1.751	0.272, 0.554
$a_2/\lambda, b_2/\lambda$	1.376, 1.669	1.037, 1.457	–	0.608, 1.743
$a_3/\lambda, b_3/\lambda$	–	1.836, 1.956	–	1.797, 2.079
d/λ	1.773	2.512	1.946	2.351
H/λ	0.957	1.192	0.335	0.430
η_0	96.9%	88.2%	100%	100%
η_1	96.0%	88.9%	100%	100%
E_0	0.8%	1.0%	0.2%	0.9%
E_1	3.0%	1.3%	0.3%	0.9%

The polarization-sensitive node illustrated in Fig. 4 is a passive device that passes through two TM-polarized, uncorrelated input beams incident at angles $\pm\theta$, and swaps the corresponding TE-polarized beams. By controlling the polarization state of the input beams using e.g. twisted nematic liquid crystal cells, this component can be used like more conventional active nodes based on acousto-optic or electro-optic Bragg diffraction. With a symmetric lamellar profile, a component acting as described above is obtained if we choose the image window $\mathcal{W} = \{-1, 0\}$, and the goal distributions $\mathcal{P} = \{0, 1\}$ and $\mathcal{P} = \{1, 0\}$ for

TM and TE-polarized input waves, respectively. The solution

$$\{d/\lambda, H/\lambda, c/\lambda\} = \{0.724, 1.720, 0.140\} \tag{10}$$

at $\theta = 27.4°$ gives the efficiencies $\{0.02\%, 98.5\%\}$ and $\{95.9\%, 0.05\%\}$. On the other hand, a reflection-type node with parameters

$$\{d/\lambda, H/\lambda, c/\lambda\} = \{1.368, 0.391, 1.061\} \tag{11}$$

and $\theta = 21.45°$ functions almost ideally.

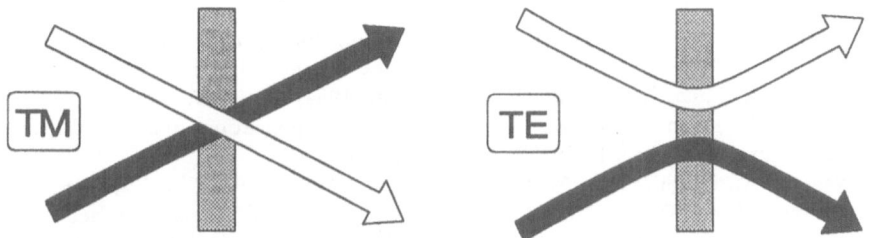

Fig. 4. The operation of a polarization-sensitive node that either passes through or swaps two mutually uncorrelated input beams incident at $\pm\theta$, depending on their state of polarization

5 Experimental Results

The sensitivity of the resonance domain designs to the fabrication errors can be estimated by introducing random perturbations to the optimized parameters. The deviations from the optimized groove structure increase the reconstruction error. Numerical simulations show that the inaccuracy of the transition points should not exceed $\lambda/20$ for satisfactory performance. Therefore, the present-day microlithographic fabrication methods restrict the demonstrations to infrared wavelengths.

We realized two reflection-type resonance domain fan-out gratings for $\lambda = 10.6\mu$m. The fabrication method [9] guarantees almost perfect reflectivity at this wavelength. The reconstruction errors of the fabricated beamsplitters with fan-out to six and seven were, 11% and 15%, compared to the design values 2.3% and 3.5agree roughly with the numerical analysis of fabrication errors, assuming that the transverse fabrication accuracy is 0.2μm. For normal incidence (and odd fan-out), the zeroth order is particularly sensitive to the fabrication errors of the groove depth. In our experiment, the relief depth error for the one to seven fan-out grating was about 1%, and the zeroth order was consequently weaker than the others. Excluding the zeroth order, a reconstruction error of 8% was measured for the remaining orders.

References

1. Streibl, N., Brenner, K.-H., Huang, A., Jahns, J., Jewell, J., Lohmann, A. W., Miller, D. A. B., Murdocca, M., Prise, M. E., and Sizer, T.: Digital optics. Proc. IEEE **77** (1989) 1954–1969.
2. Goodman, J. W.: Introduction to Fourier optics. McGraw-Hill, San Francisco (1968).
3. Vasara, A., Noponen, E., Turunen, J., Miller, J. M. and Taghizadeh, M. R.: Rigorous diffraction analysis of Dammann gratings. Opt. Commun. **81** (1991) 337–342.
4. Vasara, A., Noponen, E., Turunen, J., Miller, J. M., Taghizadeh, M. R., and Tuovinen, J.: Rigorous diffraction theory of binary optical interconnects. Holographic Optics III: Principles and Applications, Morris, G. M., ed. Proc. SPIE **1507** (1991) 224–238.
5. Petit, R., ed.: Electromagnetic theory of gratings. Springer, Berlin (1980).
6. Maystre, D.: Rigorous vector theories of diffraction gratings. Progress in Optics XXI Wolf, E., ed. North-Holland, Amsterdam (1984), pp. 1–67.
7. Gaylord, T. K. and Moharam, M. G.: Analysis and applications of optical diffraction by gratings. Proc. IEEE **73** (1985) 894–937.
8. Vasara, A., Taghizadeh, M. R., Turunen, J., Westerholm, J., Noponen, E., Ichikawa, H., Miller, J. M., Jaakkola, T., and Kuisma, S.: Binary surface-relief gratings for array illumination in digital optics. Appl. Opt. **31** (1992) 3320–3336.
9. Noponen, E., Vasara, A., Turunen, J., Miller, J. M., and Taghizadeh, M. R.: Synthetic diffractive optics in the resonance domain. J. Opt. Soc. Am. A **9** (1992) 1206–1213.
10. Miller, J. M., Turunen, J., Taghizadeh, M. R., Vasara, A., and Noponen, E.: Rigorous modal theory for perfectly conducting lamellar gratings. Third International Conference on Holographic Systems, Components, and Applications. Conf. Proc. **342**. Institute of Electrical Engineers, London (1991) 99–102.
11. Lesem, L. B., Hirsch, P. M., and Jordan J. A.: The kinoform: a new wavefront reconstruction device. IBM J. Res. Dev. **13** (1969) 150–155.
12. Kirkpatrick, S., Gelatt, C. D., and Vecchi, M. P.: Optimization by simulated annealing. Science **220** (1983) 671–680.
13. Turunen, J., Vasara, A., Westerholm, J., Jin, G., and Salin, A.: Optimisation and fabrication of grating beamsplitters. J. Phys. D: Appl. Phys. **21** (1988) S102–S105.
14. Killat, U., Rabe, G., and Rave, W.: Binary phase gratings for star couplers with high splitting ratio. Fib. Int. Opt. **4** (1982) 159–167.

The Design of Quasi-Periodic Fourier Plane Array Generators

A. G. Kirk, A. K. Powell, and T. J. Hall

Department of Physics, King's College, London, The Strand, London WC2R 2LS, UK

A technique for the reduction of the peak intensity of noise which arises from Fourier plane array generators is introduced. This is achieved by designing the array generator as a quasi-periodic computer generated hologram. In this way an even distribution of noise is obtained while the required diffraction orders are left as tightly sampled spots. Using the method of generalised error diffusion several different quasi-periodic 4×4 array generators are designed. A 14.5 dB reduction in peak noise is obtained for a hologram with 8×8 quasi-periods. Diffraction efficiency and array uniformity are not adversely affected by this technique.

1 Introduction

A Fourier plane array generator (FPAG) is a periodic diffractive element which when illuminated with a plane beam produces an array of equal intensity diffraction orders in the far field [1]. This is shown in Fig. 1. A single period of this element is represented by the function h which has a period width L so that the array generator f can be described by

$$f(x,y) = h(x,y) \otimes \mathrm{comb}\left(\frac{x}{L}, \frac{y}{L}\right) \tag{1}$$

where the comb function represents the (in principle) infinite periodicity of the element. The diffracted amplitude is thus $F = \mathcal{F}(f)$ where $\mathcal{F}$ represents the Fourier transform. Therefore the diffracted amplitude is

$$F(u,v) = H(u,v)\mathrm{comb}(Lu, Lv) \tag{2}$$

where $H = \mathcal{F}(h)$ represents the diffraction envelope from a single period of the element which is sampled by the comb function. The diffraction orders thus appear as sharply defined spots. These devices are used within optoelectronic processing systems for the generation of beamlet arrays which can act as free-space optical data buses [2,3]. In addition they can act as parallel data routing devices when used in a multiple imaging system [4,5,6].

For large arrays the best uniformity is achieved by the use of computer generated holograms (CGHs) [7]. Electron beam microlithography may be used to fabricate binary CGHs with very high space-bandwidth products with very

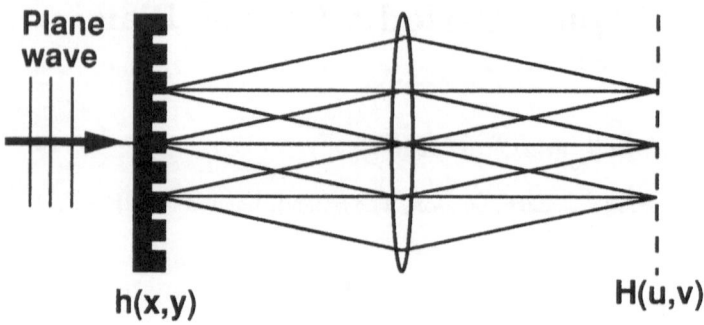

Fig. 1. The operation of a Fourier plane array generator

good accuracy [8,9]. It is however much more difficult to obtain multilevel holograms with the same degree of accuracy. Many different techniques have been investigated for the design of binary CGHs. These include inverse optimization techniques [9,10,11,12], error diffusion [13,14,15], phase retrieval techniques [16] and neural network algorithms [17]. Regardless of the technique used any design algorithm for binary holograms inevitably results in the appearance of errors in the reconstruction plane in addition to the desired amplitude. The choice of algorithm will, however, have an important effect upon the location and the relative magnitude of these errors.

Within optoelectronic processing systems there are 4 important figures of merit for the effectiveness of a CGH as an array generator. These are the efficiency η which measures the total percentage of the illuminating intensity which is directed into the fanout spots, the standard deviation σ of the intensities in the fanout spots, the signal to noise ratio S and a parameter which we call S_m which is a measure of the ratio between the mean signal intensity and the maximum noise intensity. For most applications it is only important to measure S and S_m in the immediate vicinity of the fanout spot array as higher diffraction orders are lost from the system or do not fall upon photodetectors. The measure S_m is more important than the simple SNR measure S because for many systems it is the peak noise intensity which limits the performance of the system. It is thus important that S_m should be as large as possible and therefore any noise around the fanout spot array should be distributed as evenly as possible.

The original binary FPAG was the Dammann grating [4]. Because this type of element is designed as a one dimensional grating which is then crossed to produce a separable two-dimensional structure any noise orders are also separable. This results in the appearance of characteristic lines of noise around the edge of the fanout array. Because this noise is highly structured and is confined to a small number of orders the measure S_m is very small for a Dammann grating. This is greatly improved when two-dimensional design techniques [8-16] are used because a non-separable hologram allows more freedom in the possible locations of the noise orders. The noise is shared among more orders and so the peak noise value is reduced and hence S_m is increased. In this paper we investigate a technique by which the maximum noise intensity by be reduced still further.

2 Method

2.1 Quasi-periodicity

Due to the periodic replication of the hologram the diffracted amplitude described in equation 2 is sampled on a square grid which has dimensions determined by the hologram size. This results in sharply defined diffraction orders. Any noise will also be sampled on this grid. If these noise orders were 'smeared out' while the signal orders remained tightly sampled then the signal to peak noise ratio S_m would be increased. This is shown in Fig. 2. This places two conflicting demands upon the the hologram. In order to obtain sharply defined diffraction spots for the fanout array then the hologram must be periodically replicated. In contrast a smooth noise distribution of noise is obtained from an aperiodic function where no sampling occurs. There is thus a requirement for a quasi-periodic hologram. Such a hologram is composed of several almost identical replications which produce the required pixellated array but which have sufficient differences to give a smooth distribution of noise.

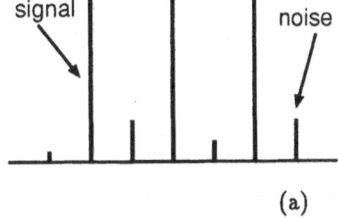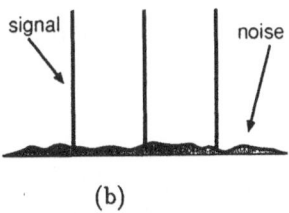

(a)   (b)

Fig. 2. Representation of signal and noise for a periodic CGH (a) and the desired effect of a more even distribution of noise (b)

At first sight it may be thought that this could be achieved by designing a single period of the array generator and then operate some 'quasi-periodicity' algorithm on it which produces several almost identical copies. It is however difficult to envisage how such an algorithm would operate in practise. A more straight-forward approach is to design the hologram as a single function which is as large as the equivalent periodically replicated hologram. For example if a hologram with a period size of L is to be designed and which has $R \times R$ replications then the equivalent quasi-periodic hologram would have a single period of size $RL \times RL$. In order to maintain the relationship between the separation of the fanout spots and the size of the hologram the fanout spots in the quasi-periodic hologram must be separated by R times more than for the original $L \times L$ hologram.

2.2 Generalised error diffusion

It is apparent that in order to obtain a quasi-periodic hologram it is necessary to use a design algorithm which scales well to large problems. This makes the error diffusion algorithm attractive. Error diffusion was originally introduced by Hauck and Bryngdahl [13] as a half-toning algorithm for hologram binarization and has since been extended to the design of multi-phase level holograms [14]. Here we use the recently introduced generalised error diffusion algorithm (GED) [15]. The basic principle of this algorithm is as follows:

A binary hologram h is to be designed which produces a diffracted amplitude H. The required diffracted amplitude F would be obtained from some (non-binary) real amplitude f. The total binarisation error E is thus

$$E = \sum_{uv} P_{uv} |G_{uv}|^2 \tag{3}$$

where $G = F - H$ and P is a filter function which acts to give a spatial weighting to the relative importance of errors in the reconstruction plane. If this equation is transformed into the Fourier plane then after some manipulation [15] we can obtain an expression for the change in error ΔE which would be obtained if a hologram pixel at location $h(r, s)$ was flipped to its opposite state. This is

$$\Delta E = 4h(r, s)[F(r, s) + (p', g)] \tag{4}$$

where $g = \mathcal{F}(G)$ and p' is obtained by taking the Fourier transform of the filter P and then removing the central component. If p' is limited in extent then the choice of whether to flip a pixel depends only upon the state of the neighbouring pixels and upon the local value of f. The function p' thus acts as an error diffusion mask and the location of the diffused errors is determined exactly by P.

For on-axis designs such as the array generator it is useful to specify P to be a sinc^2 function which has its first zeros at the edges of the zero-order reconstruction. Any errors will thus be diffused towards the edges of the reconstruction. These are then substantially reduced because a binary hologram which is made up of square pixels has a sinc^2 diffraction envelope which suppresses the diffused errors.

2.3 Design of quasi-periodic holograms

The technique was demonstrated by first designing a normal binary fanout holo-gram by GED. This had 64 × 64 pixels and was designed to diffract light with high efficiency into the central 4 × 4 odd orders. This was then periodically replicated 2 × 2 times in order to obtain a hologram with 128 × 128 pixels. This is shown in Fig. 3(a). Figure 3(b) shows the simulated diffraction pattern obtained from this hologram. Little noise can be seen on this plot but when the intensity was thresholded at $I_{max}/200$ where I_{max} was the maximum intensity in the fanout array then noise was observed between the pixels and is shown in Fig. 4. The 2 × 2 replication has resulted in a 2 × 2 sampling so that the isolated noise orders may clearly be seen between the fanout orders. The GED design algorithm was then used to design a quasi-periodic hologram with 128 × 128 pixels which would produce the same 4 × 4 fanout array. For this holo-gram however the fanout pattern was specified on a 4 × 4 lattice rather than the 2 × 2 lattice of the original odd-order hologram. This resulted in the hologram with 2 × 2 quasi-periods which is shown in Fig. 5(a). The simulated diffraction pattern from this hologram is shown in Fig. 5(b). This intensity was thresholded at the same level of $I_{max}/200$ and the thresholded intensity pattern is shown in Fig. 6(a). It can be seen that the noise is greatly reduced and that in order to make it more visible it was necessary to threshold the intensity at $I_{max}/1000$. This is shown in Fig. 6(b). It can be seen from this that the noise has been smeared out over all the available pixels rather than being sampled as was the case for the periodic hologram.

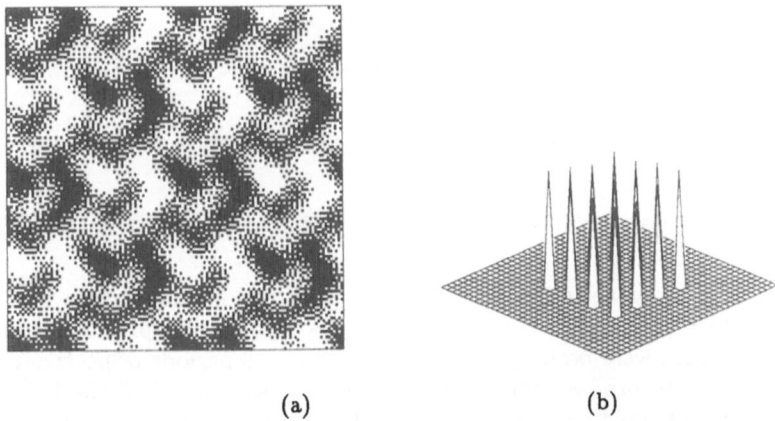

(a)   (b)

Fig. 3. 2 × 2 replications of a 64 × 64 pixel binary hologram (a) which produces the 4 × 4 odd order fanout (b)

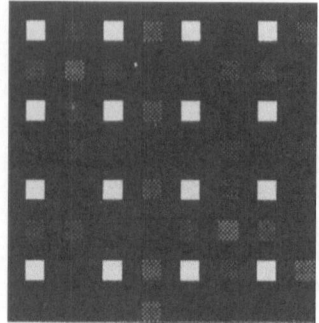

Fig. 4. Fanout array generated by periodic hologram, thresholded at $I_{max}/200$. The sampled noise can clearly be seen

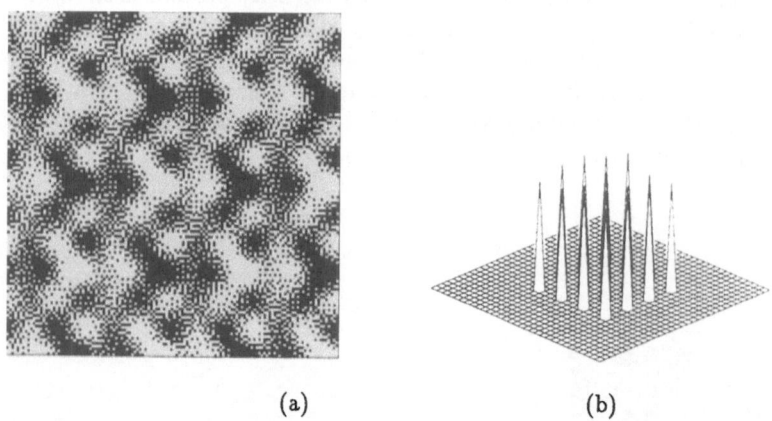

(a)   (b)

Fig. 5. A 128 × 128 pixel quasi-periodic binary hologram (a) which produces the 4 × 4 odd order fanout (b)

In order to obtain quantitative results several quasi-periodic holograms were designed. In addition to the 128 × 128 pixel hologram a 256 × 256 pixel and 512 × 512 pixel hologram were designed with 4 × 4 and 8 × 8 periods respectively. Both were designed to produce the same 4 × 4 fanout as before. All the holograms were then replicated appropriately so that they had a total of 512 × 512 pixels. The diffraction efficiency η and standard deviation σ were calculated, together with the values for the total signal to noise ratio S and the ratio of the mean signal intensity to maximum noise intensity S_m. Both S and S_m were calculated only within the area of the fanout array.

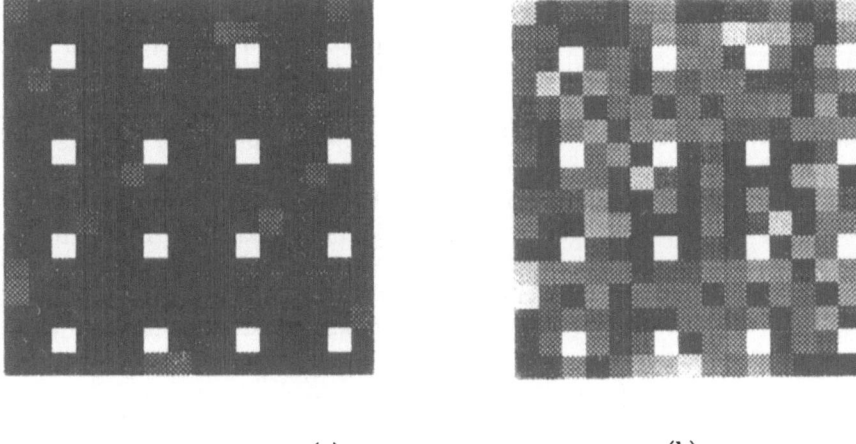

(a)   (b)

Fig. 6. Fanout array generated by periodic hologram, thresholded at $I_{max}/200$ (a) and at $I_{max}/1000$ (b). The noise is more evenly distributed than for the periodic hologram

3 Results

The performance of the various holograms is shown in the table below.

Size of hologram	No. of quasi-periods	$\eta(\%)$	$\sigma(\%)$	S	$S_m(\times 10^3)$
64×64	1×1	74.1	4.4	480	0.53
128×128	2×2	72.7	4.9	505	1.30
256×256	4×4	70.6	1.7	378	4.72
512×512	8×8	72.8	3.3	507	14.75

The efficiency values shown in this table include power lost by diffraction into all higher orders due to the finite size of the square binary pixels. Hence $\eta = \eta_{DFT} \times 0.81$ where η_{DFT} is the diffraction efficiency calculated by from the discrete Fourier transform only. It can be seen from these results that maximum noise was reduced to less than 7×10^{-5} times the value of the mean signal strength for the 512×512 pixel hologram. This is equivalent to an almost 30-fold increase in the value of S_m when compared to the single period 64×64 pixel design. From this table it can also be seen that the reduction in maximum intensity has not been gained at the expense of diffraction efficiency, accuracy or the total signal to noise ratio.

4 Discussion

The reduction in maximum noise intensity is achieved by increasing the number of locations available for occupation by the noise. It might therefore be expected that the degree of noise reduction would be proportional to the increase in the space available to it. For example a hologram with 2×2 quasi-periods has 2^2 times more locations at which noise can be placed when compared with a normal periodic hologram and so a four-fold reduction in maximum noise intensity should be expected. Fig. 7 shows the relationship between S_m and the number of hologram pixels $N \times N$. The dotted line shows the predicted increase while the solid line shows the observed increase. It can be seen that S_m increases only half as rapidly as expected and so indicates that the noise is not smeared out to the fullest possible extent. This is due to random fluctuations in the distribution of noise. If these fluctuations could be reduced it would be possible to obtain a two-fold improvement in the performance of this technique. It is however, difficult to envisage a simple method by which this could be achieved with the current design algorithms.

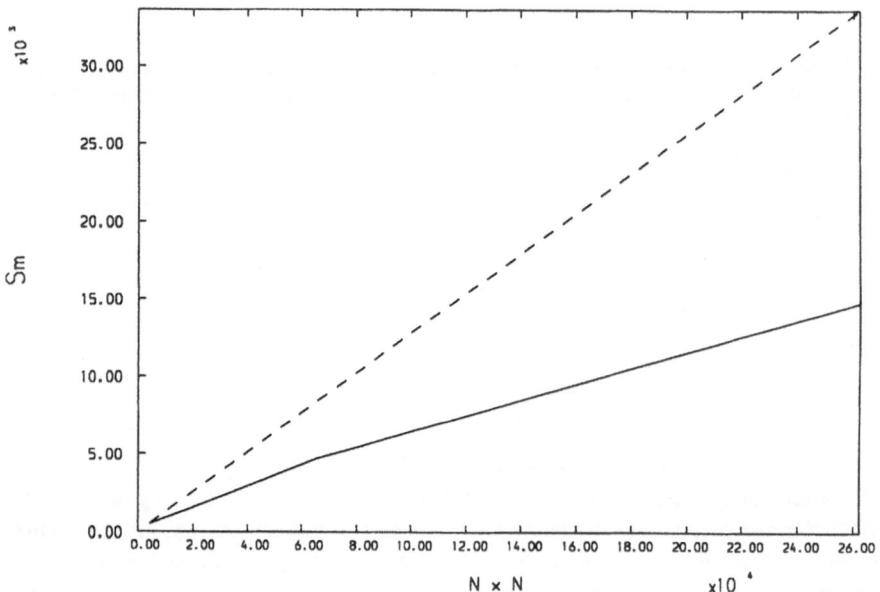

Fig. 7. Ratio of mean fanout intensity to maximum noise in array S_m as a function of the total number of hologram pixels $N \times N$. The solid line shows the experimental results while the dashed line shows the result which would be observed if the noise was evenly distributed

Although this technique may not achieve the best possible results it remains an effective way of dramatically reducing the maximum noise intensity in the fanout pattern. This improvement is gained at the expense of the increased time required to design a hologram. The 512×512 quasi-periodic hologram required approximately 30 minutes of CPU time to design. However when compared with the time and expense required for the fabrication of high performance CGHs then this is not an important factor. Consequently if a hologram is to be plotted as a periodically replicated structure then it will be profitable to design it using this quasi-periodic technique.

The holograms which have been described in this paper have not been fabricated and the results shown are simulation only. There is however some reason to argue that quasi-periodicity may give the such holograms a greater resistance to fabrication errors than traditional periodic holograms. In particular for holograms which are fabricated by high resolution electron beam microlithography the accuracy which is achieved is often limited by the proximity effect [18]. This results in regions of the hologram being over-exposed due to the proximity of neighbouring exposed regions. As this is a function of the local hologram pattern then for periodic holograms this effect is identical for every hologram period. The same error is thus introduced to each period and so the effect of these errors will add constructively. For the quasi-periodic hologram the proximity effect in each quasi-period will be different. The errors will therefore differ and so will not add together in this way.

5 Conclusions

Within many optoelectronic processing systems it is often the maximum noise intensity generated by a CGH which limits the performance of the system rather than the overall SNR. A method has been introduced which significantly reduces this by designing a CGH as a quasi-periodic structure. This allows an even distribution of noise while leaving the signal orders as sharply defined points. Several quasi-periodic holograms were designed by the method of generalised error diffusion and a 28 times reduction in the maximum noise intensity was observed. The diffraction efficiency and standard deviation between fanout orders were not affected by the use of this technique. In addition the total SNR remained approximately constant, thus proving that the observed reduction in maximum noise intensity was due to the action of the quasi-periodic design technique.

6 Acknowledgements

Andrew Kirk would like to thank British Aerospace plc for sponsorship which supports his research.

References

1. Streibl, N.: Beam shaping with optical array generators. J. Mod. Opt. **36** (1989) 1559.
2. McCormick, F. B., Tooley, F. A. P., Cloonan, T. J., Brubaker, J. L., Lentine, A. L.: S-SEED-based photonic switching network demonstration. In Photonic Switching 1991, Technical Digest Series, OSA, 44.
3. Kirk, A. G., Imam, H., Bird, K., Hall, T. J.: The design and fabrication of computer generated holographic fanout elements for a matrix-matrix interconection scheme. Proc. SPIE **1574** (1991) 121.

4. Dammann, H., Görtler, K.: High-efficiency In-line Multiple Imaging by Means of Multiple Phase Holograms. Opt. Commun. **3** (1971) 312.

5. Weible, K. J., Pedrini, G., Xue, W., Thalman, R.: Optical interconnection of a neural network associative memory using diffraction gratings. Jpn. J. Appl. Phys. **29** (1990) L1301.

6. Kirk, A. G., Hall, T. J., Crossland, W. A.: A compact and scalable free-space optical crossbar. In Holographic Systems, Components and Applications. IEE Publication No. 342 (1991) 137.

7. Lee, W.-H.: Computer-generated holograms: techniques and applications. Progress in Optics XVI, ed. Wolf, North Holland 1978.

8. Lu, X., Wang, Y., Wu, M., Jin, G.: The Fabrication of a 25×25 Multiple Beam Splitter. Opt. Commun. **72** (1989) 157.

9. Dames, M. P., Dowling, R. J., McKee, P., Wood, D.: Design and fabrication of efficient optical elements to generate intensity weighted spot arrays. Appl. Opt. **30** (1991) 2685.

10. Seldowitz, M. A., Allebach, J. P., Sweeney, D. W.: A synthesis of digital holograms by direct binary search. Appl. Opt. **26** (1986) 2789.

11. Feldman, M. R., and Guest, C. C.: Iterative Encoding of high-efficiency holograms for generation of spot arrays. Opt. Lett., **14** (1989) 479.

12. Turunen, J., Vasara, A., Westerholm, J.: Stripe-geometry two-dimensional Dammann gratings. Opt. Commun. **74** (1989) 245.

13. Hauck, R., Bryngdahl, O.: Computer generated holograms with pulse density modulation. J. Opt.Soc. Am. A **1** (1984) 5.

14. Weissbach, S., Wyrowski, F., Bryngdahl, O.: Digital phase holograms: coding and quantization with an error difusion concept. Opt. Comm. **72** (1989) 37.

15. Kirk, A. G., Powell, A. K., Hall, T. J.: A generalisation of the error diffusion method for binary computer generated hologram design. submitted to Optics Communications (1992).

16. Fienup, J. R.: Phase retrieval algorithms: a comparison. Appl. Opt. **21** (1982) 2758.

17. Just, D., Ling, D. T.: Neural networks for binarising computer generated holograms. Optics Commun. **81** (1991) 1.

18. Thompson, L. F., Wilson, C. G., Bowden, M. J.: Introduction to Microlithography. American Chemical Society, Seattle, 1983.

Microlenses in PMMA with High Relative Aperture Fabricated by Proton Irradiation Combined with Monomer Diffusion

K.-H. Brenner, M. Frank, M. Kufner, S. Kufner, and M. Testorf

Angewandte Optik, Physikalisches Institut der Universität Erlangen-Nürnberg, W-8250 Erlangen, FRG

1 Motivation

A miniaturization of refractive optical elements is interesting for various applications including 3D-integrated optical systems. In these concepts the typical functions of microlenses like image formation and collimation of light are required. In most fabrication techniques for microlenses there is a trade off between the lens size and the achievable range of focal length. A new fabrication method for microlenses with high relative apertures over a wide range of diameter sizes in polymethyl methacrylate (PMMA) by irradiation with a high energy proton beam and diffusion of monomer vapor in the irradiated domains is reported in [1].

2 Fabrication Process

The main principle of the lens fabrication process is based on the fact that an irradiation of PMMA with a high energy ion beam reduces the molecular weight by splitting the polymer chains and thus changes the diffusion properties of the material. Thus there are two processing steps, an irradiation and a subsequent diffusion (Figs. 1 and 2).

By irradiation of PMMA with a high energy proton beam through a structured metal mask domains with reduced molecular weight are produced. For the fabrication of microlenses metal masks with circular apertures have to be used.

After irradiation the structured substrate is placed into an atmosphere of monomer vapor. The diffusion of monomer vapor causes the volume of the irradiated domains to expand. By surface tension these volumes form lenslike shapes.

Figure 3 illustrates the effect of the processing steps on the molecular structure of the PMMA matrix.

3 Parameters

An important point in this context is the classification of the process parameters with respect to their effect to the shape of the lens surface [2]. The two separate

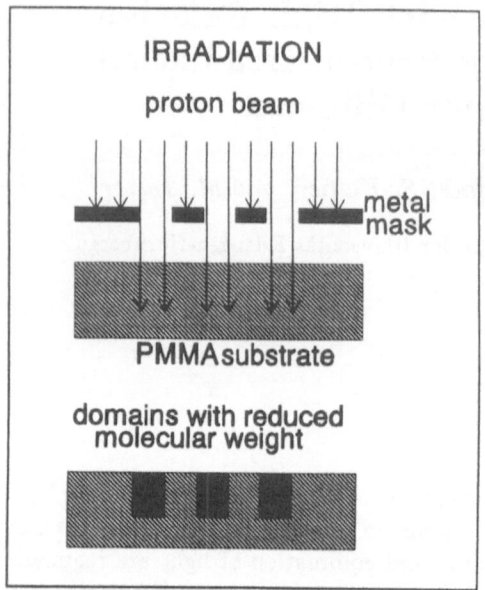

Fig. 1. Irradiation process

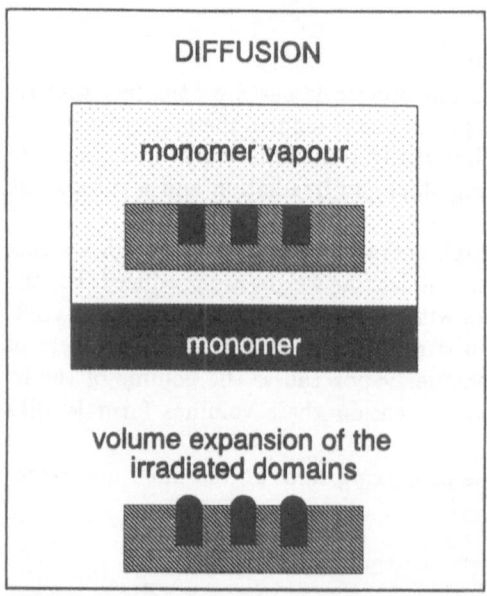

Fig. 2. Diffusion process

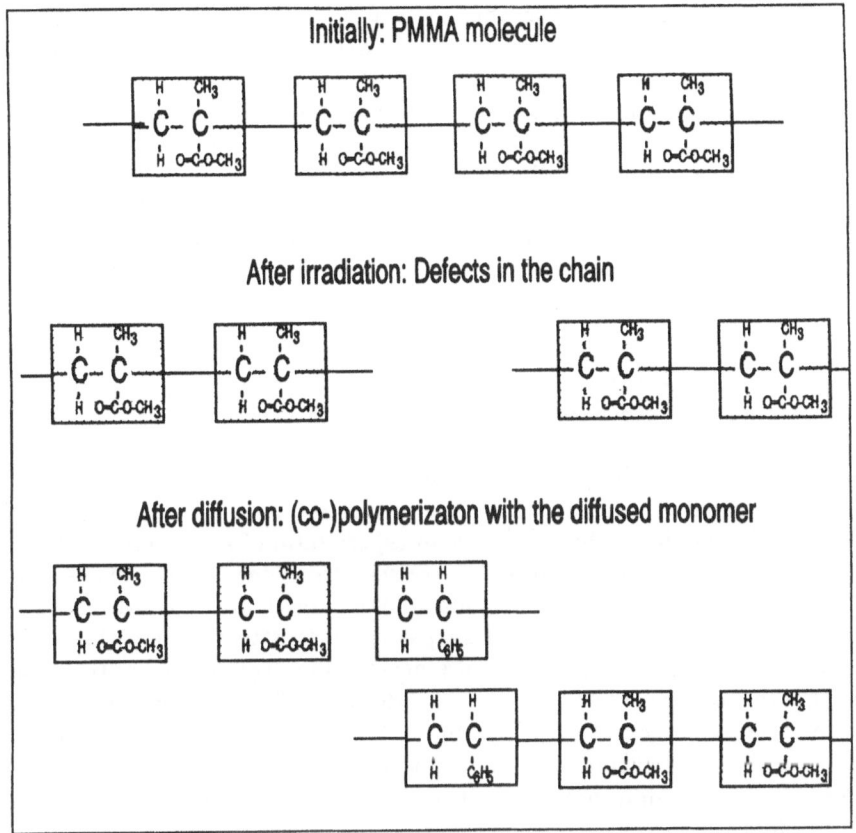

Fig. 3. Effect of processing steps on the molecular structure of the PMMA matrix

processing steps suggest a corresponding classification of the parameters. In the irradiation process the proton energy determines the penetration depth of the beam into the substrate, the dose deposition is responsible for the degree of radiation damage in the PMMA matrix. In the diffusion process the volume expansion can be controlled by the diffusion time and temperature.

3.1 Parameters of the Irradiation Process

The purpose of the irradiation process is to create well defined domains in a PMMA substrate with reduced molecular weight and thus with different diffusion properties. As an effect of the stopping range, the almost constant energy loss and the small lateral straggling of the incident protons it is possible to consider the average molecular weight to be approximately constant over the irradiated volume.

Two factors are important in this context, first the shape of the volume penetrated by the proton beam and second the average molecular weight within this volume.

The application to microlenses suggests a choice of volume shapes with rotational symmetry. Metal masks with circular apertures and orthogonal incidence of the proton beam result in cylindrical penetration volumes. The aspect ratio of these volumes can be varied easily by controlling the penetration depth. The penetration depth in PMMA is determined by the energy of the proton beam, this is the first parameter to discuss.

Proton Energy To determine how the aspect ratio of an irradiated domain influences the final surface shape an experiment was performed with an increasing proton energy and all other conditions constant. This causes an increasing penetration depth of the proton beam and an increasing aspect ratio of penetration depth/mask aperture. The penetration depth is varied in a range between 63 μm and 480 μm with a constant mask aperture of 300 μm.

The experiment shows that a minimum aspect ratio of about 1:2 (depth to diameter of the penetrated volume) is necessary to achieve convex shapes. In the case of lower aspect ratio the absolute volume growth over a relatively large area is not enough to form spherical shapes by surface tension. On the other hand there is a saturation effect if we use penetration depths of more than the diameter of the mask aperture. Then, even with short diffusion times, the volume expansion gets too large to be confined by surface tension and the convex shape cannot be maintained. Therefore an irradiation of volumes deeper than the lens diameter restricts the diffusion time but provides no advantages with respect to design freedom.

In view of the application to microlenses the aspect ratio of the irradiated cylindrical volume has an optimum range interval. Convex shapes can be achieved best with aspect ratios between 1:2 and 1:1.

Dose Deposition The radiation damage resulting in a reduction of the average molecular weight can be controlled by the dose deposition. Since the molecular weight affects critically the diffusion properties of the irradiated substrate it is interesting to analyze the effects of varying dose deposition. The irradiation process of PMMA with a proton beam allows a structuring of the substrate with domains of different molecular weight. The average molecular weight within a considered domain can be controlled by dose deposition. The range of the deposited dose is limited by two properties of the used material. Since the PMMA substrate heats up by absorbing kinetic energy a deposition of too high doses destroys the substrate. In this case damages like bubbles can be observed. On the other hand a minimum dose deposition is necessary as well. A minimum difference of molecular weight must be created to cause a volume expansion in the irradiated domains without affecting the substrate shielded by the mask.

Experimentally the variation of the dose deposition can be done in different ways. The linear relationship between dose deposition and the product of

radiation time, ion current and energy allows a choice of the control parameter. The variation of the radiation time with all other irradiation conditions
constant is most practicable for the experiment. The PMMA substrate was irradiated through a metal mask with 500 μm diameter circular apertures with
a current of 13.5 nA and a proton energy of 5 MeV resulting in a penetration
depth of 300 μm. For the diffusion process a diffusion temperature of 90 °C and
a diffusion time of 30 min were used. The radius of the surface shapes resulting
from spherical fits are shown in Fig. 4 for a range of dose deposition between
3 kJ/cm^3 and 11 kJ/cm^3. A dose deposition of less than 3 kJ/cm^3 is insufficient
for the fabrication of lenslike surface structures. The upper bound for the dose
deposition depends on the trade off between heating the material and the required time. For the used parameter set damages from a too high dose can be
observed for more than 11 kJ/cm^3. This value can be increased by cooling or dose
deposition with reduced current and correspondingly longer radiation time. In
the range between 3 kJ/cm^3 and 11 kJ/cm^3 the resulting surface shapes can be
considered approximately spherical. Within the considered range the deposited
dose allows an easy control of the focal length.

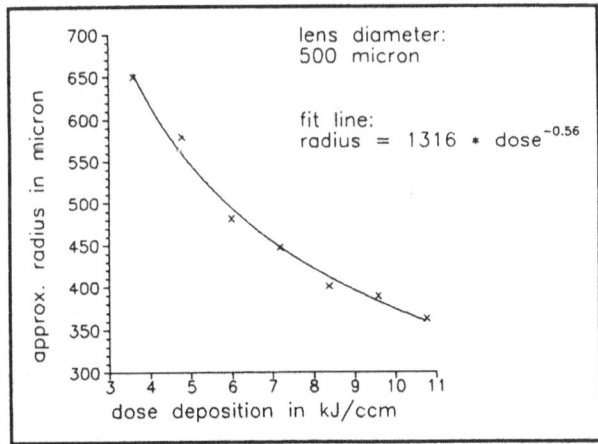

Fig. 4. Radius of spherical approximation over deposited dose

4 Parameters of the Diffusion Process

The diagram in Fig. 5 illustrates the conditions for the ranges of the diffusion
parameters. The volume expansion is too small for the fabrication of microlenses,
if the surface tension cannot form a spherical shape, on the other hand it is too
large, if it can no longer be confined by surface tension.

Comparing the irradiation and the diffusion parameters the experimental
results show that we have to meet the optimum conditions for the diffusion

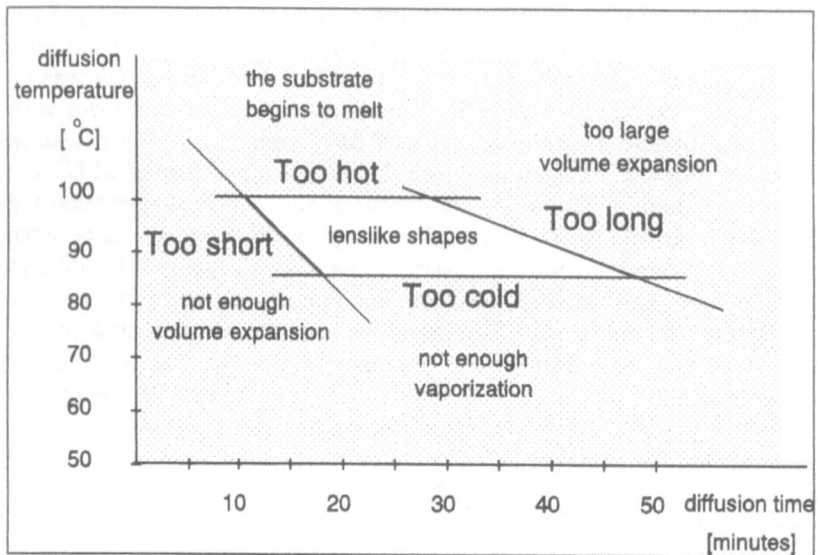

Fig. 5. Ranges of the diffusion process parameters

process, whereas the irradiation process includes the significant parameters for the control of lens properties like focal length, lens diameter and surface shape.

5 Results

The numerical apertures achieved in our experiments are in a range from 0.2 to more than 1. An example of a microlens with high relative aperture is shown in Fig. 6. The lens diameter is 300 μm and geometrical calculations result in a focal length of less than 300 μm. Thus the relative aperture is even larger than one.

The lens diameters we realized experimentally are from 50 μm to 1 mm, the availability of metal masks with sufficient thickness is the only limit for the achievable size.

The fabrication of lens arrays is possible either by separate irradiation of each lens in a step and repeat process or by irradiation through a mask with an array structure. The latter was used for the fabrication of the lens array shown on the scanning electron microscope photo in Fig. 7. The picture shows a part of a 20 × 20 lens array of 500 μm diameter lenses with a spacing of 840 μm, a focal length of 620 μm and thus a relative aperture of 0.8.

We have investigated the imaging properties of microlenses fabricated with the reported method. The used test patterns are shown in Fig. 8, the diameter of the Siemens-star in (a) is 225 μm and the spatial frequencies of the line grids (b) are from top to bottom 16, 32, 64, 128 and 256 line pairs per mm. The corresponding 1:1 images formed by a PMMA microlens with 500 μm diameter

Fig. 6. Example of a microlens with high relative aperture

and 620 μm focal length are shown in Fig. 9. The images in Fig. 9(b) show that each of the line grids is resolved by the 500 μm diameter lens. The 256 line pairs correspond to a spatial period of 4 μm or a line width of 2 μm. Thus the resolution of this lens is better than 2 μm.

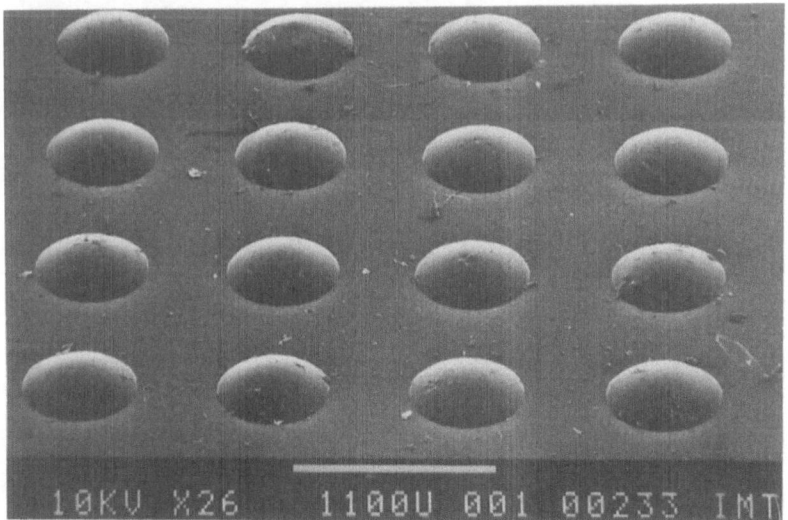

Fig. 7. Scanning electron microscope photograph of a lens array

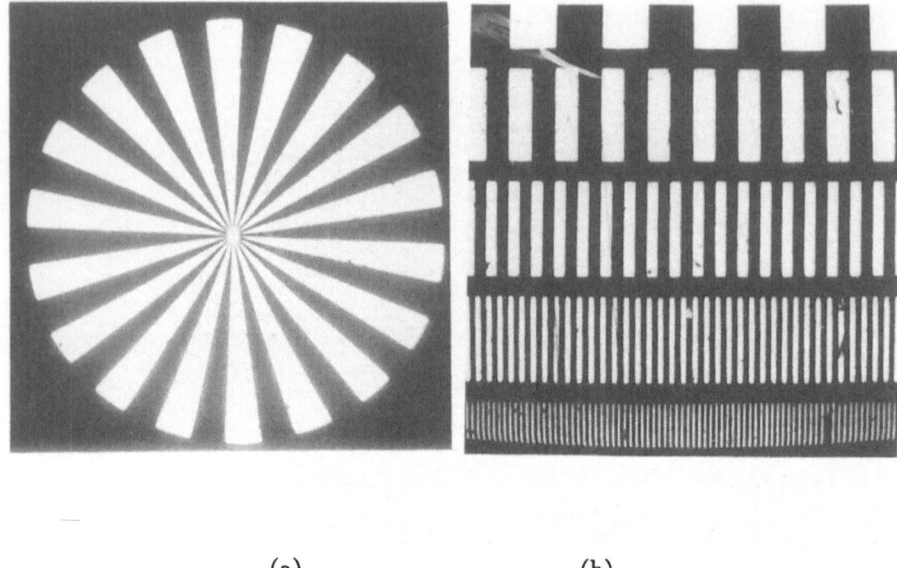

(a)   (b)

Fig. 8. Test patterns

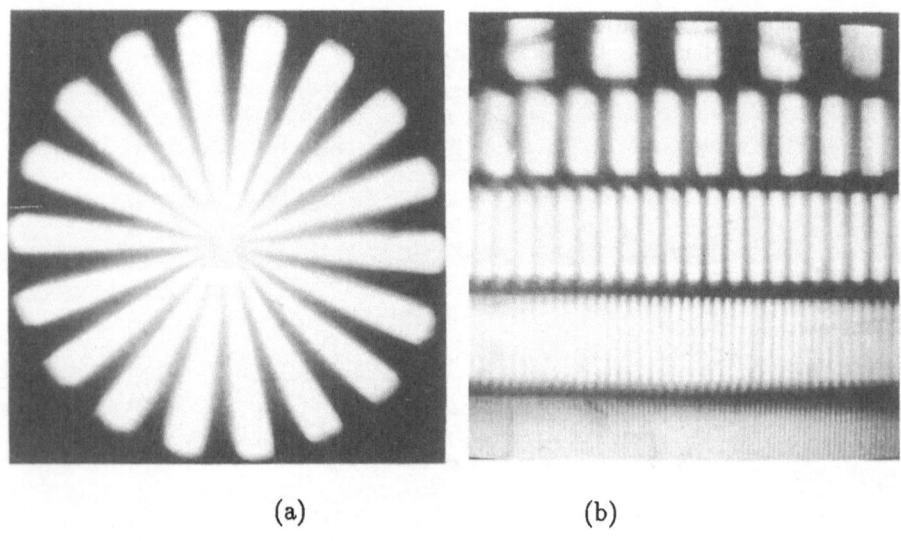

(a)   (b)

Fig. 9. Images

For a determination of the phase profile we used a Mach-Zehnder interferometer. Figure 10(a) shows a plot of the two-dimensional phase distribution of the central region of one lens. A parabolic fit corresponds quite well to the measured phase profile with a standard deviation of 0.31 μm. The difference between measurement and the fit is shown in Fig. 10(b). The parabolic approximation yields a 620 μm focal length for the 500 μm diameter lens. According to these values the relative aperture is 0.8.

(a)

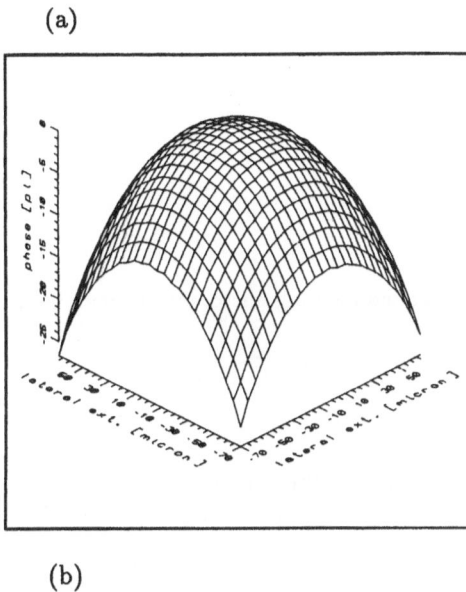

(b)

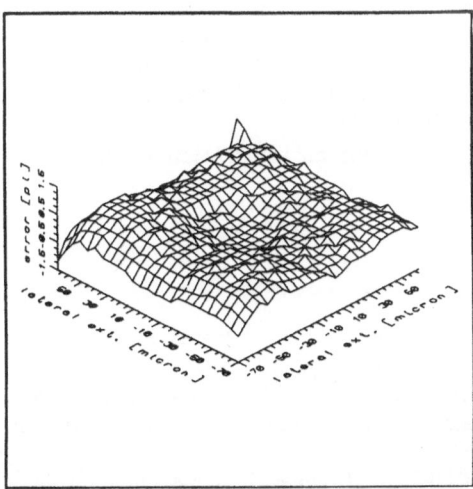

Fig. 10. (a) Interferometric phase profile measurement of the PMMA lens. (b) Deviation from the best-fit paraboloid

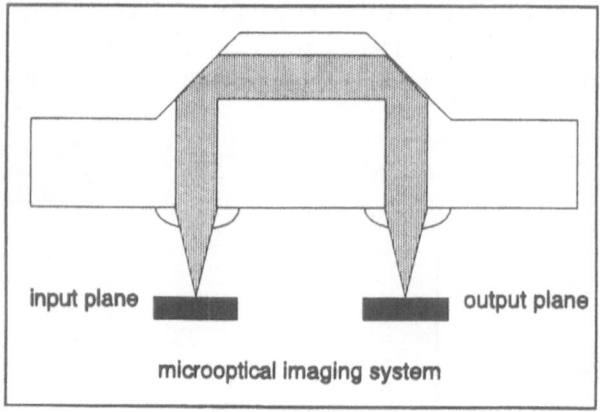

Fig. 11. Monolithic 3D integration of deflecting and focusing microoptical components

6 Summary

The surface profile lenses are purely refractive and thus the wavelength dependence is only based on dispersion. Since the lens effect is caused by a surface profile only (the change of the refractive index is negligible) the lenses can be copied easily by casting.

The used material PMMA allows a combination of different optical components fabricated by different techniques monolithically in the same substrate. Figure 11 shows an example of a three-dimensional monolithic integration of microlenses and microprims on one PMMA substrate. The fabrication of the microprisms is also based on proton irradiation [3].

These properties make the lenses suitable for an integration in three-dimensional microoptical systems.

References

1. Frank, M., Kufner, M., Kufner, S., Testorf, M.: Microlenses in polymethyl methacrylate with high relative aperture. Appl. Opt. **30** (1991) 2666–2667.
2. Kufner, M., Kufner, S., Frank, M., Moisel, J., Testorf, M.: Microlenses in PMMA with high relative aperture: A parameter study. Pure and Applied Optics, in press.
3. Kufner, S., Kufner, M., Frank, M., Müller, A., Brenner, K.-H.: 3D integration of refractive microoptical components by deep proton irradiation. Pure and Applied Optics, in press.

Fabrication of Microoptic Components by Thermal Imprinting

K.-H. Brenner, C. Doubrava, and T.M. Merklein

Angewandte Optik, Physikalisches Institut der Universität Erlangen-Nürnberg,
W-8520 Erlangen, FRG

3D-microoptic systems can be realized by a thermal imprinting processes in Poly-MethylMethAcrylate (PMMA). Hereby each microoptic component is realized by thermally imprinting a metal master into the substrate. This process, though being simple, provides high quality components and allows a mass production. Experimental results are demonstrated for an interconnection scheme.

1 Introduction

Microoptical elements are of interest for many applications such as integrated sensors, interconnections, optical computing and measurement [1]. The small size of the components and the required high positional accuracy is typically achieved by planar, lithographic fabrication technologies, favoring diffractive optical elements (DOE). There are, however, numerous applications, where the modest spectral operating range, the limited diffraction efficiency and the high angular sensitivity of DOEs is not sufficient. Light deflection by refractive effects shows none of the above mentioned limitations. For the fabrication of refractive optical components (ROE), suitable technologies exist only for the case of micro-lenses. For the case of micro-prisms, micro-mirrors or micro-beamsplitters very little has been published yet. One of the main reasons for this is that structuring depths of several hundreds of micrometers are required for these types of components. The LIGA-process [2], for example, allows a structuring of PMMA in a depth of up to several hundred micrometers·by exposure to high energy gamma-radiation.

2 Microoptical Systems Layout

The scheme of deep thermal imprinting was developed especially for the fabrication of arrays of microprisms and beamsplitters to be used in microoptic systems (Fig. 1). These systems are arranged in layers of active components, micro-lenses, micro-prisms and phase-only filters. The layer structure simplifies the alignment of the different microoptical components. The micro-prisms can act as mirrors or as beam-splitters, depending on the type of coating. It is, however, necessary that these surfaces are oriented at 45 degrees with respect to the surface, therefore requiring structures with a depth of up to 1 mm. Furthermore it is desirable that these components offer good optical quality of the surfaces.

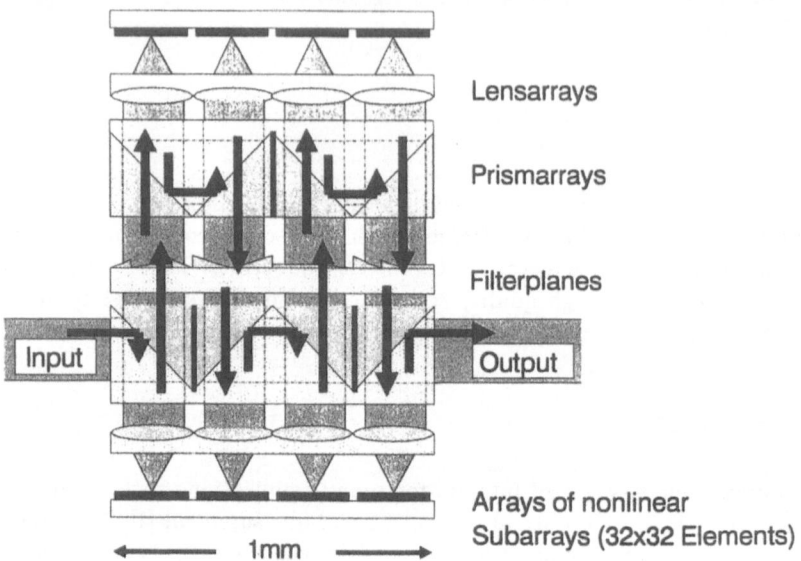

Fig. 1. Scheme for a microoptic system for digital data processing, consisting of different layers of nonlinear devices, lenses, beamsplitters and prisms

3 Deep Thermal Imprinting

For the fabrication of the tilted micro-mirrors, a metal master is heated above the glass temperature ($\approx$ 120 °C) of PMMA and is pressed into the PMMA sample. A small region of the PMMA close to the master ($\sim$ 100 μm) reaches a temperature above the glass temperature and becomes viscous. In this state the PMMA assumes the shape the surface of the master. After the master is cooled below the glass temperature at about 80 °C the PMMA matrix becomes stable again and the master is removed [3].

3.1 Fabrication of the master

The metal masters for prisms and mirrors are realized by milling and polishing of steel. More general metal masters can be realized also by electroforming of surface profiles, which have been produced by other methods. The diameter of the metal masters, we are using is typically in a range between 500 μm to 1 mm (Figs. 2, 3).

As can be seen from the measurements (Fig. 3) the accuracy of the imprints is only limited by the surface quality of the master. To test this, we also used polished YIG-crystals as imprint masters. It was possible to achieve very good results due to the excellent quality of the crystal surfaces. YIG-crystals are particularly suitable for thermal imprints because of its excellent heat conductivity properties and the high degree of hardness.

Fig. 2. Metal master (diameter 0.8 mm) and its imprint in a 5 mm thick substrate

3.2 Substrate Materials

We are presently using PMMA as the material for imprinting optical components. There are various types of PMMA available, differing in the molecular weight and the mixture with additives. Some types have been optimized with respect to low absorption and scattering over a wide spectral range. In our applications the path of light in PMMA is usually very short (several mm). Consequently the optical absorption and scattering properties of the PMMA substrate do not affect the imaging quality significantly. The thermal properties, however, affect the speed of imprinting and the quality of the imprinted profiles. The temperature of the phase transition of PMMA depends on the molecular weight,

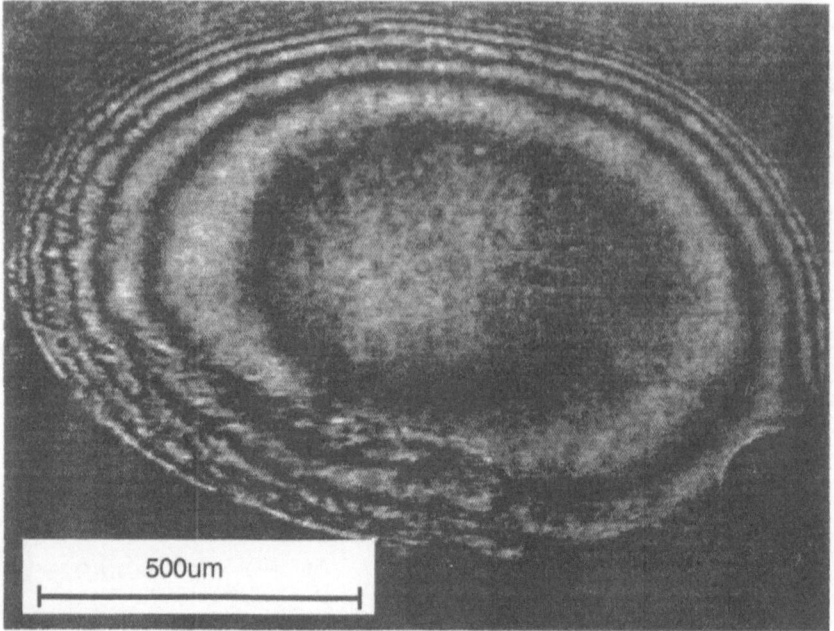

Fig. 3. Interferogram of the polished metal master

the degree of cross linkage between the molecules and on the purity. Favorable for this method are pure samples, however the thermal stability has to be determined individually [4,5]. Some additives in PMMA are not suitable for this process because they can be damaged by chemical reactions occurring already at temperatures below the glass temperature.

3.3 The Imprinting Process

For the positioning and imprinting of the master into the substrate we have used precision positioning stages, driven by step motor drives under computer control with an accuracy of 1 μm. Until now the accuracy of position, inclination θ and rotation ϕ of the imprint is limited by the mounting accuracy of the master ($\delta x \approx \delta y \approx \delta z \approx 10$ μm, $\theta \approx \phi \approx 0.1°$). This accuracy of the master can be improved to $\delta x \approx \delta y \approx \delta z \approx 1$ μm, $\theta \approx \phi \approx 0.01°$ by using optical measurements as aid for the adjustment. The heating of the metal master to a temperature above the glass temperature (< 120 °C) of PMMA is done with a temperature controlled electrical heater. The vertical movement for pressing the die into the PMMA substrate is also under computer control. Only a small region of the PMMA close to the master reaches a temperature above the glass temperature and is deformable. If the temperature of the master is only slightly above the glass temperature, the PMMA gets viscous without chemical damage occurring.

The displaced material flows along the die to the surface of the substrate. After the master is cooled down below the glass temperature or, more precisely, below the Martens temperature ($\approx 80\,°C$), which indicates the limit of thermal stability of PMMA, the master is removed. Thermal expansion of PMMA is not a critical parameter because only a small region (some $100\ \mu m$) around the master is notably heated. The relative thermal expansion ($\Delta l/l \approx 70 \times 10^{-6}/°C$) results in a lateral shrinking of approx. $1\ \mu m$ of the imprinted elements. Birefringence due to mechanical stress was not observable.

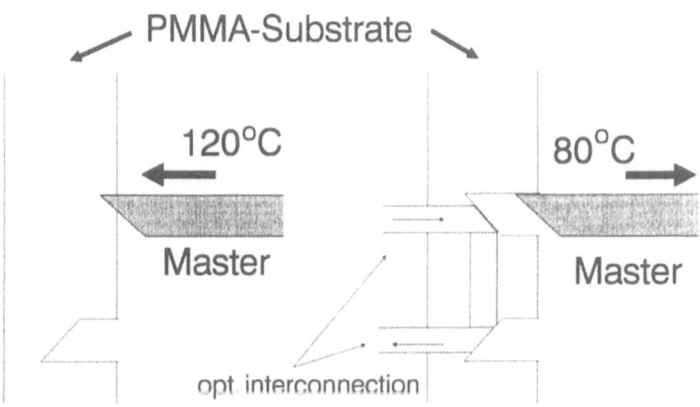

Fig. 4. Deep thermal imprinting of optical interconnections

4 Measurement of the Optical Quality of the Imprints

A special interferometer was implemented for comparing the imprinted structures with the masters (Figs. 5, 6). With the master in one arm of the interferometer and the imprint in the other arm, the interferogram shows the height difference between the two (Fig. 6). The interferometer was adjusted with a slight tilt. Therefore fringes can be seen. The figure indicates that the imprint is an almost perfect copy of the master.

For an absolute measurement of the imprinted microprisms we have used a Mach-Zehnder-Interferometer (Figs. 7, 8). In this setup the deflection of the transmitted light after the microprism (object) is compensated with a macroscopic prism.

5 Microoptical Interconnections

As an application of the proposed method we have realized a special interconnection pattern. The perfect shuffle plate (Fig. 9) demonstrates the potential of

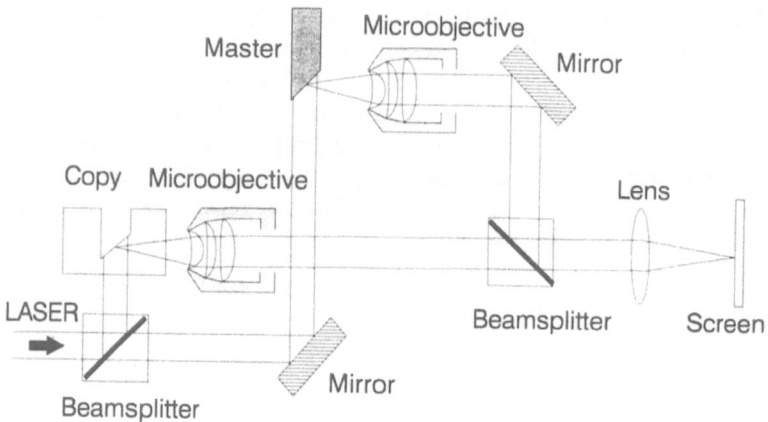

Fig. 5. Interferometer for a comparative measurement of the metal master and an imprinted microprism

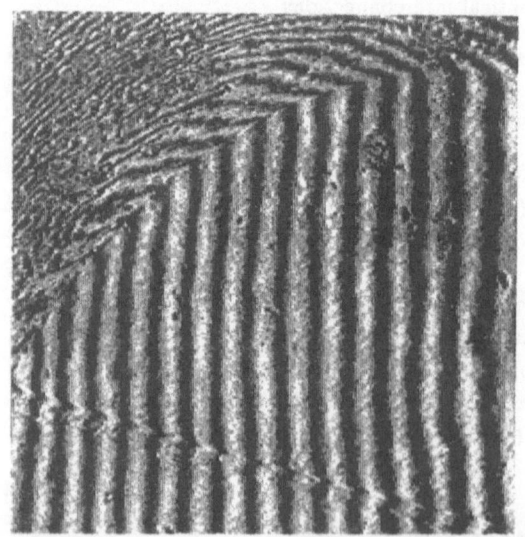

Fig. 6. Interferometric comparison between a master and its imprint using the interferometer in Fig. 5

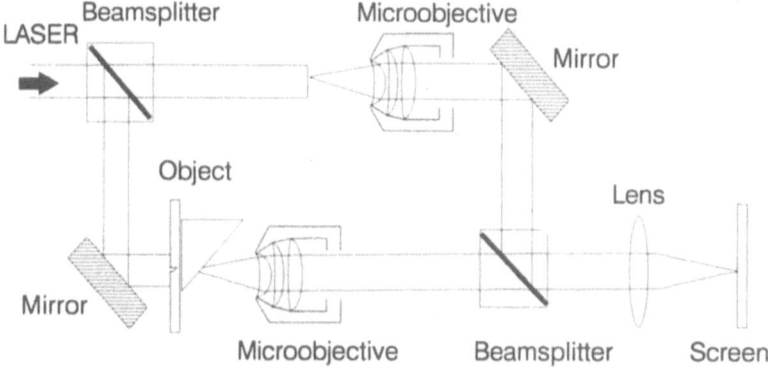

Fig. 7. Mach-Zehnder interferometer for absolute measurements of imprinted microprisms

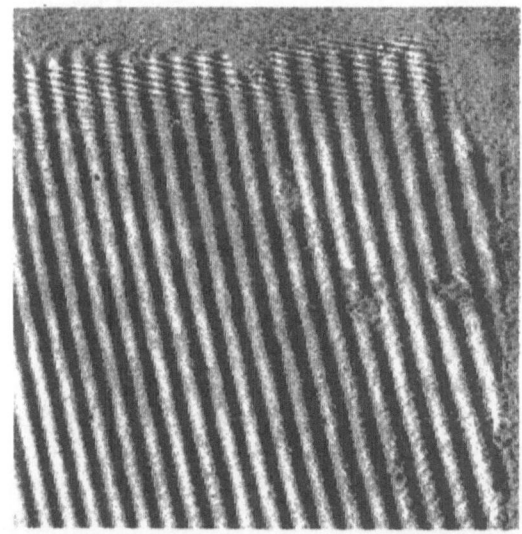

Fig. 8. Measurement with the interferometer described in Fig. 7. The microprism was realized by imprinting the surface of a YIG-crystal into a PMMA-sample

the proposed method. A perfect shuffle interconnection interlaces the first half
of the input channels regularly with the second half. For the special case of four
inputs, as in Fig. 9, the perfect shuffle leaves the edge channels unchanged and
interchanges the inner channels. The light enters from the left side, is permuted
and exits at the right side.

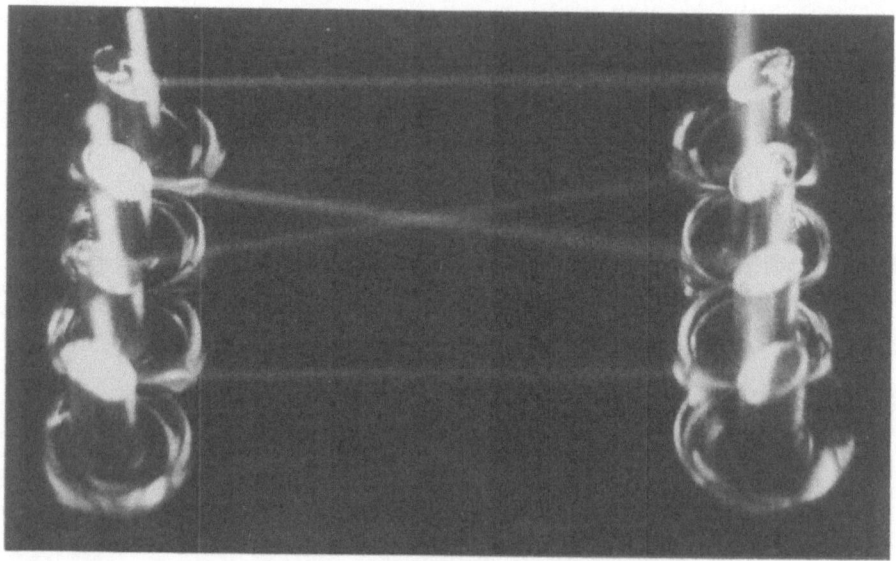

Fig. 9. Perfect shuffle plate realized by thermal imprinting of microprisms. The master
was rotated to adjust the interconnection angle between the input and the output
positions

6 Conclusion

The method, developed for the fabrication of microoptic components is attractive by its simplicity and accuracy. Especially the possibility of a free design of
the components and their locations opens a wide range of optical applications. In
particular optical inter-connections and microoptic elements for digital optical
data processing systems can be realized cost efficiently. Due to the refractive or
reflective nature of the imprints, the angular sensitivity and the spectral sensitivity for transmitted or reflected light are significantly smaller as compared to
diffractive interconnections. A mass production of imprinted structures is also
possible by electroforming. To this end, the substrates can be coated with silver
(Ag) or nickel (Ni) by chemical vapor deposition (CVD), resulting in a conducting surface. The thickness and stability of the metal layers is increased by

electroforming of nickel. The PMMA substrates are then dissolved e.g. by acetone. Consequently a negative copy of the imprinted component arrangement can be achieved. This copy can be used later for thermally imprinting the whole scheme in one step. A reactive or thermal injection molding [2] may cause problems due to the shrinking during polymerization. With deep thermal imprinting, however, the distances can be maintained sufficiently.

References

1. Jahns, J., Däschner, W.: Optical cyclic shifter using diffractive lenslet arrays, Opt. Comm. **79** (1990) 407-410.
2. Hagmann, P., Ehrfeld, W.: Fabrication of microstructures of extreme structural heights by reaction injection molding. In International Polymer Processing IV **3** (1989), 188-195.
3. Brenner, K.-H., Doubrava, C., Merklein, T. M.: Fabrication of microoptic components by thermal molding. Appl. Opt. (in press).
4. Nitsche, R., Wolf, K. A.: Kunststoffe Struktur und Physikalisches Verhalten der Kunststoffe. Springer Verlag, Berlin (1962).
5. Vieweg, R., Esser, F.: Kunststoff-Handbuch Band 9 'Polymethacrylate'. Carl Hanser Verlag, Munich (1975).

Invariant Pattern Recognition: Towards Neural Network Classifiers

G. Lebreton,[1] E. Marom,[2] N. Konforti,[2] and D. Mendlovic[2]

[1] University of Toulon, BP132, F-83130 La Garde cedex, France
[2] University of Tel Aviv, Ramat Aviv, 69978 Tel Aviv, Israel

1 Introduction

From the many studies on invariant pattern recognition, the coauthors have selected single harmonic filters [1,2]. A real-time optical implementation of such filters was demonstrated on photographic input images [3], but its 10-F length yielded a poor signal-to-noise ratio, making it unusable with the limited dynamic range and flatness of a light valve for real-time video input. For real applications, the other difficulty is the usual requirement of all invariances simultaneously: scale and projection can be processed in a single filter [1], but a pre-processing is necessary for the rotation before any object recognition, since the projection and scale invariant filter do not work on rotated objects.

The research program presented here is intended to solve these two major problems :

1. implement a real-time processor with complex filter generation
2. utilize a hybrid neural network image classification for rotation pre-processing before object recognition.

A theoretical research has led to the feasibility of a hybrid opto-electronic processor, and its realization is now in progress.

The first part presents the novel architecture developed for that purpose. The second part describes the active optical components actually experimented for the final system. The third part shows first results using fixed invariant filters to test the real-time optical correlator, before the implementation of the hybrid neural network.

2 Hybrid Processor Architecture for an Associative Memory with Complex Images

An associative memory architecture is the only way to solve the two main difficulties for a real- time invariant pattern recognition: first is to correct for the orientation of an unknown object before its detection with projection and scale invariant filters; second is to improve the signal-to-noise ratio, always poor with fully invariant filters which utilize a very small part of the spectral information from the distorted object.

Many optical implementations have been proposed to implement associative memories, and some of them have been experimentally demonstrated, but usually for 1-D inputs, which require a 2-D memory. For image processing, where a 4-D memory is necessary, the utilization of volume holograms has been proposed, but their practical utilization in real-time is not realistic with the existing technology. The basic idea here is to utilize a double correlator at video rates, with computer-driven integration and feed-back. D.Psaltis et al. [4,5] have demonstrated that the outer product associative memory

$$\hat{h}(x,y) = \int T(x,y,\xi,\eta)\hat{f}(\xi,\eta)d\xi d\eta \ , \ \text{where} \ T(x,y,\xi,\eta) = \sum_{m=1}^{M} h_m(x,y)f_m(\xi,\eta)$$

(1)

[with output $\hat{h}(x,y)$, input $\hat{f}(\xi,\eta)$, m^{th} memory $f_m(\xi,\eta)$ and associate output $h_m(x,y)$ identical to $f_m(\xi,\eta)$] could be expressed as

$$\hat{h}(x,y) = \sum_{m=1}^{M} \Gamma_{f_m f}(0,0)f_m(\xi,\eta)$$

(2)

where $\Gamma_{f_m f}(0,0) = \int f_m(\xi,\eta)\hat{f}(\xi,\eta)d\xi d\eta$ is the correlation sampled at the origin.

But to maintain the shift invariance provided by the Fourier transform in a coherent optical correlator, the weights in the sum must be replaced with the convolution with a non-linear (e.g. quadratic) image of the correlation output over a restricted 2-D aperture:

$$\hat{h}(x,y) = \sum_{m=1}^{M} |\Gamma_{f_m f}(\xi,\eta)|^2 * f_m(\xi,\eta)$$

(3)

This can be implemented with the double-correlator architecture shown on Fig. 1, where a reflective hologram on a BGO crystal gives the desired squared correlation which is back convolved with the memory images recorded on the first hologram.

Half-part of the liquid-crystal light valve, between crossed polarizers for intensity modulation, is imaged on the second-half to multiply the modulus of each addressed with its phase term, either to enable the complex inputs required for the invariant filters, or only to correct for the phase defects related to the non-uniform flatness of the light valve.

The dynamic range of the photo-thermoplastic memory (about 64 grey levels) enables to record successively several inputs, and at least two filters can be written vertically at the same time with the actual device (320×230 pixels) but another one (640×460) will permit simultaneously 6 images of 128×128 pixels. The system is thus suitable to utilize a set of distortion-invariant filters in a neural network classification for noise robustness.

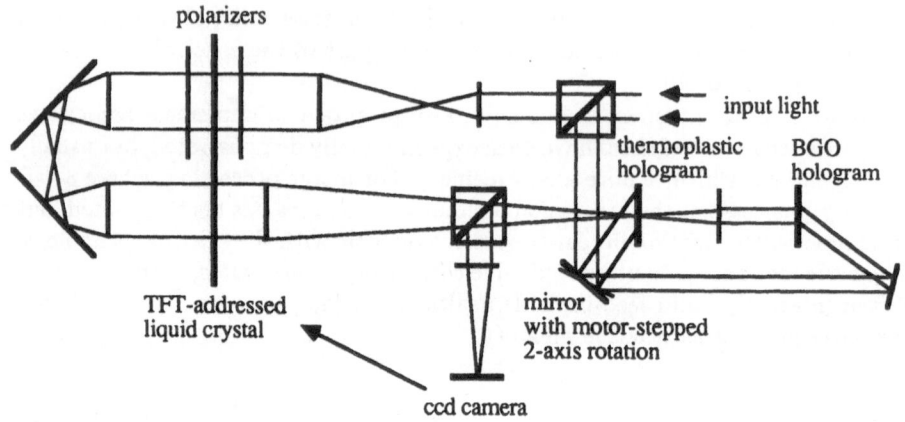

Fig. 1. Double-correlation optical associative memory with complex input

The CCD output image is thresholded by the computer before sending back to the optical system; this feedback loop can be done at video rate, since the photo-thermoplastic memory keeps fixed during the iterations and as long as the same set of filters is adequate.

A future objective will be to utilize the system for a tree classification, where a few successive sets of filters might provide a very general image classifier.

A rotation pre-processing before recognition can be implemented in a similar way, if we consider how the human vision is usually able to decide, even for unknown objects, that they are oriented in a given direction. The same result can be expected with an associative memory made with 36 rotated replicas of a simple object of the type shown on Fig. 2.

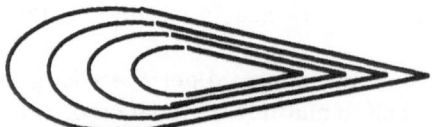

Fig. 2. Example of pattern for orientation classification

But at least for a restricted set of objects to be detected one can compute an appropriate pattern with frequency components extracted from a few number of distorted objects. This part of the program is just starting now, with still no results available.

3 Active Components for the Optical Part of the System

The BGO is a standard crystal configuration for holography, with a 10×12mm aperture. The NRC holocamera is utilized now with its standard photothermoplastic plate, but new thermoplastic layers made by the CNET will be used in the following. These new emulsions present two advantages. First, they have demonstrated a much longer life time: 1500 to 3000 cycles instead of 100 to 300 cycles for the NRC plates. But in addition the research in progress is expected to enable writing with near-infrared wavelength, which will make them attractive to implement programmable interconnects with laser diodes. The electronically addressed light valve (with Thin Film Transistors) is a prototype with 320×240 pixels, from the French CNET (Lannion). The dynamic range is linear over 64 levels, as shown on Fig. 3.

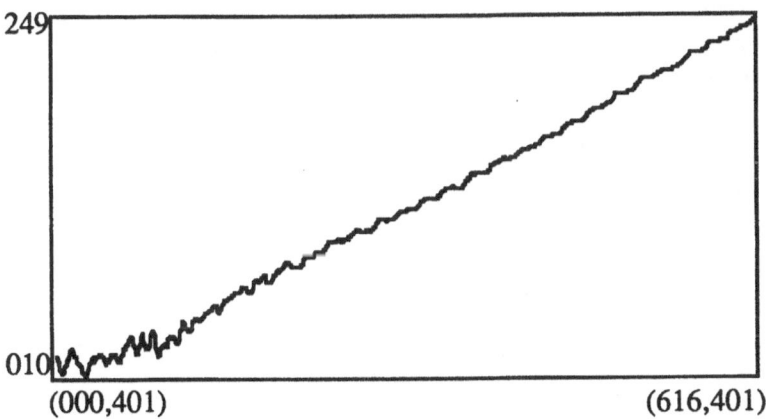

Fig. 3. Line profile of a grey-scale on the LCLV

This grey-level capability is illustrated by a white light image on Fig. 4, followed with a picture of the light valve. Shown on Fig. 5 is the experimental apparatus (the Argon laser is not visible).

4 First Correlation Results Using Video Input with Fixed Invariant Filters

The purpose of the experiments, as a first step, was to utilize the invariant filters developed at the University of Tel Aviv as fixed masks, with a video input, to adapt all parts of the future processor. These computer-generated filters utilize a spatial carrier frequency for phase encoding. A grating with the same spatial carrier had to be placed in front of the LCLV (Liquid crystal light valve); the

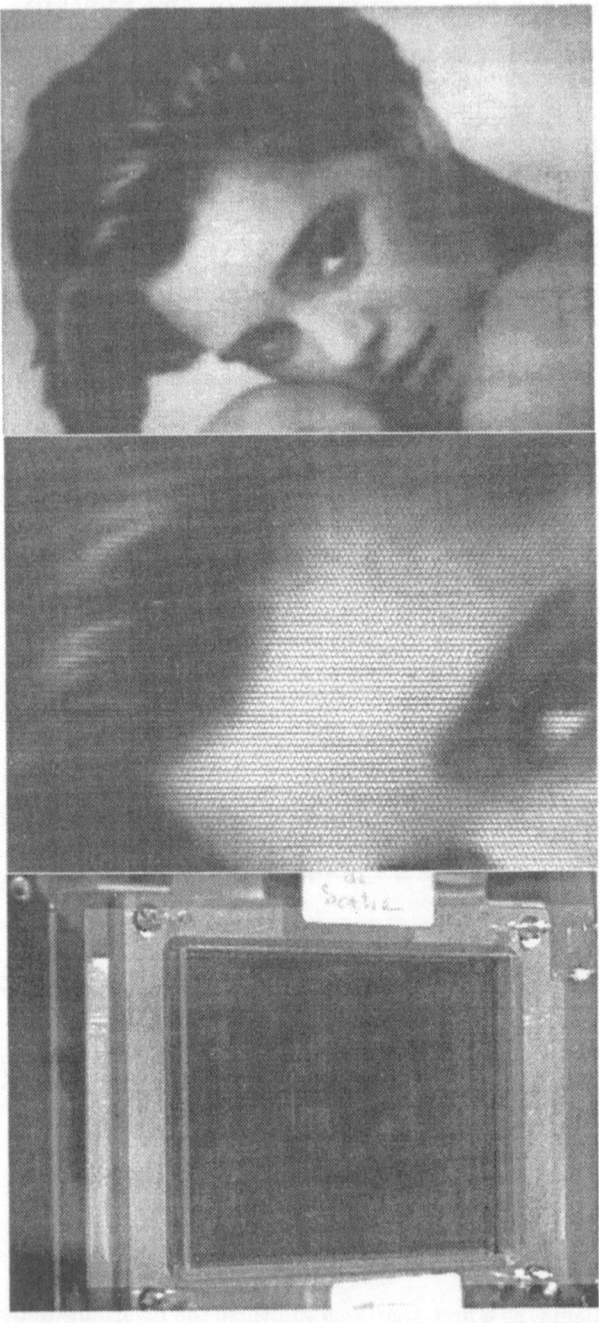

Fig. 4. Liquid crystal light valve from C.N.E.T.: front view and grey image, with enlargement showing the pixels

Fig. 5. Views of the experimental setup

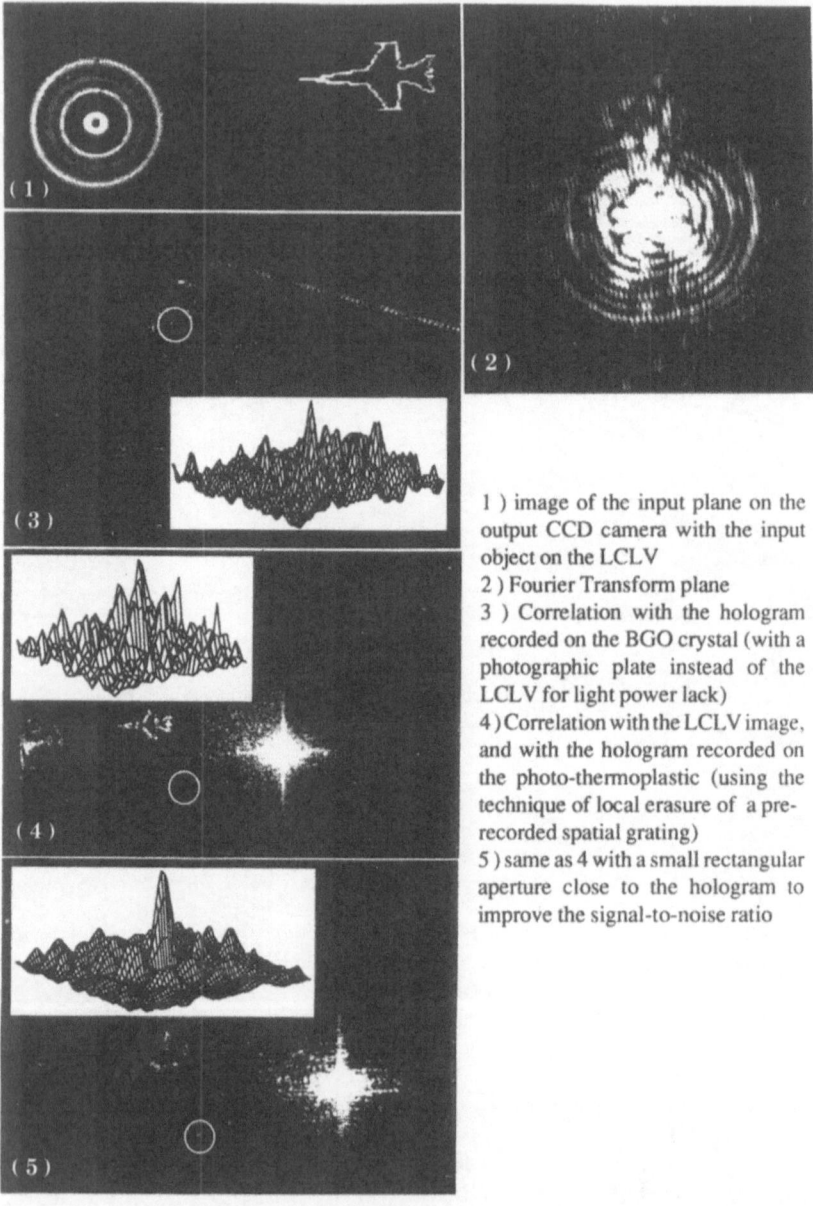

1) image of the input plane on the output CCD camera with the input object on the LCLV

2) Fourier Transform plane

3) Correlation with the hologram recorded on the BGO crystal (with a photographic plate instead of the LCLV for light power lack)

4) Correlation with the LCLV image, and with the hologram recorded on the photo-thermoplastic (using the technique of local erasure of a pre-recorded spatial grating)

5) same as 4 with a small rectangular aperture close to the hologram to improve the signal-to-noise ratio

Fig. 6. Joint-correlator results with a computer-generated rotation invariant complex filter on photographic plate. (1) Image of the input plane on the output CCD camera with the input object on the LCLV. (2) Fourier-transform plane. (3) Correlation with the hologram recorded on the BGO crystal (with a photographic plate instead of the LCLV for light power lack). (4) Correlation with the LCLV image, and with the hologram recorded on the photo-thermoplastic (using the technique of local erasure of a prerecorded spatial grating). (5) Same as 4 with a small rectangular aperture close to the hologram to improve the signal-to-noise ratio

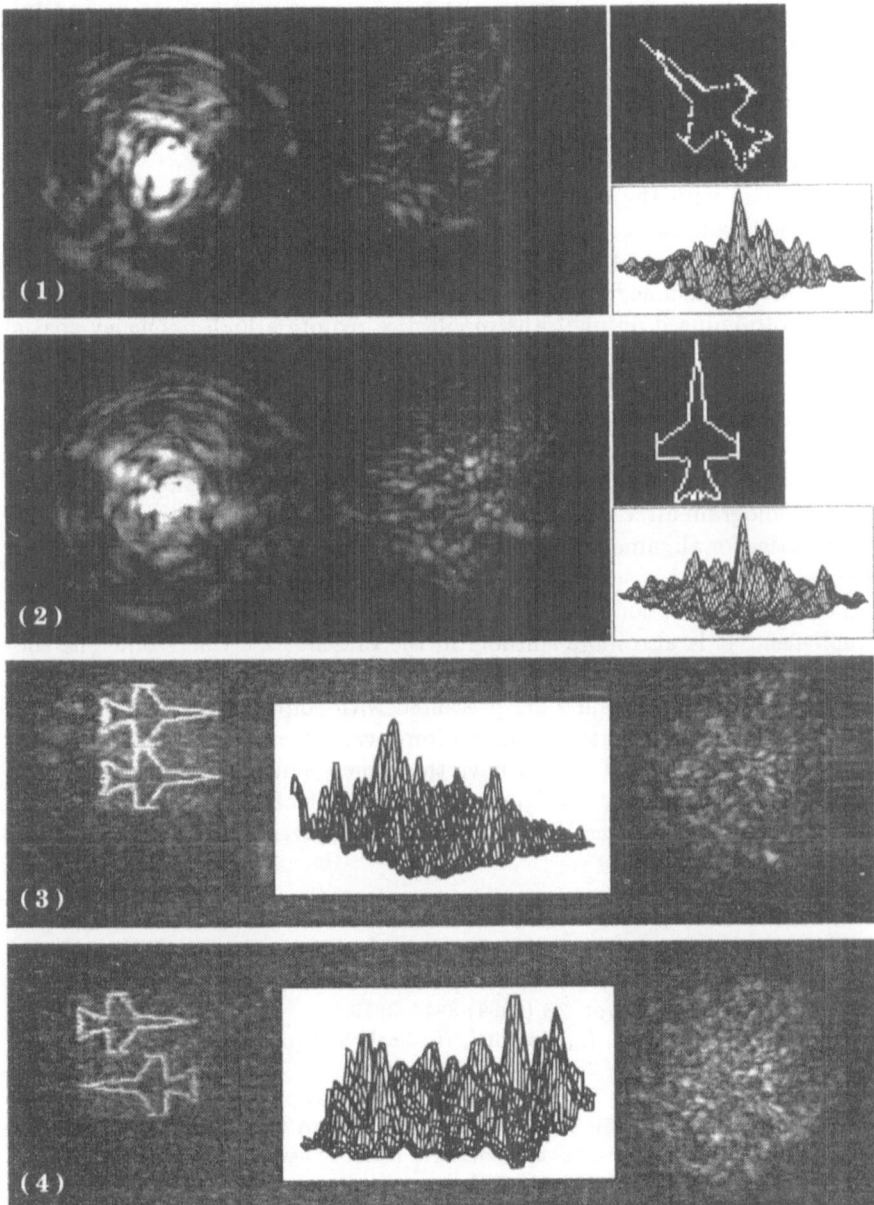

Fig. 7. Matched filter results with the photo-thermoplastic single exposure standard mode and LCLV images, using a rotation invariant filter (3-D plot shows saturation on 4)

frequency was fixed to twice the spatial frequency of the pixels on the LCLV. These two masks were written on photographic plates.

Several types of invariant filters using harmonic components have been prepared (described in [1,2,3]), the most important for our final experiment being the projection and scale invariant filter. All these filters are matched to a 64×64 bits binary image of the contours from a DF18 aircraft.

To compare with the results obtained at Tel Aviv with photographic holograms, a joint- correlator architecture has been experimented first. This was also an opportunity to extend a previous work at GESSY [6] on the utilization of photothermoplastic holographic cameras for image recording. The surface deformation photo-induced on photothermoplastics require a high frequency spatial carrier, about 800 mm^{-1}. Image record was demonstrated [6] by local erasure of a pre-exposed grating by the image. Here, this two-exposure method is utilized to perform a quadratic recording in the hologram plane, necessary for the joint-correlator configuration. But the main advantage of a joint-transform versus a matched-filter set-up, to avoid the difficulty for accurate repositioning of a photographic hologram after chemical processing, was lost in the present case, while the difficulties for alignment between the LCLV and the photographic filter were severe.

The matched-filter configuration (VanderLugt correlator) was then more easy to implement. It is also more suitable in the present case where different successive images on the LCLV can be processed with the same recorded hologram of the filter. These first results are presented with comments on Figs. 6 and 7. The poor signal-to-noise ratio should be improved when the complex filters will be recorded also on the LCLV, to have the same point spread function as the processed image. However, this confirms the interest of an associative memory architecture for a better classification with noisy images.

References

1. Mendlovic, D., Marom, E., Konforti, N.: Improved rotation or scale invariant matched filter. Appl. Opt. **28** (1989) 3814–3819.
2. Mendlovic, D., Marom, E., Konforti, N.: Scale and projection invariant pattern recognition. Appl. Opt. **28** (1989) 4982–4986.
3. Mendlovic, D., Marom, E., Konforti, N.: Real-time optical generation of circular or Mellin radial-harmonic filters. J. Opt. Soc. Am. A **7** (1990) 225–230.
4. Paek, E.G., Psaltis, D.: Optical associative memory using Fourier transform holograms. Opt. Eng. **26** (1987) 428–433.
5. Psaltis, D., Hong, J.: Shift-invariant optical associative memories. Opt. Eng. **26** (1987) 10–15.
6. Lebreton, G., Bamler, R., Glünder, H., Platzer, H.: Imaging on thermoplastic films: a new recording technique for a real-time coherent light valve. Appl. Opt. **24** (1985) 450–453.

Holographic Interconnect Components for Optical Processing Systems

A. C. Walker, M. R. Taghizadeh, E. J. Restall, B. Robertson, and J. M. Miller

Department of Physics, Heriot-Watt University,
Riccarton, Edinburgh EH14 4AS, UK

Both computer-generated and optically-recorded holograms play important roles in the current generation of optical computing demonstrator systems. This paper reviews the various types of holographic interconnect components that have been used in a series of optical parallel digital processing experiments at Heriot-Watt University. Both high-level and low-level space-invariant fan-out elements are covered, along with components for space-variant interconnect systems.

1 Introduction

Holographic and diffractive optical elements have been finding widespread use in parallel optical computing demonstration experiments [1,2,3]. Three basic functions can be identified: (i) high-level fan-out (one to hundreds) of a single beam to provide the multiple inputs required to power logic planes containing arrays of optical logic elements [4]; (ii) low-level fan-out (one to a few) to provide regular interconnection patterns between elements within processing arrays (e.g. between nearest-neighbours) [5]; and (iii) irregular (space-variant) interconnects across processing arrays. The following sections discuss the various approaches to developing the necessary components for these applications.

2 High-level Space-Invariant Fan-Out

The most effective approach to converting a single laser beam into a regularly-spaced array of equal-irradiance light spots uses a periodic two-dimensional phase-grating in the Fourier plane of a lens (see Fig. 1). The detailed form of the grating, which can be computer calculated, depends on the precise approach chosen. However, in all cases, the aim is to introduce a diffractive sub-structure within each basic period of the grating, so that the envelope function that determines the distribution of power across the two-dimensional array of diffracted orders takes the desired form. The simplest approach, developed by Dammann and Görtler [7], assumes a binary surface-relief phase sub-structure: calculated, for a one-dimensional grating, to produce a top-hat envelope function with equal-irradiance diffractive orders up to the desired arrays size and near-zero power in the higher orders. Such a 1-D array generator, etched as regions of either 0 or

π phase change into the surface of a suitable transparent substrate (e.g. fused silica), can exhibit an efficiency of $\approx 80\%$ (incident power into the desired orders), while by simply crossing two such gratings a full 2-D array generator with $\approx 65\%$ efficiency can be obtained.

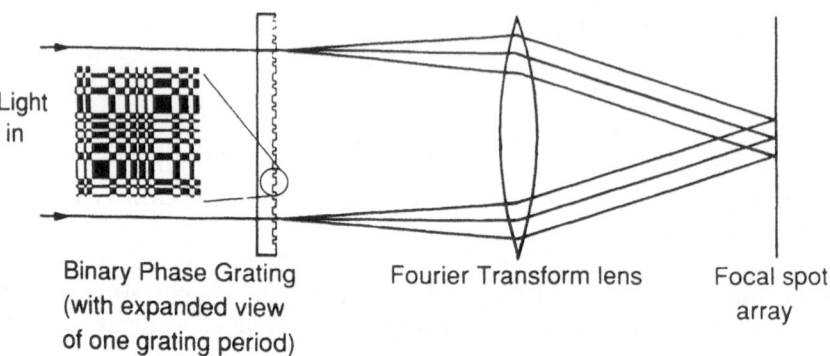

Light
in

Binary Phase Grating   Fourier Transform lens   Focal spot
(with expanded view   array
of one grating period)

Fig. 1. Schematic showing how a diffractive phase element (in this example a 2-D binary Dammann grating) can be used to generate an array of equal power focussed light spots from a single input beam.

One problem with this approach is that the etch depth must be very accurately controlled to avoid excessive power remaining in the zero-order – which forms the central spot of the desired array pattern. This difficulty can be avoided by calculating a grating structure which suppresses all the even orders. A square (or rectangular) array of spots can then be formed by the odd-orders only – leaving any unwanted zero-order power separated from the desired output pattern. Good uniformity can then be achieved: e.g. for a 16×16 array generator only $\pm 2\%$ variation in power across the full 256 beams.

To obtain a better efficiency, it is possible to calculate a full 2-D grating structure directly, instead of using crossed 1-D gratings. Such "non-separable" designs are more computationally intensive but offer significantly higher efficiency: e.g. $\approx 75\%$ for a fan-out element with 1% uniformity.

To improve efficiency further, and to consequently reduce the unwanted power in the higher orders (which in some circumstances could cause problems), it is necessary to move away from binary phase gratings to more complex structures. The ideal structure is the kinoform, which by using a continuous phase profile can achieve, theoretically, close to 100% efficiency. Of course, such a structure is difficult to realise perfectly in practice, but two approaches have been investigated. The first exploited the spatially-filtered output from a specially designed binary grating to record a volume-phase optical element holographically. This used the ability of the holographic recording medium (dichromated gelatin) to produce continuous phase variations. Good efficiencies (e.g. > 90%) have been demonstrated using this "hybrid kinoform" approach [6]. The alternative method is to fabricate a multi-level surface-relief phase grating as an approximation to the ideal continuous profile. This has been demonstrated using a set of four masks to make a 1-D, 32-beam array generator based on a 16-level phase grating. Details of these techniques are presented in the accompanying paper by Taghizadeh et al. [9].

3 Low-level Space-Invariant Fan-Out

As an example of a low-level fan-out device, we discuss the production of a nearest-neighbour interconnect (NNI) for digital image processing applications based on holographic optical elements (HOEs) recorded in dichromated gelatin (DCG). We have previously fabricated a transmission HOE for the visible which performs such an interconnection – details can be found in reference [10]. The interconnect used a space-invariant volume phase fan-out hologram to diffract light in the required directions (see Fig. 2).

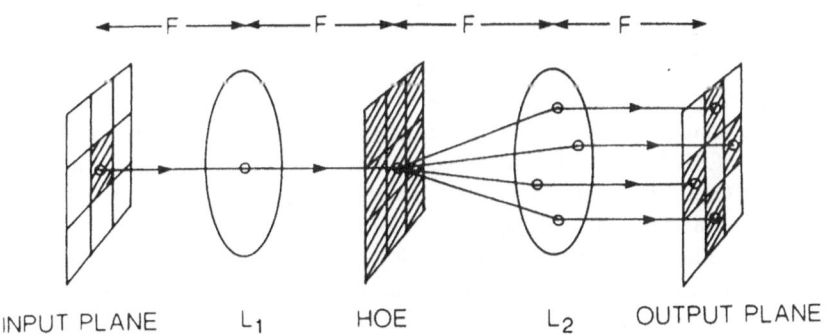

Fig. 2. Schematic showing a transmission HOE being used in a $4 - f$ imaging system to provide a 1-to-4 nearest-neighbour interconnect.

This approach to a fan-out interconnect was dropped as the basis for an NNI hologram because in the digital image processing system (O-CLIP [3]) in which it was to be used, the array of pixels is coherently illuminated. The interconnect attempts to fan-in nearest neighbour signal beams into the same spatial mode. For this to be done with acceptable efficiency and good uniformity they would have to interfere constructively at the HOE. If this interference condition is not met at all times across the whole plane of the HOE (which in all probability will be the case, due to either vibration and/or slight misalignment of optics), then optical power will remain in the zero-order beam of the hologram and will not be interconnected. Two solutions can be identified to overcome this fan-in problem. Firstly, if the acceptance areas of the devices (to which the signals are being directed) are sufficiently large, then incoming beams can be focussed as spatially-separated, and therefore non-interfering, adjacent spots. Alternatively, the signal beams can be spatially separated at the HOE so that signals fanning in together only overlap at the output plane. Interference effects then simply modulate the irradiance within the focussed spots, rather than direct power elsewhere. This requires the hologram to be aperture divided. Thus, for a 1-to-4 NNI the hologram is spatially split into four regions, each containing a separate grating which will interconnect light into the correct direction. The HOE remains a Fourier plane device and retains its spatially invariant nature.

The inevitable penalty for adopting the aperture division approach is that in halving the aperture of each redirecting grating the f-number of the interconnected signal beam is doubled, resulting in a larger diffraction limited spot size for each signal in the output plane. This is exactly equivalent to the alternative spot-separation approach described above and is a direct result of the constant brightness theorem, which, including the inherent fan-out loss, fixes a maximum theoretical efficiency, in terms of irradiance averaged over the spot area, of 1/16th. This results from a combination of factor of four for the 1-to-4 fan-out and a further factor of four for the doubled spot sizes. These problems are general to all forms of incoherent spatially invariant interconnection schemes. Even if interferometric precision could be guaranteed with coherent addition there are other problems associated with multiplexed fan-out HOEs, outlined in reference [5], which will still limit the realisable efficiency to a comparable degree.

In addition to incorporating aperture division into the NNI design, it was also concluded that it would be advantageous to use a reflection HOE geometry. Transmission HOEs have a relatively narrow Bragg acceptance angle compared to reflection HOEs and consequently an optical system that performs the interconnect using a transmission HOE will be long, (e.g. metres) because of the need to maintain low field angles with long focal-length lenses. Reflection HOEs can have approximately twice the angular bandwidth (measured from diffraction maximum to minimum) compared to a transmission HOE modulated to the same level. In addition, this peak of efficiency in the angular spectrum is very much squarer for a reflection HOE; typically four to five times at full-width-half-maximum for the same modulation depth. Implementing the NNI in reflection allows the overall system to shrink to a more practical and compact size.

The reflection NNI HOE that was produced (in DCG) had a bandwidth at full-width-half-maximum of 60 nm, corresponding to an angular bandwidth of about ±18 degrees, which easily meets the requirements of the image processing system for which it was designed. It had an efficiency in excess of 99% and an angular response of ±4 degrees with no measurable loss of efficiency (less than 0.1%). When tested with a full 16 × 16 array of input signals, the interconnected array was visually promising, however, the spot positioning showed small errors (corresponding to < 0.37 mrad deviations) due to imperfect control of the recording conditions. Nonetheless, this work has shown that such a spatially invariant interconnect is viable and that, as a result of analysing various techniques, an optimum approach has been identified.

4 Irregular Space-Variant Interconnects

In recent years there has been an increasing interest in free-space optical interconnection networks suitable for use in communication and computing systems. These include several multistage architectures, for example the perfect shuffle, cross-over, Banyan and butterfly networks, all of which require the spatial permutation of an array of (signal carrying) light beams. We have developed such space-variant interconnects, recorded in dichromated gelatin (DCG), utilising its high space-bandwidth product, efficiency and uniformity. They provide point to point on-axis interconnection with a moderately-high packing density and can be manufactured for any wavelength from the visible to the near infra-red.

The interconnects are implemented using coupled holographic array elements. The basic unit, illustrated in Fig. 3, is a doublet made up of a holographic microlens array and a planar grating redirection array, with the two holographic structures aligned and cemented together.

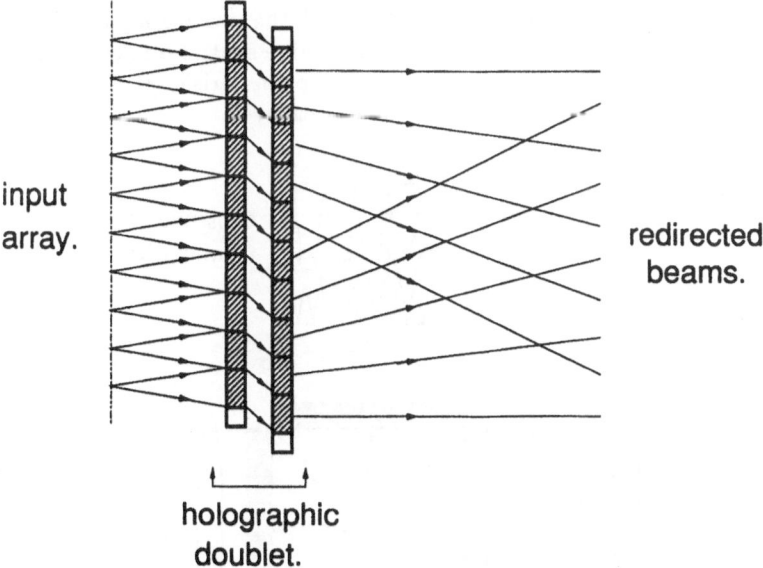

input array.

redirected beams.

holographic doublet.

Fig. 3. A doublet-HOE structure corresponding to half of a point-to-point (on axis) spatially variant optical interconnect.

The lens stage collimates an array of (signal carrying) light beams of a specified divergence f-number into an array of plane waves at a fixed carrier angle. (This off-axis nature satisfies the spatial frequency requirements for an efficient volume hologram.) The redirecting stage receives the array of plane waves at the fixed carrier angle and redirects them in free space in accordance with the spatial permutation required by a specific interconnect. These two off-axis HOE arrays are designed to have a common carrier angle, so that when they are cemented together (with an index matched adhesive) they behave as a single on-axis element. A second similar doublet completes the interconnect, taking the re-arranged signals and refocussing them onto the next device plane. With a symmetric interconnect, a reflection geometry, with a second pass through the input doublet, can provide a compact alternative.

The collimating/focussing and the redirecting functions are divided between the two separate HOEs forming each doublet, both of which being optimised for use at the operation wavelength. Shifts between the recording and replay wavelengths are straightforward to implement for planar gratings and shifted focussing elements have been demonstrated by Redmond et al. [10]. The compound doublet automatically works at the wavelength for which the individual elements are designed.

Of all the space-variant interconnects, the full perfect shuffle is the most flexible. It can be built up from two sub-stages, a perfect shuffle and a crossover, as shown in Fig. 4. Note that, in this scheme, a 1-to-2 fan-out is provided by the polarising beam splitter; the crossover is conveniently implemented in a reflection geometry; and quarter-wave plates are used to ensure the correct beam paths between the input and output planes. Although the two overlapping interconnected arrays in the output plane fan-in to the same spatial modes, because they are orthogonally polarised no interference problems arise. At the output plane a full perfect shuffle interconnection stage is achieved.

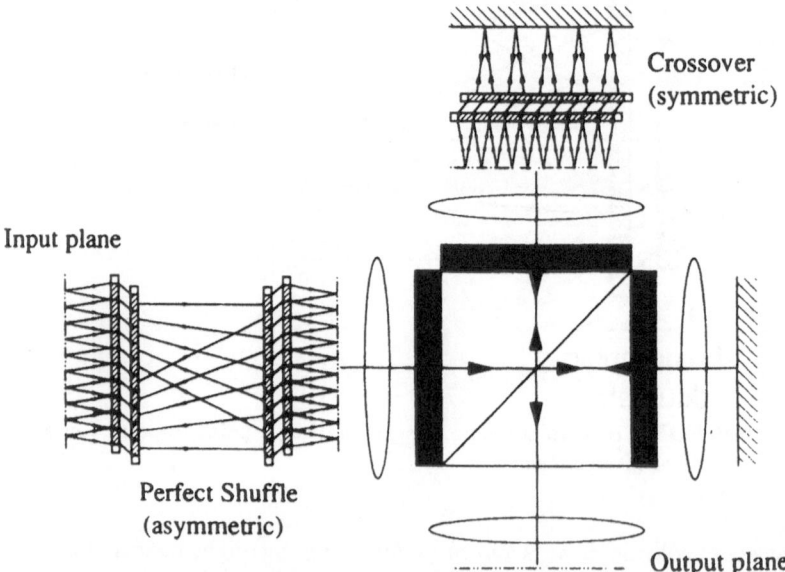

Fig. 4. Schematic of a full-shuffle interconnect. Half the shuffled input goes directly to the output plane (via the mirror on the right) and the other half via the crossover.

A demonstration of a symmetric interconnect working in this reflective orientation (at 514 nm) is shown in Fig. 5 [11]. This example shows the first stage of a two-dimensional Banyan interconnect. The input image was a 16 × 16 square array, masked diagonally. The effect of this interconnect is to swap diagonally opposed quadrants of the image, as shown correctly implemented in Fig. 5. The interconnect HOE had a facet spacing of 200 μm, was designed for use with $f/5$ light beams and had an interconnection distance of 5 cm. The total efficiency of such an interconnect is about 80% once sealed and anti-reflection coated.

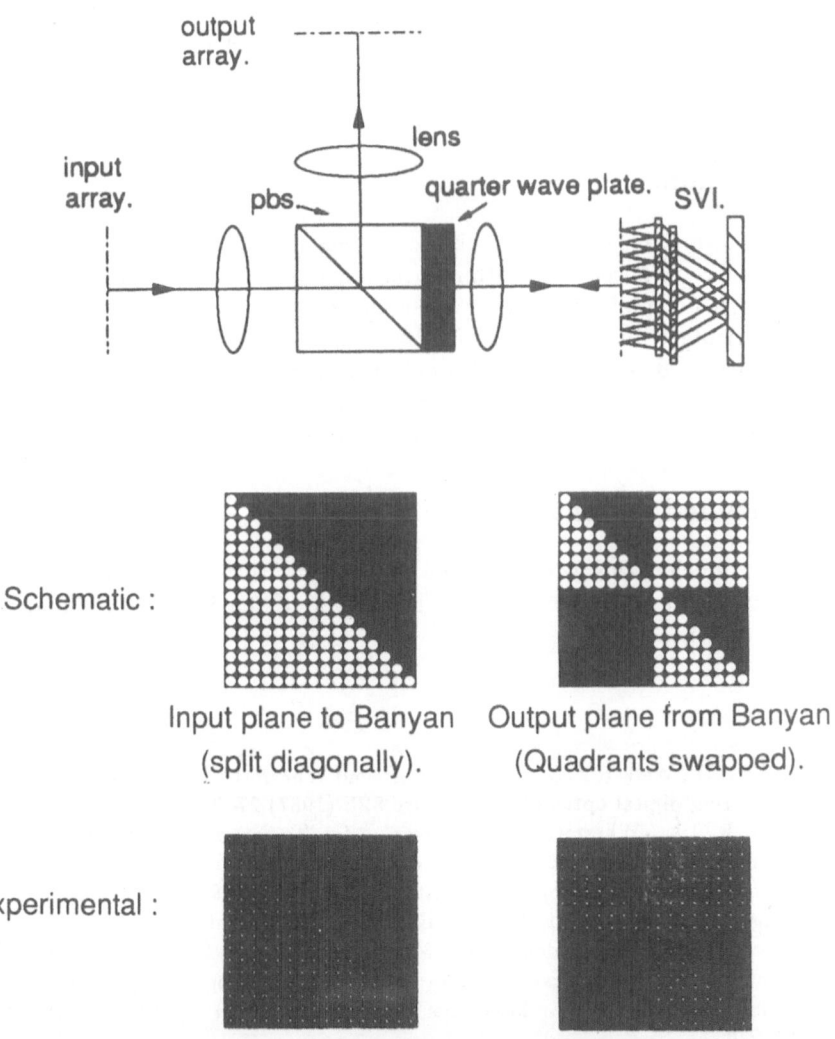

Fig. 5. The top diagram shows the layout used to demonstrate a reflective (symmetric) 2-D Banyan interconnect (1st stage). Below an example of its operation, first schematically and finally the result recorded from the experimental implementation.

Crosstalk and scalability problems need to be addressed before optical interconnects of this type can be considered as a practical way of creating complex,
routing systems. For example, the microbeams being routed are typically 200
μm in diameter and will undergo a certain amount of diffractive spreading in
the interconnection distance. This leads to crosstalk and a drop in throughput
efficiency, while alignment and fabricational tolerances can lead to a further
degradation in interconnect uniformity. Many paper studies, based on Gaussian
beam optics, have been made concerning these problems, for example McCormick
et al. [12], but none have been conclusive as to the ultimate limits. Preliminary
studies indicate that the positional and focal-length uniformity tolerances of
the collimating/focussing element of the doublet is a critical factor. If this is
the case then it may be that surface-relief diffractive or refractive microlenses,
rather than volume holographic components, will be better suited to making
such interconnects.

5 Conclusions

We have shown that holographic/diffractive techniques can satisfy a wide range
of interconnect requirements within experimental optical computing systems.
This remains a rapidly developing research area, with many new advances in
sight – particularly in the context of surface-relief diffractive optical elements. It
can be anticipated that the optical processing systems of the future will rely on
holographic fan-out/fan-in components of this sort, closely integrated into the
compact optomechanical assemblies that are also currently under development.

6 Acknowledgement

Support for much of this work has come through the UK Science and Engineering
Research Council, partly within the framework of the Scottish Collaborative
Initiative on Optoelectronic Sciences.

References

1. Smith, S.D., Walker, A.C., Tooley, F.A.P. and Wherrett, B.S.: The demonstration
   of restoring digital optical logic. Nature 325 (1987) 27–31.
2. Craig, R.G.A., Wherrett, B.S., Walker, A.C., Tooley, F.A.P. and Smith, S.D.:
   The optical cellular logic image processor: implementation and programming of
   a single channel digital optical circuit. Appl. Opt. 30 (1991) 2297–2308.
3. Craig, R.G.A., Wherrett, B.S., Walker, A.C., McKnight, D.J., Redmond, I.R.,
   Snowdon, J.F., Taghizadeh, M.R., MacKinnon, G. and Smith, S.D.: First programmable digital optical processor: optical cellular logic image processor. In
   Optics for Computers: Architectures and Technologies. Proc. SPIE 1505 (1991)
   264–270.
4. McKnight, D.J., Redmond, I.R., Walker, A.C., Taghizadeh, M.R., Buller, G.S.,
   Mathew, J.G.H. and Smith, S.D.: Parallel optical digital data transfer between
   bistable logic arrays. Opt. Comp. Proc. 2 (1991) 137–144.

5. Restall, E.J., Redmond, I.R. and Walker, A.C.: A nearest neighbour spatially invariant reflection interconnect in dichromated gelatin Proc. Third Int. Conf. on Holographic System Components and Applications IEE Conference Vol. **342** (1991) 40–44.

6. Robertson, B., Restall, E.J, Taghizadeh, M.R. and Walker, A.C.: Space-variant holographic optical elements in dichromated gelatin. Appl. Opt. **30** (1991) 2368–2375.

7. Dammann, H. and Görtler, K.: High-efficiency in-line multiple imaging by means of multiple phase holograms. Opt. Commun. **3** (1971) 312–315.

8. Robertson, B., Turunen, J., Ichikawa, H., Miller, J.M., Taghizadeh, M.R. and Vasara, A.: Hybrid kinoform fanout holograms in dichromated gelatin. Appl. Opt. **30**, (1991) 3711–3720.

9. Taghizadeh, M.R., Turunen, J., Ichikawa, H., Miller, J.M., Robertson, B., Blair, P., Ross, N., Vasara, A., Byckling, E., Jaakkola, T., Noponen, E. and Westerholm, J.: Binary, multilevel and hybrid holographic optical array illuminators. These proceedings.

10. Taghizadeh, M.R., Redmond, I.R., Robertson, B., Walker, A.C. and Smith, S.D.: High efficiency holographic optical elements for all-optical digital computing. In Holographic Optics II: Principles and Applications. Ed. G.M. Morris. Proc. SPIE **1136** (1989) 265-274.

11. Restall, E.J., Robertson, B., Taghizadeh, M.R. and Walker, A.C.: Two-dimensional non-local multi-facet holographic interconnects in dichromated gelatin. In Proc. Third Int. Conf. on Holographic System Components and Applications in IEE Conference Vol. **342** (1991) 127–131.

12. McCormick, F.B., Tooley, F.A.P., Cloonan, T.J., Sasian, J.M. and Hinton, H.S.: Microbeam optical interconnections using microlens arrays OSA Proc. Photonic Switching, H.S. Hinton and J.W. Goodman eds. (OSA Washington DC.) Vol. **8** (1991) 90–96.

Binary, Multilevel, and Hybrid Holographic Optical Array Illuminators

M. R. Taghizadeh,[1] J. Turunen,[1] H. Ichikawa,[1] J. M. Miller,[1]
B. Robertson,[1] P. Blair,[1] N. Ross,[1] A. Vasara,[2] E. Byckling,[2]
T. Jaakkola,[2] E. Noponen,[2] and J. Westerholm[2]

[1] Department of Physics, Heriot-Watt University,
Riccarton, Edinburgh EH14 4AS, UK
[2] Department of Technical Physics, Helsinki University of Technology,
SF-02150 Espoo, Finland

1 Introduction

Since the recognition of the importance of optical array illuminators [1] in the realization of parallel digital optical processors, a wide variety of optical components have been designed and demonstrated that convert a single laser beam into a regularly-spaced array of $M \times N$ equal-intensity light spots. Space-variant array illuminators, which form the spot array in a Fresnel plane or an image plane of the aperture are in general easy to design and fabricate, but they require uniform plane-wave illumination that is difficult to provide. Space-invariant array illuminators generate the spot array in the Fourier plane and are therefore rather immune to the exact shape of the incident beam, but they are more difficult to design and require tight fabrication tolerances.

Here we compare three techniques of realizing space-invariant optical array illuminators by means of diffraction gratings with a computer-synthesized periodic structure. The spot separation and the compression ratio in the Fourier plane can then be controlled straightforwardly by choosing a suitable grating period and the size of the illuminating beam, respectively. The relative intensities of the diffraction orders are, however, highly nonlinear functions of the grating structure, even in the domain where the paraxial scalar diffraction theory is valid.

We discuss first binary and multilevel surface-relief gratings fabricated directly on fused silica by microlithographic methods involving electron-beam written masks and reactive ion etching. Additionally, we apply the method of hybrid holography [2] to record (in dichromated gelatin) the object wave produced by spatially filtering the output of an electron-beam written binary-amplitude master grating. The potential and the limitations of these techniques are compared quantitatively, using 32×1 array illuminators as an illustration.

2 Binary Array Illuminators

Dammann pioneered the use of binary synthetic diffractive optical elements to produce equal-intensity spot arrays [3]. Because of the inherent inversion symmetry of the power spectrum of a binary grating, the generation of highly desirable array sizes $M \times N = 2^K \times 2^L$ (K, L integers) requires that all even orders are extinguished by utilization of the phase-profile symmetry

$$\phi(x, y) = \phi(x + d_x/2, y + d_y/2) = \phi(x, y + d_y/2) + \pi = \phi(x + d_x/2, y) + \pi \ , \quad (1)$$

where d_x and d_y denote the grating periods in x and y directions. Using this symmetry, we have designed and fabricated both separable and non-separable binary array illuminators. The separable designs were obtained using the method of simulated annealing as presented in [4]; some of the results are collected in Table 1. The non-separable profiles were designed assuming the stripe-geometry of [5], but using polygonal instead of rectangular apertures to provide additional degrees of freedom [6]. Some of these results are collected in Table 2. These efficiencies could be increased slightly by using a larger number of stripes. The main conclusion is that the non-separable designs are particularly valuable for low fan-out ($< 10 \times 10$), while for large fan-out the gain in efficiency is about 10%.

Table 1. One-dimensional and crossed two-dimensional binary grating designs. Here η_{1D} is the one-dimensional efficiency and $\eta_{2D} = \eta_{1D}^2$ is the efficiency of the crossed design

N	η_{1D}	η_{2D}	N	η_{1D}	η_{2D}
2	0.811	0.657	64	0.812	0.659
4	0.707	0.499	128	0.804	0.646
8	0.762	0.521	256	0.808	0.653
16	0.816	0.666	512	0.810	0.656
32	0.824	0.679	1024	0.806	0.650

Table 2. Non-separable two-dimensional binary grating designs: η is the efficiency and P is the number of stripes per half-period

$M \times N$	η	P	$M \times N$	η	P
2×2	0.657	any	4×2	0.747	8
4×4	0.775	12	8×4	0.763	16
8×8	0.756	20	16×8	0.756	20
16×16	0.762	40	32×16	0.753	40
32×32	0.745	75	64×32	0.745	80

A large number of binary array illuminators have been by fabricated on fused silica by reactive ion etching, using photolithography to transfer the pattern of an electron-beam-written mask into photoresist. The transverse accuracy of the mask is 0.1 μm and the etch depth accuracy is about 10 nm.

3 Multilevel Array Illuminators

The diffraction efficiencies of optical array illuminators can be increased by relaxing the phase quantization constraints, which leads to the kinoform technique [7]. The symmetry of Eq. (1) can still be used but it is no longer necessary for obtaining even fan-out. The results given in Tables 3 and 4 for separable and non-separable kinoform array generators, respectively, were obtained using methods based on [8] and [9]. The diffraction efficiencies given here are for unconstrained phase profiles. If only $K = 2^N$ equally spaced levels are allowed, the efficiency is given approximately by $\eta_K = \eta \sin^2(\pi/K)/(\pi/K)^2$. Comparing the results of Tables 3 and 4, it is clear that the non-separable designs again have a significant advantage for small arrays, but the difference in efficiency is rather negligible for large arrays ($> 10 \times 10$).

Table 3. One-dimensional and crossed two-dimensional phase grating designs with unrestricted phase profiles

N	η_{1D}	η_{2D}	N	η_{1D}	η_{2D}
2	0.811	0.657	16	0.970	0.941
4	0.917	0.841	32	0.971	0.943
8	0.959	0.920	64	0.965	0.931

Table 4. Non-separable two-dimensional multilevel phase grating designs with unrestricted phase levels

$M \times N$	η	$M \times N$	η
2×2	0.916	4×2	0.917
4×4	0.949	8×4	0.909
8×8	0.920	16×8	0.915

Several multilevel gratings have been fabricated using three or four electron-beam-written masks and reactive ion etching steps to generate eight or sixteen relief depth levels, respectively, in fused silica. A mask alignment accuracy of 0.5–1 μm and an etch depth accuracy of about 10 nm were achieved routinely; the diffraction efficiencies of the multilevel gratings were found to be within 2–7% of the design values.

4 Hybrid Array Illuminators

Merging the techniques of synthetic and optical holography [2] allows one to construct optical array illuminators that reconstruct a nearly phase-only wavefront without the need to fabricate the actual surface relief structure. This is achieved by spatially filtering the output of a binary amplitude mask, which can be fabricated with a high degree of accuracy by direct-write electron-beam lithography. Binary-phase wavefronts can be reconstructed by filtering out the zeroth order only [10], while continuous-phase wavefronts are derived by passing through only one diffraction order (containing the desired array) of a (pulse width and density modulated) carrier grating [9]. Using the latter method, it is in principle possible to record optical volume holograms that reconstruct the array at the efficiency predicted in Tables 3 and 4. In addition, a focusing lens can be integrated in the element.

5 Experimental Results

To provide a quantitative comparison of the relative merits of the three methods of realizing optical array illuminators described above, we consider three array illuminators with comparable parameters. The fan-out is 32×1, and a period is 1 mm. We present the results in the form of line scans across the array images in Fig. 1.

Figure 1 clearly shows the relative intensities of the higher, undesired diffraction orders outside the desired array. These are much more prominent in Fig. 1a than in Figs. 1b and 1c, reflecting the higher efficiency of the kinoform designs. The array uniformities can be estimated roughly from the line scans: the exact values were measured by detecting the integrated intensity of each spot in the array. The measured uniformities are $\pm 5\%$, $\pm 8\%$, and $\pm 11\%$ for the binary, 16-level, and hybrid elements, respectively.

The compression ratio (spot spacing/spot size) is above 4 in every case (although this can not be seen from the scans because of the finite slit width). The fact that the compression ratio of the array in Fig. 1a appears much better than that of the kinoforms follows from the use of symmetry (1), which means that the effective period of the binary grating is only 0.5 mm.

6 Discussion and Conclusions

When comparing the relative merits of the three different fabrication techniques, one must consider (in addition to the complexity of design which is of little concern as long as only regular fan-out is required) the complexity of the fabrication procedure, its suitability to large-scale production, and the ease of use of the resulting elements.

The binary gratings are the easiest to fabricate and, along with the multilevel elements, the easiest to use since they reconstruct the array on-axis. Of these two, the multilevel approach offers higher efficiency but large-scale fabrication requires low-cost replication techniques. Hybrid holograms can be replicated by optical holography but the present recording material (dichromated gelatin) requires careful processing.

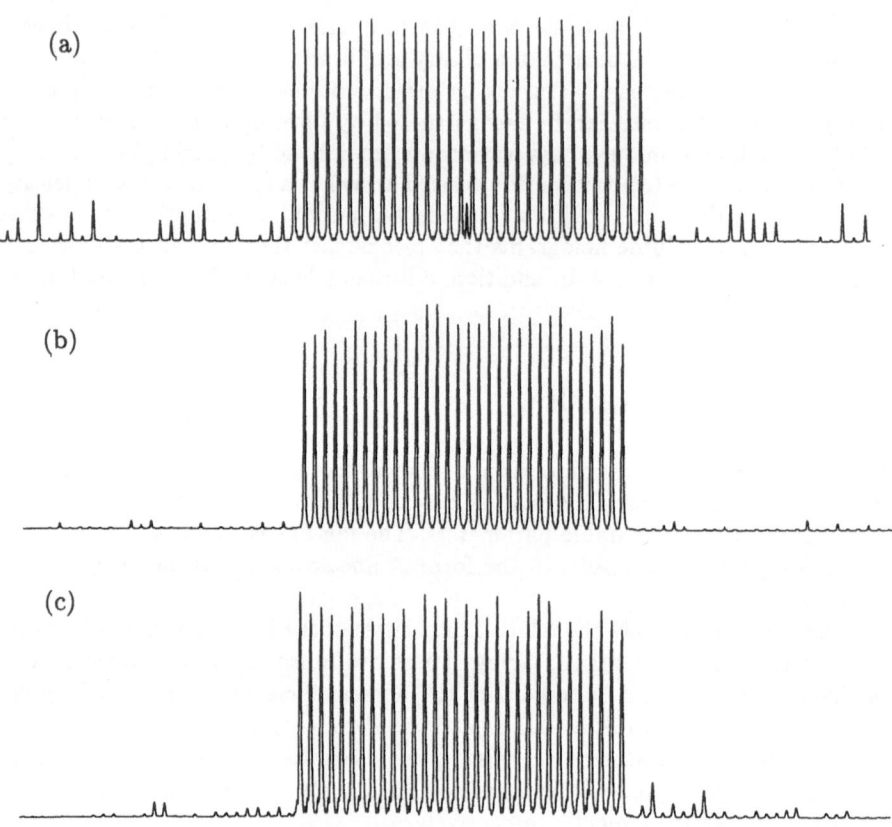

Fig. 1. (a) The array generated by a binary 32×1 array illuminator (b) The array generated by a 16-level 32×1 array illuminator (c) The array generated by a hybrid 32×1 array illuminator

References

1. Streibl, N.: Beam shaping with optical array generators. J. Mod. Opt. **36** (1989) 1559–1573.
2. Bartelt, H., and Case, S. K.: High-efficiency hybrid computer-generated holograms. Appl. Opt. **21** (1982) 2886–2890.
3. Dammann, H., and Görtler, K.: High-efficiency in-line multiple imaging by means of multiple phase holograms. Opt. Commun. **3** (1971) 312–315.
4. Turunen, J., Vasara, A., Westerholm, J., Jin, G., and Salin, A.: Optimisation and fabrication of grating beamsplitters. J. Phys. D: Appl. Phys. **21** (1988) S102–S105.
5. Turunen, J., Vasara, A., and Westerholm, J.: Stripe-geometry two-dimensional Dammann gratings. Opt. Commun. **74** (1989) 245–252.
6. Vasara, A., Taghizadeh, M. R., Turunen, J., Westerholm, J., Noponen, E., Ichikawa, H., Miller, J. M., Jaakkola, T., and Kuisma, S.: Binary surface-relief gratings for array illumination in digital optics. Appl. Opt. **31** (1992) 3320–3336.

7. Lesem, L. P., Hirsch, P. M., and Jordan, J. A.: The kinoform: a new wavefront reconstruction device. IBM J. Res. Dev. **13** (1969) 150–155.
8. Turunen, J., Vasara, A., and Westerholm, J.: Kinoform phase relief synthesis: a stochastic method. Opt. Eng. **28** (1989) 1162–1167.
9. Robertson, B., Turunen, J., Ichikawa, H., Miller, J. M., Taghizadeh, M. R., and Vasara, A.: Hybrid kinoform fanout holograms in dichromated gelatin. Appl. Opt. **30** (1991) 3711–3720.
10. Robertson, B., Taghizadeh, M. R., Turunen, J., and Vasara, A.: High-efficiency, wide-bandwidth optical fanout elements in dichromated gelatin. Opt. Lett. **15** (1990) 694–696.

Part III

II-VI-Compound Nonlinearities

Wide Bandgap II-VI Light Emitting Devices

B. C. Cavenett, K. A. Prior, S. Y. Wang and J. Simpson

Department of Physics, Heriot-Watt University, Edinburgh EH14 4AS, UK

The prospect of wide bandgap light emitting diodes (LEDs) and lasers has been transformed by recent advances in p-doping in materials such as ZnSe grown by molecular beam epitaxy. We report the growth of p-n ZnSe junctions on GaAs substrates using iodine as the n-type dopant and nitrogen from a plasma discharge source as the p-type dopant. LEDs have been fabricated using a gold contact to the p-type layer and blue CW emission has been observed under forward bias. Stripe geometry laser structures have been fabricated and blue stimulated emission has been observed for the first time at low temperatures.

1 Introduction

The II-VI semiconductors have been investigated for many years because of the potential for wide bandgap laser and optoelectronic devices which would find applications in printers, video discs and displays [1]. However, until recently, it has not been possible to dope these semiconductors both n- and p-type, because of self-compensation processes which have limited the concentration of the active dopant species [2]. For example, in the case of ZnSe which has a bandgap of 2.7 eV and so can potentially be used for blue light emitting devices, the residual conductivity is always n-type and free carrier concentrations above 10^{18} cm^{-3} can be achieved with a wide variety of dopants such as Ga, In, Cl and I.

Potential dopants for p-type conductivity come from the groups I and V of the periodic table and over a period of many years most of these elements have been examined in both bulk crystals and epitaxial layers. Only lithium showed some potential with doping levels of acceptors up to 8×10^{16} cm^{-3} being achieved [3] but this atom is very mobile and can diffuse, particularly during any form of heat treatment involved in processing. Also, since lithium interstitial atoms act as donors, auto-compensation can occur making the material unsuitable for practical devices.

A major breakthrough was made by Ohkawa et al. [4] and Park et al. [5] who both used a commercial (Oxford Applied Research) plasma source for nitrogen doping and although the active species of nitrogen has not yet been identified, acceptor levels $(N_A - N_D)$ of 10^{18} cm^{-3} have been achieved [6]. This breakthrough quickly led to the fabrication of laser structures and in 1991 the 3M company demonstrated the first II-VI laser diode operating in pulsed mode at 77 K [7]. Since then, room temperature pulsed operation and 77 K CW operation have been achieved [8]. These lasers have been based on highly efficient recombination within an undoped CdZnSe quantum well set at the centre of a ZnSe or ZnSSe waveguide but the emission is in the blue-green spectral region.

We describe in this paper the fabrication of ZnSe p-n junctions using iodine from a novel electrochemical source and nitrogen from a plasma source and the demonstration of blue stimulated emission from a stripe geometry laser structure.

2 Experimental Results

Epitaxial layers of ZnSe have been grown on GaAs substrates using a VG Semicon 288 Molecular Beam Epitaxy system. Conventional Knudsen cell sources were used for zinc and selenium with a beam equivalent pressure ratio of Se:Zn in the range 1 to 4 as measured by a movable ion gauge. Layers of less than the critical thickness of approximately 1500 Å are pseudomorphic and at low temperatures only the free exciton split by the compressive strain is observed in the photoluminescence indicating the high purity of the layers. Typical strain relaxed layers of $2 - 3$ μm were grown with growth rates of approximately 0.5 μmh^{-1} for a substrate temperature of 280 °C as measured by an optical pyrometer.

The iodine doping was achieved using an electrochemical cell as shown in Fig. 1. A pellet of AgI is compressed between a gauze platinum anode and a silver cathode. The Ag$^+$ ions are mobile when the cell is heated to above 146 °C and so a flux of iodine is produced which is proportional to the current flowing through the cell. The n-type material has been assessed by photoluminescence and details of spectra for thick films are given in Fig. 2 where in (a) the results for an undoped thick film are shown. Both free and donor bound exciton emissions are observed, each split by the residual strain due to the difference in expansion coefficients between ZnSe and GaAs. Curves (b) and (c) show the strong (D°, X) emission for layers doped with iodine at concentrations of 10^{17} cm^{-3} and 10^{18} cm^{-3} respectively. The carrier concentrations have been determined by conventional Hall Effect as well as electrochemical C-V profiling using a Biorad profiler with a NaOH/NaSO$_3$ electrolyte. The material can be profiled to unlimited depth and the measurements have confirmed that high quality uniformly doped material can be grown.

Nitrogen from an Oxford Applied Research plasma source has been used to dope ZnSe p-type. The photoluminescence shows that a high degree of compensation takes place at carrier concentrations of $\approx 5 \times 10^{17}$ cm^{-3} and, although it is possible to incorporate up to 10^{19} cm^{-3} atoms of nitrogen, the $(N_A - N_D)$ is limited to 10^{18} cm^{-3} [6].

We have fabricated p-n junctions of ZnSe grown on n$^+$GaAs substrates and using the iodine and nitrogen dopants a 1.5 μm n-type layer was grown followed by a 1.5 μm p-type layer. The carrier concentrations were measured by electrochemical C-V profiling and as can be seen in Fig. 3 the n-type layer has $(N_D - N_A) = 10^{18}$ cm^{-3} and the p-type layer has $(N_A - N_D) = 3 \times 10^{17}$ cm^{-3}.

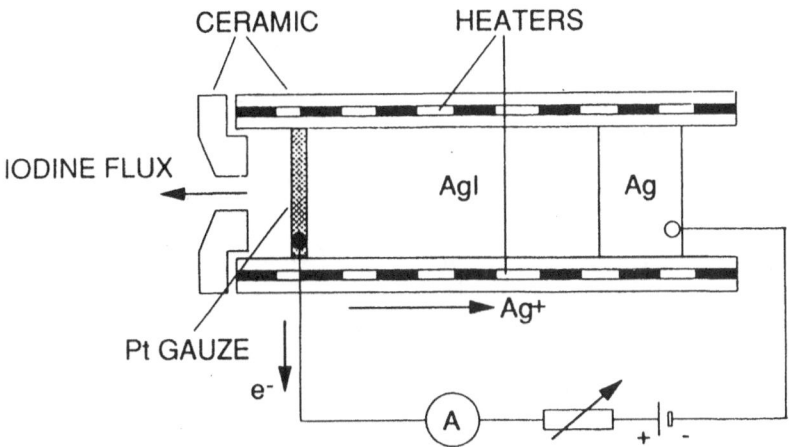

Fig. 1. Electrochemical iodine source with flux proportional to the cell current

The spike at the interface in the CV profile is an artifact of the measuring technique.

Light emitting diode structures were fabricated by evaporating gold dot (0.5 mm) contacts onto the p-layer and cleaving 1 mm × 2 mm devices. Under forward DC or pulsed bias blue emission has been observed at temperatures as low as 4 K, though at room temperature the CW emission also shows a broad deep level emission in the region associated with native defects such as zinc vacancies. The room temperature blue electroluminescence using pulsed excitation is shown in Fig. 4 and it can be seen that the deep level recombination is weak.

Laser diode structures have also been fabricated using a stripe gold contact 0.2 μm wide on the p-type layer. These structures were cleaved to 1 mm in length and excited with 0.5 μsec pulses in forward bias. At low currents (15 mA) broad spontaneous emission was observed from the cleaved end of the laser structure and this is shown in Fig. 5 curve (a). As the current is increased stimulated emission in the form of spikes dominated the output and the onset of these spikes at 50 Acm^{-2} was taken as the threshold for stimulated emission. The spectrum above the threshold is shown in Fig. 5 curve (b) where it can be seen that strong emission is obtained from 430-470 nm with a clear cutoff at 470mm. Details are given elsewhere [10].

In general, a measurement of the light output intensity versus the applied current shows a threshold corresponding to laser emission. However, in our ZnSe devices the lack of carrier confinement results in a strong emission from the ZnSe/GaAs interface and, in fact, as the current through the device increases the stimulated emission saturates and the interface emission at 800 nm increases rapidly.

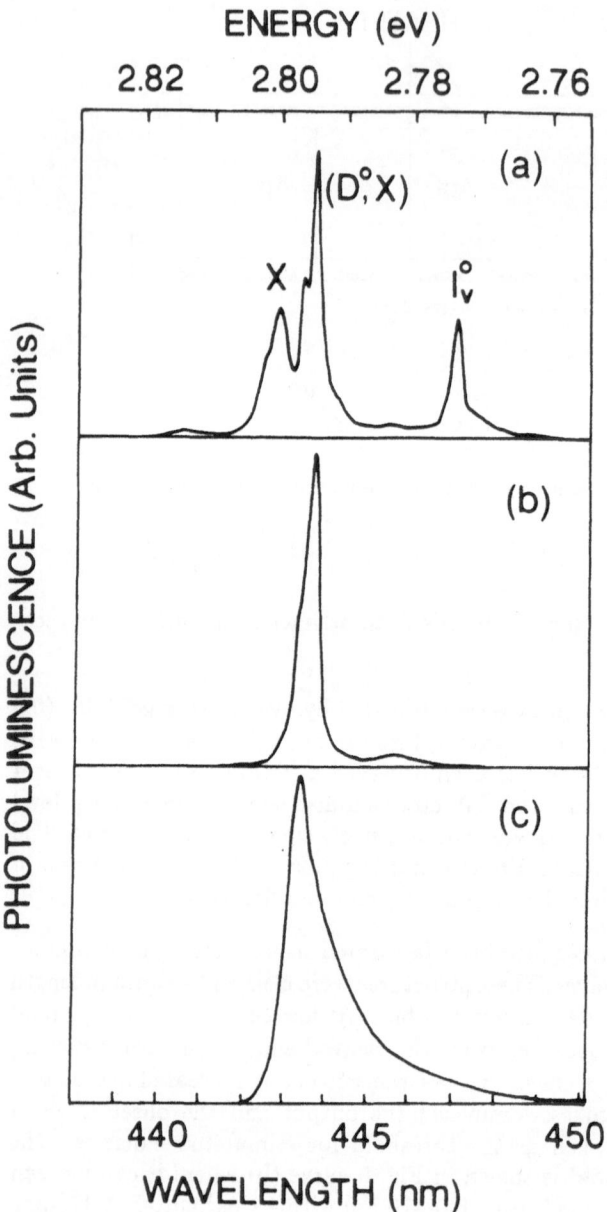

Fig. 2. Photoluminescence spectra at 4.2 K of (a) undoped ZnSe (b) iodine doped ZnSe ($n = 10^{17}$ cm^{-3}) (c) iodine doped ZnSe ($n = 10^{18}$ cm^{-3})

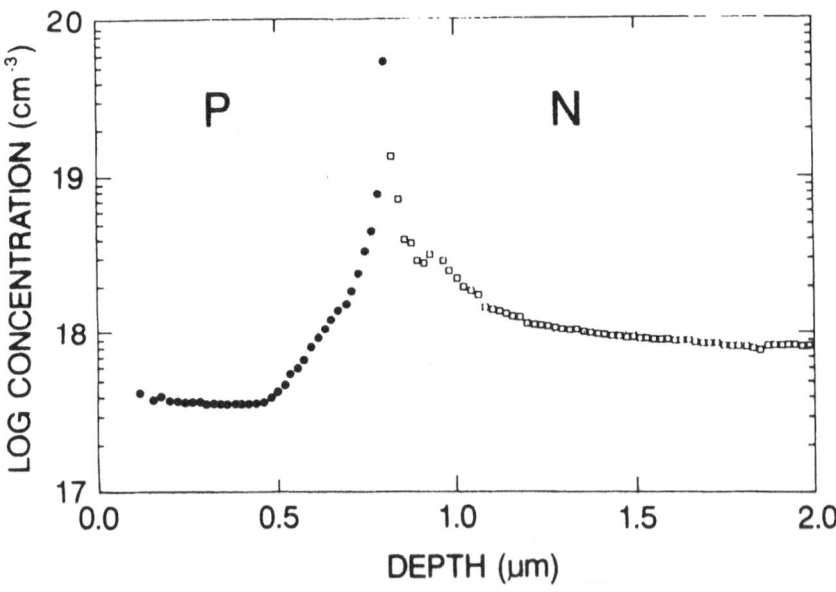

Fig. 3. Electrochemical C-V profile of carrier concentration through a ZnSe p-n diode structure

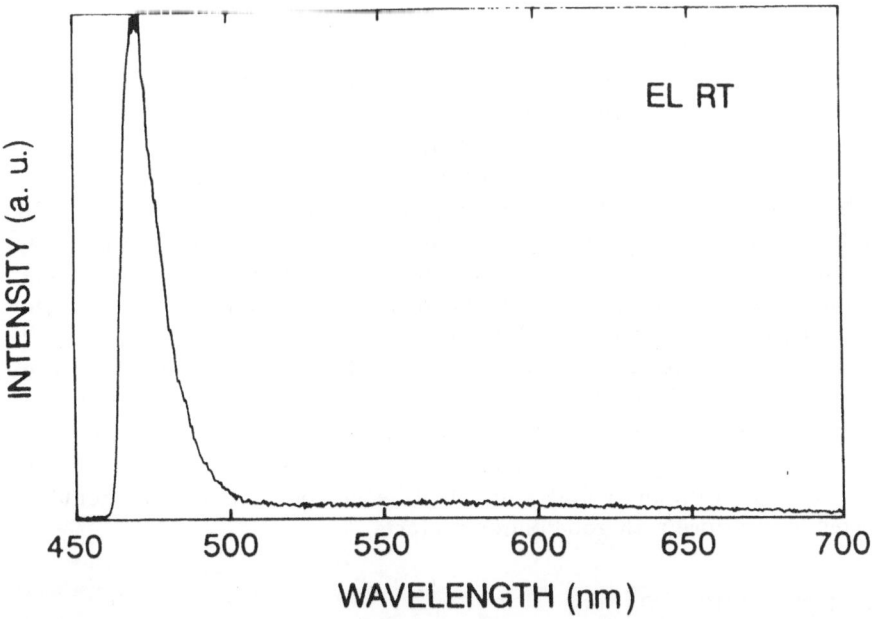

Fig. 4. Room temperature electroluminescence from a ZnSe p-n diode structure

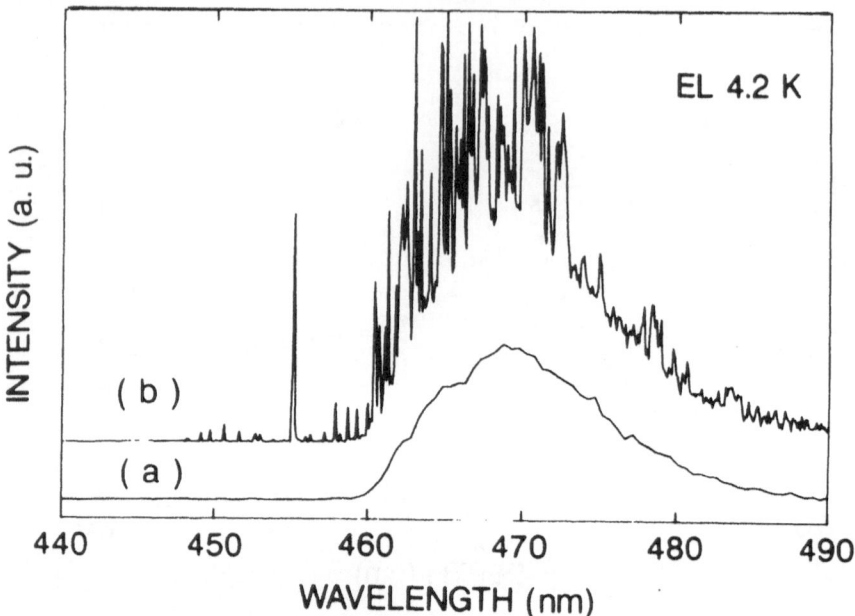

Fig. 5. Emission at 4.2 K from a stripe geometry ZnSe homojunction laser (a) Below threshold (15 mA) (b) Above threshold (150 mA)

3 Discussion

Nitrogen doping of ZnSe has allowed the possibility of device fabrication centred around p-n junctions. These include LEDs, laser diodes and a variety of modulator structures. At this stage of the research it is not clear what is the active species of nitrogen produced in the discharge. This could be atomic nitrogen or excited N_2^* dimers. However, it is clear that at incorporation levels, as measured by SIMS, of greater than 10^{16} cm^{-3} a compensating donor is formed which limits the active number of nitrogen acceptors to mid 10^{17} cm^{-3}. This process is clearly observed in the photoluminescence where, in contrast to the data shown in Fig. 2, as the nitrogen incorporation is increased the (A°, X) emission broadens and a strong DAP emission appears which eventually dominates the spectrum [11]. This problem of doping contributes to the difficulties in forming an ohmic contact with the p-type ZnSe. Metals with large work functions such as Pt and Au have had most success but in both cases a large Schottky barrier is formed which provides a high series resistance in a forward biased device such as a laser diode. In fact, typical "turn-on" voltages are 20 V instead of a few volts as would be expected for a device with good ohmic contacts.

4 Conclusions

Blue emitting devices can be fabricated by the growth of ZnSe on GaAs by molecular beam epitaxy. Both n- and p-type material can be characterized by electrochemical C-V profiling and both p-n LEDs and laser diode structures can be fabricated. The latter show that blue laser emission can be obtained in a device with poor optical and electrical confinement.

Acknowledgements

We are grateful for support from SERC and VG Semicon Ltd.

References

1. Bhargava, R. N.: J. Crystal Growth **117** (1992) 894.
2. Kukimoto, H.: In Growth and Optical Properties of Wide-Gap II-VI Low-Dimensional Structures. NATO ASI series B, **200**, Eds. T.C. McGill, C.M. Sotomayor Torres and W. Gebhardt (Plenum, New York, 1989) 119.
3. DePuydt, J. M., Haase, M. A., Cheng, H., Potts, J. E.: Appl. Phys. Lett. **55** (1989) 1103.
4. Ohkawa, K., Karasawa, T., Mitsuyu, T.: Jpn. J. Appl. Phys. **30** (1991) L152.
5. Park, R. M., Troffer, M. B., Rouleau, C. M., DePuydt, J. M., Haase, M. A.: Appl. Phys. Lett. **57** (1990) 2127.
6. Qiu, J., DePuydt, J. M., Cheng, M., Haase, M. A: Appl. Phys. Lett. **59** (1991) 2992.
7. Haase, M. A., Qiu, J., DePuydt, J. M., Cheng, H.: Appl. Phys. Lett. **59** (1991) 1272.
8. Jeon, H., Ding, J., Nurmikko, A. V., Xie, W., Grillo, D. C., Kobayashi, M., Gunshor, R. L., Hua, C. G., Otsuka, N.: Appl. Phys. Lett. **60** (1992) 2045.
9. Wang, S. Y., Haran, F., Simpson, J., Stewart, H., Wallace, J. M., Prior, K. A., Cavenett, B. C.: Appl. Phys. Lett. **60** (1992) 344.
10. Wang, S. Y., Hauksson, I., Simpson, J., Adams, S. J. A., Wallace, J. M., Kawakami, Y., Prior, K. A., Cavenett, B. C.: Appl. Phys. Lett. **61** (1992) 506.
11. Hauksson, I., Simpson, J., Wang, S. Y., Prior, K. A., Cavenett, B. C.: Appl. Phys. Lett., in press.

Optical Nonlinearities of CdS for Optical Addressing

J. Oberlé,[1] *B. Kippelen,*[1] *A. C. Walker,*[2] *and A. Daunois*[1]

[1] Institut de Physique et Chimie des Materiaux de Strasbourg, Unité mixte 380046, CNRS - ULP - EHICS, Groupe d'Optique Nonlineaire et d'Optoelectrique, 5, rue de l'Université, F-67084 Strasbourg Cedex, France
[2] Department of Physics, Heriot-Watt University, Riccarton, Edinburgh EH14 4AS, United Kingdom

Electronic dispersive optical nonlinearities of CdS platelets are investigated by wave mixing and pump-probe experiments. These nonlinearities enabled the realization of a picosecond single-wavelength Fabry-Pérot type logical gate that exploits polarisation dichroism.

1 Introduction

CdS exhibits strong electronic optical nonlinearities at room temperature in the spectral vicinity of the absorption edge. The nonlinearities are related to a band filling effect (mainly due to the filling of the electronic states of the conduction band) which causes a bleaching of the absorption and corresponding refractive index changes [1,2,3,4]. In order to characterize the magnitude of the dispersive nonlinearities of CdS platelets, we have performed degenerate four-wave mixing experiments (DFWM) under stationary excitation, and pump-probe measurements with picosecond pulses. We have determined values up to $n_2 = 10^{-7}$ cm^2/W where n_2 is the Kerr refractive index characterizing the nonlinearity. As the magnitude of the nonlinearity depends on the carrier density, we have also deduced the nonlinear cross section σ_{eh}, which is more suitable to characterize dispersive nonlinearities. We have obtained values up to $\sigma_{eh} = 10^{-20}$ cm^3. Good agreement has been found between the values measured by these two different experimental techniques.

In addition to these characterization experiments, we have studied a Fabry-Pérot based optical logic gate. The use of the absorption dichroism of CdS is proposed as a method of making a directly cascadable single-wavelength logic device, operating on the picosecond timescale.

2 Degenerate Four-wave-mixing Experiments

The wave-mixing experiments are carried out, at 300 K, in a two beam self-diffraction arrangement using the 514 nm line of an Argon laser, acousto-optically modulated. Both beams are focussed to a spot of 40 μm diameter producing a grating with a period of 5 μm. At this photon energy (2.41 eV), the sample absorption is such that the diffraction process results from a phase grating [4]. The magnitude of the nonlinearity (Δn) is then related to the first-order diffracted beam intensity I_1 by [3]:

$$I_1 = I(1 - R)\left(\frac{\pi d_{\text{eff}}}{\lambda}\Delta n\right)^2 \exp(-\alpha d) , \qquad (1)$$

where I is the incident intensity, α is the linear absorption, $R = 0.2$ is the reflectivity coefficient of the platelet of thickness d and effective thickness d_{eff} given by:

$$d_{\text{eff}} = \frac{1 - \exp(-\alpha d)}{\alpha} . \qquad (2)$$

The CdS samples have been fabricated by vapor phase transport and consist of platelets with thicknesses ranging from 0.5 to a few tens of microns. Their crystalline structure is of the wurtzite type with the c axis in the plane of the platelet.

The variation of the frequency and pulse duration of the acousto-optic modulator allowed us to distinguish between thermal and purely electronic contributions to the nonlinearity. It could be shown that at repetition rates greater than 1 kHz the wave-mixing signal saturates for average incident powers of about 1 mW. This is due to an increase of absorption related to sample heating. For example, Fig. 1(a) shows the average diffracted power versus the average incident power obtained by varying the pulse duration τ_p with a constant delay between pulses ($\Delta t = 1$ ms) and a constant peak power. At low power, the slope of 1 (in logarithmic scale) confirms the electronic origin of the nonlinearity. The peak, observed for an average incident power of 1 mW in Fig. 1(a), is attributed to a thermal shift of a Fabry-Pérot fringe. Fig. 1(b) shows the diffracted power obtained by varying the peak power of the pulses with a constant pulse duration (200 ns) and a constant delay between pulses ($\Delta t = 60$ μs). The cubic dependence of the first-order diffracted power, observed in Fig. 1(b), is typical of a Kerr-like dispersive nonlinearity of electronic origin.

To avoid any thermal effects, the values shown in Tab. 1 are obtained, for different samples, at low peak-power densities (≈ 10 kW/cm^2). The n_2 parameter is given by $\Delta n/T_{\text{max}}$, where I_{max} is the peak-power at the interference maximum. The nonlinear cross-section ($\sigma_{\text{eh}} = \Delta n/N$) is deduced using the carrier density calculated in stationary regime using a carrier life-time of 1 to 3 ns which we determined by time resolved luminescence.

The increase of the nonlinearity with decreasing sample thickness, displayed in Tab. 1, is not fully understood at the present time.

3 Picosecond Pump-probe Experiments

In these experiments, the nonlinear refractive index changes are measured directly by the shifts of the Fabry-Pérot transmission peaks of either free-standing samples or samples placed in a resonator. The experimental set-up is shown in Fig. 2. The light source is a distributed feedback dye laser (DFDL) pumped by a

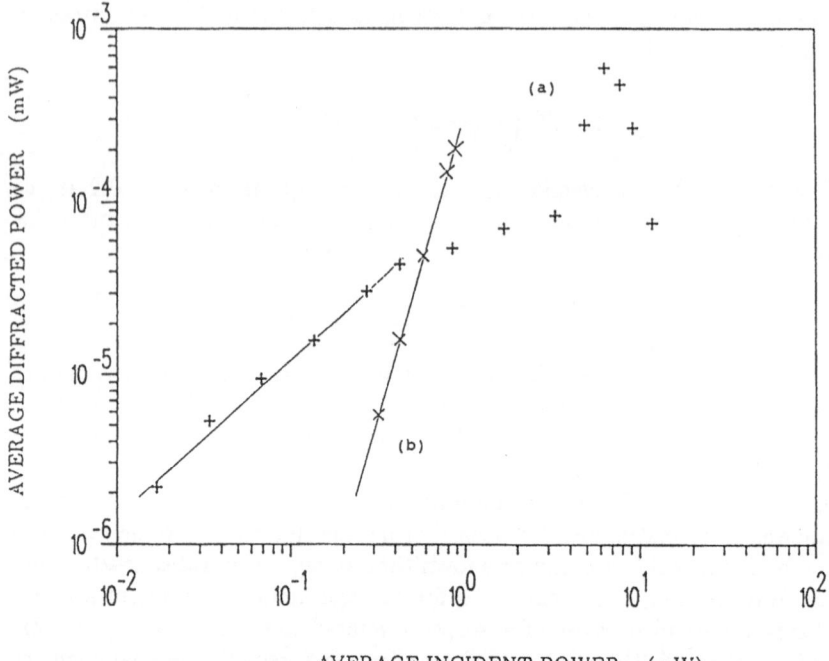

Fig. 1. Average first-order diffracted power as a function of the average incident power. (a) The incident power is varied by changing only the pulse duration. (b) The variation of the incident power results here from a variation of the peak-power.

mode-locked Nd^{3+}:YAG laser. The dye laser uses Coumarin 500 in methanol as the active medium and the emission consists of pulses of 10 ps duration tunable between 490 and 530 nm. The spectrally broad continuum (probe) is produced in a second dye cell containing also Coumarin 500 and pumped transversly by the third harmonic pulse of the Nd^{3+}YAG. To avoid background scatter, the polarisations of the pump $\mathbf{E} \perp c$ and the probe $\mathbf{E} \parallel c$ are orthogonal.

Figure 3(a) shows the shifts of a Fabry-Pérot fringe of a 4 μm thick free-standing sample induced by pump pulses of 24 and 185 MW/cm^2. The different values of n_2 and σ_{eh}, noticed in Tab. 2, are deduced from the peak shifts measured far from the saturation of the nonlinearity (at ≈ 20 MW/cm^2). The n_2 values given in this table are multiplied by the ratio T_1/τ_p in order to be compared to those obtained by DFWM experiments under stationary excitation. The nonlinear cross-section is obtained using a carrier density given by:

$$N = I(1 - R)\frac{[1 - \exp(-\alpha d)]\,\tau_p}{\hbar \omega d} \tag{3}$$

where I is the incident peak power density and τ_p the pulse duration. In Tab. 2, the values of n_2 and σ_{eh}, corresponding to the thinnest sample ($d = 0.5$ μm), were obtained with the sample placed in a Fabry-Pérot cavity formed by two dielectric mirrors of high reflectivity ($R = 0.96$).

Table 1. Results of DFWM experiments obtained for samples of different thicknesses with the optical polarisation parallel or perpendicular to the c-axis of the crystal at $\lambda = 514$ nm (2.41 eV).

Thickness (μm)	Polarization	$-n_2$ (cm^2/W)	$-\sigma_{et}$ (cm^3)
14.3	$\mathbf{E} \parallel \mathbf{c}$	10^{-8}	2×10^{-21}
14.3	$\mathbf{E} \perp \mathbf{c}$	3×10^{-8}	5×10^{-21}
4	$\mathbf{E} \parallel \mathbf{c}$	2×10^{-8}	2×10^{-21}
2.8	$\mathbf{E} \perp \mathbf{c}$	10^{-7}	2×10^{-20}
1.2	$\mathbf{E} \parallel \mathbf{c}$	10^{-7}	4×10^{-20}

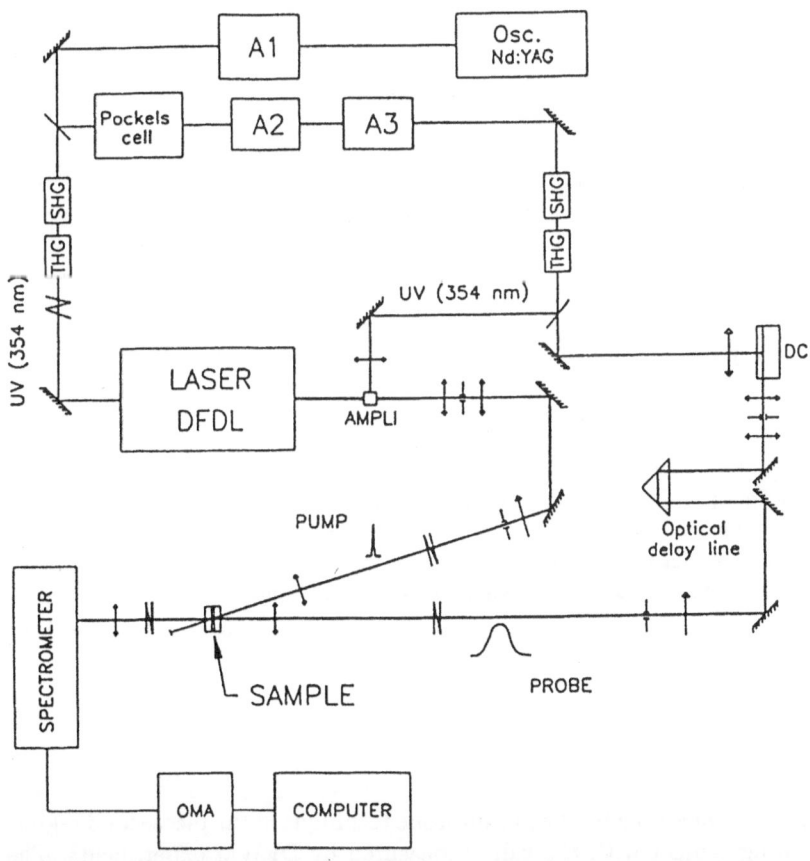

Fig. 2. Experimental set-up : A1, A2, A3: amplifiers; SHG, THG: second and third harmonic generators; AMPLI: dye laser amplifier; DC: dye-cell; OMA: optical multi-channel analyser.

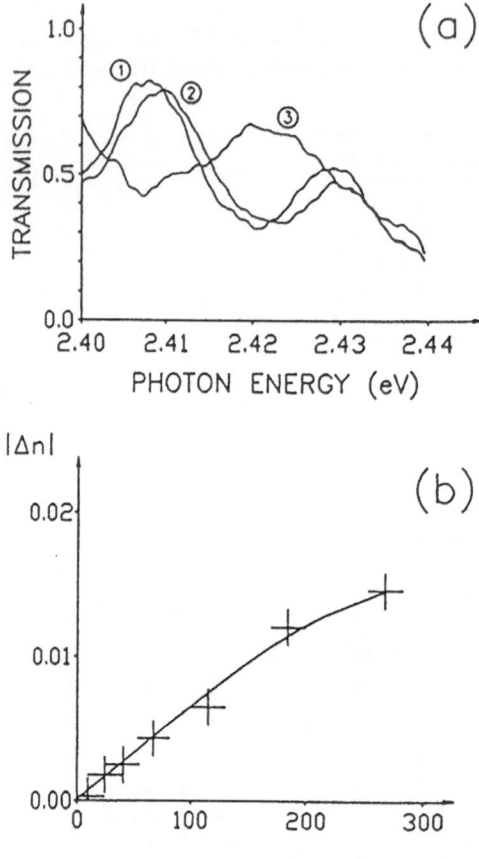

Fig. 3. (a) Probe beam transmission spectra of a 4 μm thick platelet; without (1) and with excitation ($E_{ex} = 2.421$ eV): (2) $I_{ex} = 24$ MW/cm^2, (3) $I_{ex} = 185$ MW/cm^2. (b) Variation of the refractive index as a function of excitation irradiance, deduced from data provided by (a). The curve is a polynomial fit to the data.

These values measured by the pump-probe technique in the picosecond regime are in good agreement with the values measured by DFWM experiments. The highest value of n_2 (10^{-7} cm^2/W) we have found is in agreement with the value given by Spiegelberg et al. [3]. However it should be noticed that our highest value for σ_{eh} (10^{-20} cm^3) is a factor of ~ 10 smaller than those quoted by these authors. This discrepancy is due to the shorter carrier lifetime used in [3].

Table 2. Results of the pump-probe experiments. These values are deduced from the probe beam transmission spectra obtained with excite pulses of ~ 20 MW/cm^2 at $\lambda = 512$ nm

Thickness (μm)	$-n_2$ (cm^2/W)	$-\sigma_{eh}$ (cm^3)
14.3	10^{-8}	10^{-21}
4	2×10^{-8}	2×10^{-21}
0.5	2×10^{-7}	10^{-20}

4 Picosecond Single-wavelength Logical Gate

Optical logic systems working on timescales greater than the relaxation time of the nonlinearity have been demonstrated [5,6]. However logical operation can also be carried out using an excite-probe technique with light pulses of duration shorter than the relaxation time of the medium [7,8,9]. But, in this short pulse regime, the achievement of a differentiel gain greater than unity, an essential requirement for the cascadability, is not possible if the excite and probe pulses are undistinguishable [10].

In our experiments (described in detail in [11]) the use of the absorption dichroism of CdS is proposed as a method of making directly cascadable devices in the short pulse regime with gain greater than unity. The device is based on a nonlinear Fabry-Pérot cavity containing a CdS platelet and using excite/probe operation. The excite pulse alters the tuning of the Fabry-Pérot and the new state is then read out by the probe pulse, before recovery to the initial state. This device operates at a single wavelength and uses the dichroism of the CdS platelet to distinguish the excite and probe pulses. The polarisation of the probe pulse (which is more energetic than the pump pulse) is parallel to the c axis of the crystal, so that it is less absorbed than the pump pulse whose polarisation is perpendicular and, therefore, strongly absorbed by the sample. Gain measurements were obtained by recording the transmission of the probe pulse with and without excitation.

The set-up used in these experiments is similar to the one represented in Fig. 2, but now the fluorescent dye-cell is removed and the DFDL output pulses are divided into two parts with orthogonal polarisations. The cavity, formed by two dielectric mirrors ($R = 0.96$), contains a 0.5 μm thick platelet.

The results shown in Fig. 4 are obtained by recording the transmission of the probe pulse ($I_p = 590$ MW/cm^2) with and without excitation ($I_{ex} = 50$ MW/cm^2) for different pulse wavelengths. Defining the irradiance gain as the change in transmitted irradiance divided by the exciting irradiance, the best gain obtained was 0.8. This relatively small value results from the shift between the two patterns of fringes associated with the two polarisations in this spectral region where the refractive index $n_\parallel$ is smaller than $n_\perp$. The performance of this device is also limited by the absorption of the probe which is not negligible so that the intensity of the probe pulse is high enough to tune the cavity itself.

Nevertheless a gain of 1.4 was obtained using a natural Fabry-Pérot fringe of a 4 μm thick platelet.

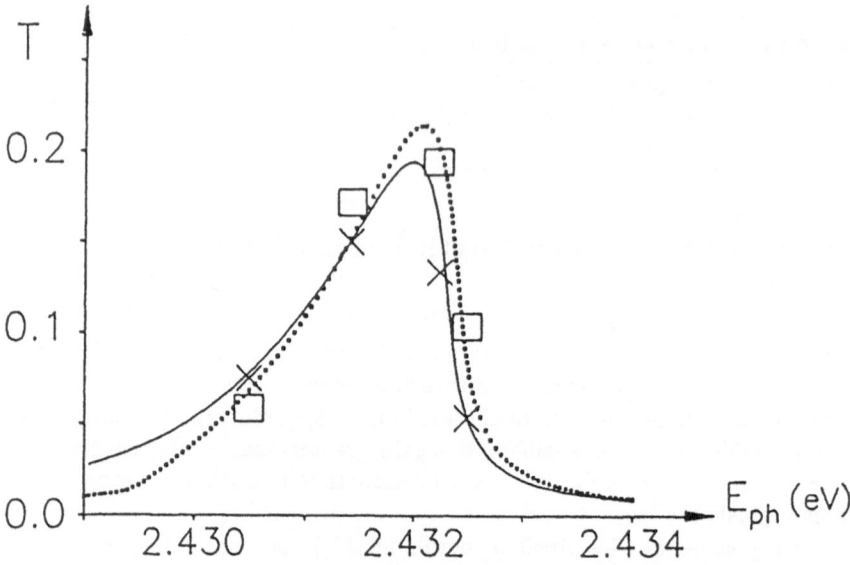

Fig. 4. Transmission of the probe pulse ($I_{\text{probe}} = 590$ MW/cm^2) as a function of photon energy. Crosses show experimentally measured values without the excite pulse and the squares with excite present ($I_{\text{pump}} = 50$ MW/cm^2). The cavity contains a 0.5 μm thick sample. The maximum irradiance gain value was measured to be 0.8. The solid and dotted lines are a numerical fit [11].

To achieve a higher gain level it is necessary to have materials which are more dichroic, i.e. the ratio of the absorption coefficients for pump and probe pulses should be larger. CdSe would appear to have more favourable dichroic properties than CdS.

5 Acknowledgements

The use of dichroic materials to overcome the limitations of two-wavelength short pulse optical logic elements was prompted by suggestions from Drs. N. C. Craft and F. A. P. Tooley. Fruitful discussions with Dr. J. Y. Bigot have stimulated the authors during this work. This study was supported by the British Council through their Alliance programme.

References

1. Swoboda, H.-E., Majumber, F. A., Lyssenko, V. G., Klingshirn, C., Banyai, L.: The electron hole plasma in CdS between 5K and room temperature. Z. Phys. B, **70** (1988) 341–348.

2. Daunois, A., Bigot, J.-Y., Oberle, J., Cherkaoui Eddeqaqi, N., Wegener, M., Witt, A., Klingshirn, C.: Spectral investigation of the switching dynamics in CdS absorptive bistable devices. in Proc. Int. Conf. on Optical Nonlinearity and Bistability of Semiconductors, Berlin (GDR), Phys. Status Solidi (b) **150** (1988) 477-481.

3. Spiegelberg, C., Kretzschmar, M., Puls, J., Henneberger, F.: Strong dispersive nonlinearity in the excitonic region of wide gap II-VI compounds at room temperature. in Proc. Int. Conf. on Optical Nonlinearity and Bistability of Semiconductors, Berlin (GDR), Phys. Status Solidi (b) **150** (1988) 769-775.

4. Spiegelberg, C., Puls, J., Henneberger, F.: Nonlinear optical properties of CdS at room temperature. A comprehensive study. in Proc. Second Int. Workshop Nonlinear Optics and Excitation Kinetics in Semiconductors (NOEKS), Bad-Stuer (GDR), Phys. Status Solidi (b) **159** (1990) 353-361.

5. Walker, A. C., Wherrett, B. S., Smith, S. D.: First implementations of optical digital computing circuits using nonlinear devices. in Nonlinear Photonics, edited by H. M. Gibbs, G. Khitrova and N. Peyghambarian (Springer Verlag, Berlin, 1990), Chap. 4.

6. Craig, R. G. A., Wherrett, B. S., Walker, A.C., Tooley, F. A. P., Smith, S.D.: The optical cellular logic image processor : implementation and programming of a single channel digital optical circuit. Appl. Opt. **30** (1991) 2297-3308.

7. Jewell, J. L., Lee, Y. H., Warren, M., Gibbs, H. M., Peyghambarian, N., Gossard, A. C., Wiegmann, W.: 3 pJ, 82 MHz optical logic gates in a room-temperature GaAs-AlGaAs multiple quantum well étalon. Appl. Phys. Lett. **46** (1985) 918-920.

8. Jewell, J. L., Scherer, A., McCall, S. L., Gossard, A. C., English, J. H.: GaAs-AlAs monolithic microresonator arrays. Appl. Phys. Lett. **51** (1987) 94-96.

0. Tai, K., Jewell, J. L., Tsang, W. T., Temkin, H., Panish, M., Twu, Y.: 1.55 μm optical logic étalon with picojoule switching energy made of InGaAs/InP multiple quantum wells. Appl. Phys. Lett. **50** (1987) 795-797.

10. Jin, R., Hanson, C., Warren, M., Richardson, D., Gibbs, H. M., Peyghambarian, N., Khitrova, G., Koch, S. W.: Room-temperature single wavelength optical latching circuits using GaAs bistable devices as logic gates. Appl. Phys. B **46** (1988) 61-67.

11. Oberle, J., Kippelen, B., Daunois, A., Grun, J.-B., Walker, A. C.: Single-wavelength pulsed optical logic based on dichroism in CdS. Opt. Commun. **90** (1992) 339-346.

Optical Nonlinearities and Switching from Excitons in II-VI Semiconductors

C. Dörnfeld, C.R. Paton, Z. Xie, J. Erland, and J.M. Hvam

Fysisk Institut, Odense Universitet, DK-5230 Odense M., Denmark

Coherent nonlinear resonances due to extended and localized excitons in CdSe and $CdSe_xS_{1-x}$ have been investigated by picosecond time resolved degenerate four-wave mixing and differential transmission experiments. Large nonlinear coefficients ($\chi^{(3)} \geq 10$ cm^2/V^2) are found, with coherence times in the picosecond range. Results on picosecond optical switching in CdSe are presented and discussed.

1 Introduction

Nonlinear optical materials for photonic switching have to meet two requirements: Low switching power and high switching speed [1]. Unfortunately, these two requirements are somewhat contradictory. Away from electronic resonances, the response times are extremely fast and basically limited only by the applied light pulses [2]. In resonances, on the other hand, the creation of real excitations may lead to the accumulation of large optical nonlinearities, at the expense of response time. It is most likely, however, that resonance enhancements of the optical nonlinearities will be needed for application in optical devices. In direct gap semiconductors, excitonic interactions in the form of state filling [3,4], exciton collisions [5], biexciton formations [6,7], etc., provide such strong resonance enhancements. The purely coherent contribution to the optical nonlinearities may exceed substantially the simultaneous incoherent contribution from the accumulation of real excitons [8-11]. Furthermore, the coherent part has a response time limited only by the dephasing time of the optical transitions [4,8-11]. Thus, if the light pulses are properly matched to the dephasing time, typically in the picosecond or sub-picosecond range, ultra-fast optical switching should be possible at a reasonably low power. We have investigated the coherent optical nonlinearities in the exciton resonances of CdSe and $CdSe_xS_{1-x}$ by degenerate four-wave mixing (DFWM) experiments. We have also performed experiments on optical bistability and switching in CdSe by transmission of nanosecond and picosecond laser pulses.

2 Third Order Nonlinearities – Degenerate Four-Wave Mixing

The lowest order optical nonlinearity, active in optical switching processes, is described by the third order nonlinear susceptibility $\chi^{(3)}$, governing the intensity dependent refractive index $n(I) \simeq n_1 + n_2 I$ through the expression $n_2 = \chi^{(3)}/\epsilon_0 n_1^2 c$, where ϵ_0 is the vacuum dielectric constant, n_1 is the linear refractive index and c is the velocity of light. In the presence of dephasing, or damping, only the real part $\text{Re}\chi^{(3)}$ enters in n_2, whereas the imaginary part gives rise to an intensity dependent absorption coefficient $\alpha(I) \simeq \alpha_1 + \alpha_2 I$, with $\alpha_2 = \omega \text{Im}\chi^{(3)}/\epsilon_0 n_1^2 c^2$ [12].

Four-wave mixing is governed by the third order nonlinear polarization

$$\mathbf{P}^{(3)}(\mathbf{k}, w) = \chi^{(3)}(\omega; \omega_1, \omega_2, \omega_3)\mathbf{E}_1(\mathbf{k}_1, \omega_1)\mathbf{E}_2(\mathbf{k}_2, \omega_2)\mathbf{E}_3(\mathbf{k}_3, \omega_3) \tag{1}$$

Transfer of significant energy to the emitted signal wave requires energy and momentum conservations, as expressed by the relations $\omega = \pm\omega_1 \pm \omega_2 \pm \omega_3$ and $\mathbf{k} = \pm\mathbf{k}_1 \pm \mathbf{k}_2 \pm \mathbf{k}_3$, where $+/-$ refer to absorbed/emitted waves. Due to the dispersion relations $\omega_i = \omega_i(k_i)$ for all the waves, these conservation laws are very restrictive and can only be fulfilled under certain conditions (phase-match). In a switching situation, all the waves are often completely degenerate, i.e. $\omega = \omega_1 = \omega_2 = \omega_3$ and $\mathbf{k} = \mathbf{k}_1 = \mathbf{k}_2 = \mathbf{k}_3$, and thereby automatically phase matched.

For spectroscopic purposes, however, an angle between the incoming waves is advantageous. In this case the nonlinear signal may appear as a collimated beam propagating in a direction with no linear signal. Thus, the linear background can to a high degree be eliminated by simple spatial filtering. In the time resolved version, there are two predominant geometries for degenerate four-wave mixing (DFWM) involving two beams or three beams, respectively.

2.1 Time Resolved Degenerate Four-Wave Mixing

With two beams, the situation is sketched in Fig. 1a. The two incident laser pulses are split off the same laser pulse (pulse length τ_L), and are impinging on the sample with a variable optical delay between them. When the two pulses arrive within the dephasing time of the optical transition, a nonlinear signal is genereated in the direction $2\mathbf{k}_2 - \mathbf{k}_1$, for pulse #1 arriving first ($\tau_{12} > 0$) as in Fig. 1a.

The time integral of the signal pulse, being detected, will for delays $\tau_{12} \gg \tau_L$ vary as $I_s \propto \exp(-a\tau_{12}/T_2)$ [8,13], where a is a constant ($1 \leq a \leq 2$) and T_2 is the dephasing time of the optical excitation (polarization) in the medium. This two-beam configuration is therefore well suited to measure dephasing times, provided of course that they exceed the laser pulse length, i.e. $T_2 > \tau_L$.

With three beams, the situation is sketched in Fig. 1b. The two first pulses arrive simultaneously, or well within the dephasing time of the material ($\tau_{12} \ll T_2$), and interfere coherently to set up a nonlinear grating in the medium. This grating can then be detected at variable time delays, τ_{13}, by diffraction of the third pulse. If also $\tau_{13} \ll T_2$, then pulse #3 will diffract off a coherent polarization grating set up by pulses #1 and #2, as in the self-diffraction case above. If, however, w is in resonance with an electronic excitation in the material a real density grating may persist in the material long after the coherent polarization grating has disappeared by dephasing. This type of experiment is therefore well suited to separate the purely coherent contribution to the optical nonlinearities from the more long-lived incoherent contributions due to a high density of excited carriers in the medium [8–11].

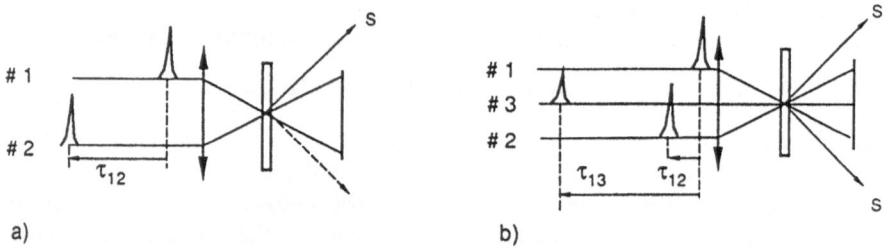

Fig. 1. Time resolved DFWM with two (a) and three beams (b), respectively

Knowing the input intensities I_1 and I_2 of the two incident beams, a good estimate of the nonlinear susceptibility can be obtained from the expression

$$\left| \chi^{(3)}(\hbar\omega) \right|^2 = \frac{4\epsilon_0 n^2 c^4}{\omega^2 I_1 I_2 I_a^2} R_s(\hbar\omega) \; , \qquad (2)$$

where $R_s \equiv I_s/I_2 \exp(-\alpha d)$ is the ratio between the transmitted signal and pump intensities, and $l_a = [1 - \exp(-\alpha d)]/\alpha$ is the nonlinear interaction length, as determined by the linear absorption length $1/\alpha$ or the sample thickness, whichever is the smaller.

2.2 Materials systems

We have performed extensive DFWM experiments to investigate the excitonic nonlinearities in wide gap II-VI semiconductors, particularly in the binary compound CdSe [8–11], but also in the mixed crystals of $CdSe_xS_{1-x}$ [14,15] where localization of the excitons takes place due to the unavoidable random compositional disorder.

At low densities, excitons are successfully treated as non-interacting Bosons. However, this assumption breaks down already at moderate densities (10^{14} cm^{-3}) due to the relatively large size of excitons (~ 50 Å) [16,17]. The resulting exciton interactions, state filling, exciton collisions, biexciton formation, etc., give rise to strong optical nonlinearities. The former resonantly enhanced around the exciton resonance, $w_x = E_x/\hbar$, where E_x is the exciton energy. The latter, i.e. the formation of biexcitons with energy E_m and Binding energy $E_m^b = 2E_x - E_m$, gives rise to new inherently nonlinear resonances just below the $n = 1$ exciton resonance [8]. Besides the fundamental transitions from the ground state to the exciton state, allowed for $\hbar\omega \geq E_x$, direct two-photon transitions (TPA) between the ground state and the biexciton state are allowed for $\hbar\omega \geq E_x - E_m^b/2$, and induced transitions (IA, M-band) between the exciton and the biexciton states for $\hbar\omega \leq E_x - E_m^b$.

A number of different groups have during the last decade been calculating the third order nonlinear susceptibility using different techniques and obtaining slightly different results [18]. The problem is that there is a large number (24) of terms, that may or may not cancel, depending on the assumptions. If population effects of the exciton and biexciton states are included, the fully resonant terms near the exciton resonance are [18]

$$\chi^{(3)} \propto \frac{\mu_{gm}^2 \mu_{xm}^2}{(\omega - \omega_x)(2\omega - \omega_m)} \left(\frac{1}{\omega - \omega_x} - \frac{1}{\omega + \omega_x - \omega_m} \right) , \qquad (3)$$

where μ_{gx} and μ_{xm} are the dipole matrix elements for the exciton and the exciton-biexciton transitions, respectively, and $\omega_i = E_i/\hbar$.

For some purposes these transitions can be approximated with a simple two-level model [8–13], including inhomogeneous broadening to take into account phenomenologically the exciton and biexciton dispersions, as well as any potential fluctuations due to real crystal inhomogeneities [8,14].

2.3 Experimental results

Simultaneous temporal and spectral information about the optical nonlinearities were obtained either by scanning the laser wavelength for a fixed delay between the incident laser pulses (excitation spectra) or by scanning the optical delay for a selected laser wavelength (correlation traces). An example is shown in Fig. 2 displaying the nonlinear excitonic resonances in CdSe at 4.2 K for different delays τ_{12} [11]. The exciton ($n = 1$ and $n = 2$) and biexciton resonances display coherent nonlinearities with a response time in the picosecond or subpicosecond range. Bound excitons and localized excitons, on the other hand, have coherence times that may extend into the nanosecond range, similar to exciton lifetimes in direct gap semiconductor.

With picosecond laser pulses, one can thus measure the real nonlinear susceptibility $\chi^{(3)}$ from forward DFWM, using (2). In CdSe as well as in the mixed crystals, we have found $\chi^{(3)} \geq 10^{-9}$ cm^2/V^2 ($\simeq 10^{-5}$ e.s.u.) [8,9,10,14] corresponding to $n_2 = 5 \times 10^{-8}$ cm^2/W. This compares very favorably with other materials. In order to obtain dispersive optical bistability, or switching, in a Fabry-Perot resonator, a refractive index change of $\Delta_n = n_2 I \simeq 10^{-3}$ is necessary [19]. With the nonlinearities investigated here, this is obtained with switching intensities $I \simeq 20$ kW/cm^2, i.e. about 0.02 μJ/cm^2 for switching with a one-picosecond pulse. If the light is focussed to a spot size of 10 μm, this in fact requires a pulse energy of only $\simeq 2 \times 10^{-14}$ J.

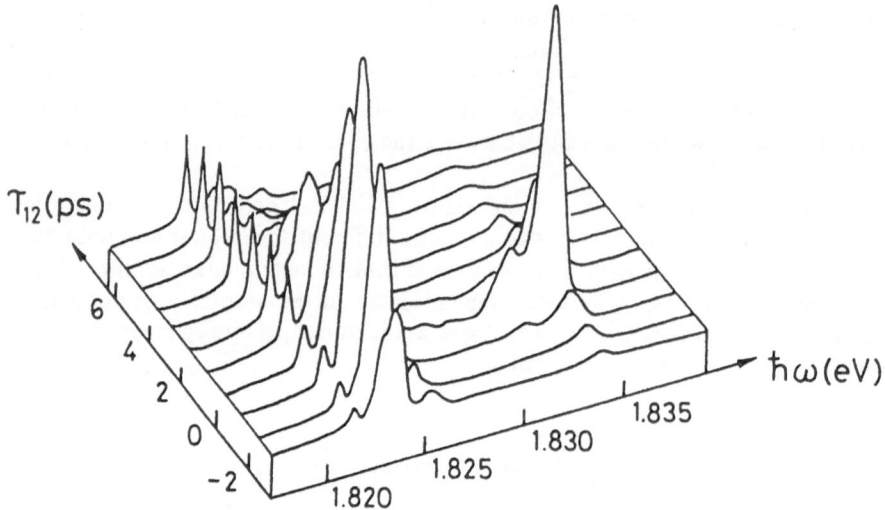

Fig. 2. Spectral dependencies of the DFWM signal for different delays τ between the incoming pulses. The incident intensity is 0.1 μJcm^{-2}

3 Picosecond Optical Switching in CdSe

Alongside the picosecond DFWM spectroscopy, we have investigated optical switching and bistability in CdSe at low temperature by transmission of picosecond and nanosecond laser pulses in the same spectral range as investigated above [20].

3.1 Nanosecond Pulses

The light pulses were provided by an excimer-pumped DCM dye laser, and the incident pulse and the transmitted pulse were recorded simultaneously by a streak camera with a 20 ps time resolution. Figure 3 shows a result of dispersive bistability at 1.820 eV in a 10 μm thick sample at low temperature (10 K). A thin metal coating (Au) is deposited on the rear surface of the as-grown sample to enhance the optical feed-back over the natural reflectivity of about 50%. The upper left traces (a) and (b) show the transmitted and incident pulse shapes $I_T(t)$ and $I_0(t)$, respectively. Trace (c) shows their ratio $I_T(t)/I_0(t)$, whereas curve (d) displays the resulting hysteresis loop I_T versus I_0.

By varying the input peak power it is certified that the apparent switching in Fig. 3 represents a real bistability, switching up at about 25 kW/cm^2 and switching down at about 15 kW/cm^2 (curve (d)). From curve (c), one can estimate switching times of 400 ps and 3 ns for up and down switching, respectively. From the switch-up time it can be seen that the coherent change in refractive index is not sufficient to switch at these power levels. A certain accumulation

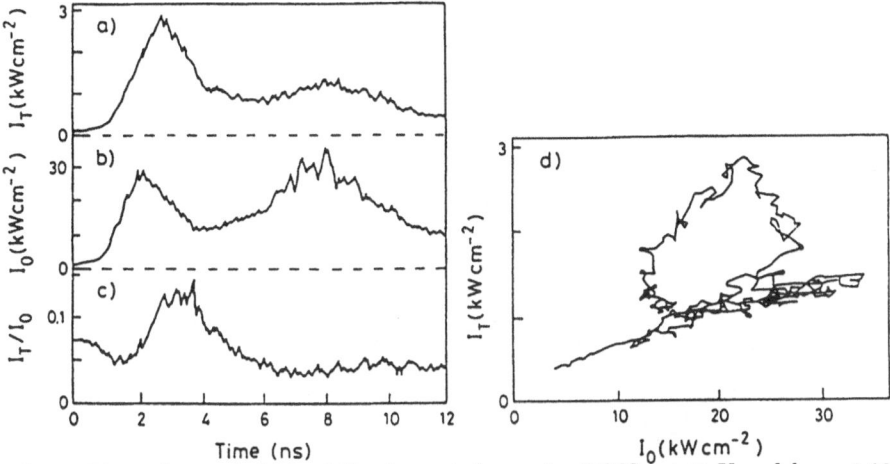

Fig. 3. Dispersive optical bistability in coated sample of CdSe at 10 K and $\hbar\omega = 1.820$ eV. I_0 and I_T are incident and transmitted intensities, respectively

of real incoherent excitons must take place before switching occurs, which again results in the slow down-switching [20].

At even higher powers, absorptive bistability was observed due to internal feed-back in the electronic system, independently on reflective coating on the sample. With increasing power, the excitonic absorption edge switches down-ward, increasing the absorption coefficient, and the transmission switches to a lower value. An example of absorptive bistability in CdSe at 5 K is shown in Fig. 4. The switching powers are high and the switching times are in the 2-3 ns range, limited by the excitonic recombination lifetime. Thus, it is clear that to obtain switching due to fast coherent nonlinearities, we have to apply laser pulses that are short and intense in order to avoid the build-up of incoherent excitons before coherent switching occurs.

3.2 Picosecond Pulses

Using the picosecond laser system, we have investigated the fast coherent nonlin-earities near the exciton resonances in thin samples of CdSe at low temperature as already shown in Fig. 2. Performing DFWM, or laser-indiced grating, experi-ments in a three-beam configuration, a direct comparison of the coherent and the incoherent nonlinearity can be made both with respect to magnitude of the non-linearity and with respect to the response times. In Fig. 5a is shown the intensity of the nonlinear signal scattered off the laser induced grating as a function of the delay of the probe delay τ. This signal is, from (2), proportional to $\left|\chi^{(3)}(\hbar\omega)\right|^2$ in the M-band resonance (exciton-biexciton transition). The fast coherent compo-nent with a response time of about 20 ps is more than two orders of magnitude larger than the incoherent component with a response time $\geq$ 500 ps. Similar results have been obtained from differential transmission experiments on thin

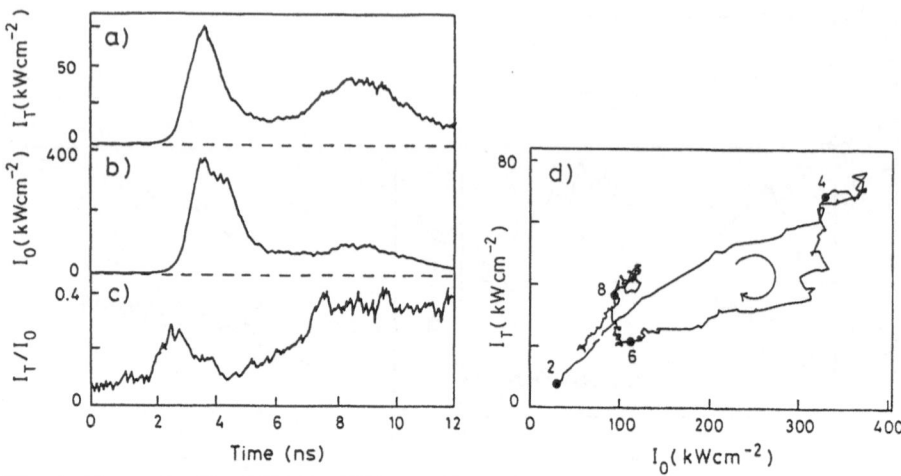

Fig. 4. Absorptive bistability in CdSe at 5 K and $\hbar\omega = 1.820$ eV. I_0 and I_T are incident and transmitted intensities, respectively

samples without any reflective coatings. Strong nonlinear absoption resonances are observed in the spectral range below the free exciton resonance as well as in the transmission window between the $n = 1$ and the $n = 2$ exciton states. The former is ascribed to two-photon absorption (TPA) to the biexciton state [21]. The corresponding induced absorption is shown in Fig. 5b again showing a fast component with a response time of about 20 ps and a slow component with a decay time of 600 ps, which is the exciton lifetime.

4 Conclusions

We have found strong nonlinear resonances of excitons in II-VI semiconductors, that could be interesting from an application point of view. The coherent nonlinear susceptibilities are sufficiently large ($\geq 10^{-9}$ cm^2/V^2) to give rise to optical switching at modest picosecond pulse energies. Thus, if the pulse length is matched to the coherent response time, a slowing-down of the switching time due to incoherent excitons can be avoided even at high repetition rates (≈ 10 GHz). We still need to demonstrate a fast coherent switching with a large contrast based on a coherent nonlinear refractive index change in a Fabry-Perot etalon with a high finesse.

It is furthermore important to find materials and structures where the excitonic nonlinearities investigated here can be exploited at room temperature. There is, however, important progress in the preparation of II-VI semiconductors with p-type doping, so that lasers and amplifiers can be produced [22], and in low dimensional quantum structures, so that excitonic nonlinearities can be observed even at room temperature [23].

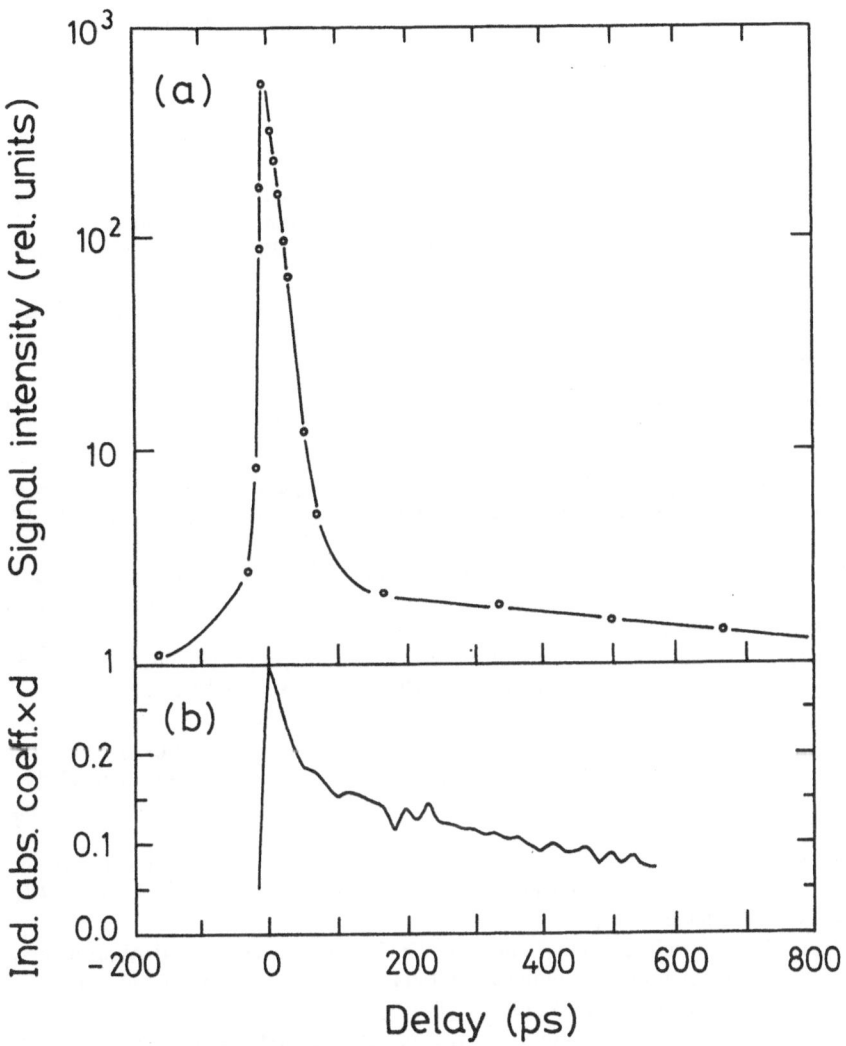

Fig. 5. Nonlinear signal $\left(\left|\chi^{(3)}(\hbar\omega)\right|^2\right)$ as a function of delay τ of the probe pulse (a), and induced absorption coefficient $(\Delta\alpha d)$ as a function of the probe delay (b)

Acknowledgements

This work was supported by the Danish Natural Science Research Council.

References

1. For recent reviews on optical switching, see e.g.: Gibbs, H.: Optical Bistability: Controlling Light by Light. Academic Press, Orlando 1985; Mandel, P., Smith,

S.D., and Wherret, B.S.: From Optical Bistability Towards Optical Computing. North Holland, Amsterdam 1987.

2. Migus, A., Antonetti, A., Hulin, D., Mysyrowicz, A., Gibbs, H.M., Peyghambarian, N., and Jewell, J.L.: One-picosecond optical NOR gate at room temperature with a GaAs-AlGaAs multiple-quantum-well nonlinear Fabry-Perot etalon. Appl. Phys. Lett. **46** (1985) 70–72.

3. Knox, W.H., Fork, R.L., Downer, M.C., Miller, D.A.B., Chemla, D.S., Shank, Gossard A.C., and Wiegmann, A.: Femtosecond dynamics of resonantly excited excitons in room temperature GaAs quantum wells. Phys. Rev. Lett. **54** (1985) 1306–1309.

4. Schultheis, L., Kuhl, J., Honold, A., and Tu C.W.: Picosecond phase coherence and orientational relaxation of excitons in GaAs. Phys. Rev. Lett. **57** (1986) 1797–1800.

5. Schultheis, L., Kuhl, J., Honold, A., and Tu C.W.:Ultrafast phase relaxation of excitons via exciton-exciton and exciton-electron collisions. Phys. Rev. Lett. **57** (1986) 1635–1638.

6. Maruani, A., and Chemla D.S.: Active nonlinear spectroscopy of biexcitons in semiconductors: Propagation effects and Fano interferences. Phys. Rev. B. **23** (1981) 841–859.

7. Masumoto, Y., and Shionoya, S.: High intensity effect on two-photon resonant, coherent Raman scattering via excitonic molecules in CuCl and CuBr. Solid State Commun. **38** (1981) 865–869.

8. Dörnfeld, C., and Hvam, J.M.: Optical nonlinearities and phase coherence in CdSe studied by transient four-wave mixing. IEEE J. Quantum Electron. **25** (1989) 904–912.

9. Hvam, J.M., and Dörnfeld, C.: Nonlinearities coherence and dephasing in layered GaSe and in CdSe surface layer, in Optical Switching in Low-Dimensional Systems, eds. H. Haug and L. Banyai, NATO ASI Series B, 194. Plenum Publishing Corp., New York 1989, pp 233–241.

10. Hvam, J.M., Dörnfeld, C., and Schwab, H.: Coherent nonlinearities due to extended and localized excitons in semiconductors. Optical Materials II, Proc. SPIE **1127** (1989) 59–64.

11. Hvam, J.M., Dörnfeld, C., and Paton, C.R.: Coherent nonlinear optical resonances in II-VI semiconductors. Nonlinear Optical Materials and Devices for Photonic Switching, Proc. SPIE **1216** (1990) 176–189.

12. Hvam, J.M.: Transient nonlinear optics in semiconductors. Nonlinear optics in solids, ed. O. Keller, Springer Ser. Wave Phen. 9. Springer, Berlin 1990, pp 213–234.

13. Yajima, T. and Taira Y.: Spatial optical parametric coupling of picosecond light pulses and transverse relaxation effects in resonant media. J. Phys. Soc. Japan **47** (1979) 1620–1626.

14. Hvam, J.M., Dörnfeld, C., and Schwab, H.: Optical nonlinearity and phase coherence in CdSe and CdSexS1-x. Phys. Stat. Sol. (b) **150** (1988) 387–391.

15. Dörnfeld, C., Noll, G., Schwab, H., Hvam, J.M., Weber, Ch., Renner, R., Gbel, E.O., Reznitsky, A., Lyssenko, V., Pendjur, S.A., Talensky, O.N., and Klingshirn, C.: Picosecond transient gratings in CdSexS1-x mixed crystals. J. Crystal Growth **101** (1990) 678–682.

16. Leonelli, R., Mathae, J.C., Hvam, J.M., Tomasini, F., and Grun, J.B.: Transient phase-space filling by resonantly excited exciton interactions in CuCl. Phys.Rev. Lett. **58** (1987) 1363–1366.

17. Haug, H. and Schmitt-Rink, S.: Basic mechanisms of the optical nonlinearities of the semiconductors near the band edge. J. Opt. Soc. Am. B, 2 (1985) 1135–1142.
18. Abram, I., Maruani, A., and Schmitt-Rink, S.: The nonlinear susceptibility of the biexciton two-photon resonance. J. Phys. C, 17 (1984) 5163–5170.
19. Wherrett, B.S.: One-electron theory of nonlinear refraction. Phil. Trans. R. Soc. Lond. A 313 (1984) 213–220.
20. Paton, C.R., Xie, Z., and Hvam, J.M.: Studies of high-speed optical switching in CdSe. J. Opt. Soc. America B 7 (1990) 1225–1230.
21. Pantke, K.-H., Erland, J., and Hvam, J.M.: Picosecond spectroscopy of exciton-biexciton transitions in CdSe. J. Crystal Growth 117 (1992) 763–767.
22. Haase, M.A., Qiu, J., DePuydt, J.M., and Cheng, H.: Blue-green laser diodes. Appl. Phys. Lett. 59 (1991) 1272–1274.
23. Chemla, D. and Miller, D.A.B.: Room-temperature excitonic non-linear optical effect in semiconductors. J. Opt. Soc. Am. B 2 (1985) 1155–1173.

Predictions of Large Optical Nonlinearities in Quantum Well Wires

S. Benner and H. Haug

Institut für Theoretische Physik, J.-W.-Goethe-Universität Frankfurt, Robert-Mayer-Strasse 8, W-6000 Frankfurt/Main, Germany

The nonlinear spectra of absorption near the band edge are calculated for quantum well wires with up to three subbands. The calculations take into account phase space filling, plasma screening and band gap renormalization due to an optically excited electron-hole plasma. Large optical nonlinearities are obtained around the exciton ground state mainly due to state-filling by the optically excited thermal electron-hole plasma, while the plasma screening effects are found to have relatively little influence. For all plasma densities n (including $n = 0$) , the free-carrier transition spectra differ strongly from those calculated with Coulomb interaction.

Recent progress in microstructuring techniques allows one to build semiconductor quantum well wires (QWW) and to study their optical properties [1-5] . These wires are usually made from GaAs/GaAlAs quantum wells by adding a further lateral confinement. The wires available up to now are not really one-dimensional, because the energy separation between the subbands is not large enough (typically 1-5 meV) . However, there is rapid improvement in the fabrication of these wires. In this paper we calculate the plasma-density dependence of the spectra of absorption and dispersion of GaAs/GaAlAs QWW's with up to three subbands. These calculations are an extension of corresponding calculations for bulk and quantum well semiconductors [6-9], which are generally in good agreement with corresponding measurements.

We start with a 2D-quantum well (with only one 2D subband), in which the electrons and holes are confined laterally by a harmonic oscillator potential with a total subband spacing Ω. In the direction of the wire axis (y) the electrons and holes can move freely. Further we assume the effective mass approximation to be valid. The wave functions of electrons and holes are supposed to have the same spatial extension which leads to spatial charge neutrality in the plane perpendicular to the wire axis. Dielectric effects due to differences in the static dielectric constants ϵ_0 of the well and the barrier material are neglected. The Fourier transform in the y-direction of the 2D-Coulomb interaction averaged with the wave functions ϕ_ℓ for the corresponding e(h) subband reads [10] ($\hbar = 1$)

$$V_{\ell,\ell'}(q) = \int dx dx' d(y - y') e^{iq(y - y')} |\phi_\ell(x)|^2 \frac{e^2}{\epsilon_0 |r - r'|} |\phi_{\ell'}(x')|^2 . \quad (1)$$

This averaged Coulomb interaction has a logarithmic divergence for small

q for particles in the same subbands (typical for 1D systems) and decreases with $1/q$ for large q as the corrseponding two-dimensional potential. Due to the presence of an electron-hole plasma which is assumed to be in a quasi-equilibrium, the bare Coulomb potential is screened. In Ref. [11] we have shown that random phase approximation (RPA) can be used to describe screening effects in quasi-one-dimensional electron systems for plasma densities having more than 1.5 particles within the length a_0 of the three-dimensional exciton Bohr radius, proving that RPA is also in one dimension asymptotically correct for high plasma densities. For lower densities it overestimates the screening. The full RPA expression for the dynamical dielectric function is given by the Lindhard formula. Here we have simplified the RPA screening by a static plasmon pole approximation [6].

As it is known from calculations for quasi-two- or three-dimensional semiconductor structures [6-9] the inclusion of the attractive Coulomb interaction between electron and hole strongly modifies the absorption spectra. A general description of the interband kinetics of the polarization can be derived within the Keldysh nonequilibrium Green's function formalism [12]. The linear Fourier component of the interband polarization P_{kl} obeys in the Hartree-Fock approximation (HFA) the following equation of motion [6]

$$i\frac{d}{dt}P_{k\ell} = \left(e^e_{k\ell} + e^h_{k\ell} - i\gamma\right) P_{kl} - \left(1 - f^e_{k\ell} - f^h_{k\ell}\right) \left[d_{k\ell}E(t) + \sum_{k'\ell'} V_{\ell,\ell'}\left(k - k'\right) P_{k'\ell'}\right]$$

(2)

where the renormalized one particle energies are given by (m is the reduced $e-h$ mass)

$$e^j_{k\ell} = \frac{E^0_g}{2} + \frac{k^2}{2m_j} + \ell\frac{m}{m_j}\Omega - \sum_{k'\ell'} V_{\ell,\ell'}\left(k - k'\right) f^j_{k'\ell'} \qquad j = e, h \ .$$

(3)

The first term on the r.h.s. of the polarization equation is the unperturbed time development as given for interband transitions. Here a phenomenological lifetime $1/\gamma$ has been included. The inhomogeneous second term shows how the polarization is driven by the field E, $d_{k\ell}$ is the interband dipole matrix element. The last term describes the influence of the attractive electron-hole interaction. The last two terms contain both the Pauli blocking factors of the inversion in terms of the Fermi occupation numbers f^j_{kl}. For zero plasma density (2) is the 1D Wannier equation which has recently been investigated in detail by Ogawa et al. [13]. They found that unlike in two- or three-dimensional systems, the Coulomb interaction reduces in quantum wires the absorption coefficient (given by the imaginary part of the polarization) for continuum states above the band edge and removes the inverse-square-root singularity of the density of states.

In Fig. 1 we give the spectra of the imaginary part of the optical susceptibility resulting from the numerical solution of (2) in the single subband case for a monochromatic test field, for various plasma densities n and a plasma temperature of 300 K. The energy broadening is assumed to be $\gamma = E_0 (\simeq 4.2)$ meV. The full lines are calculated with the above described RPA screening, while the

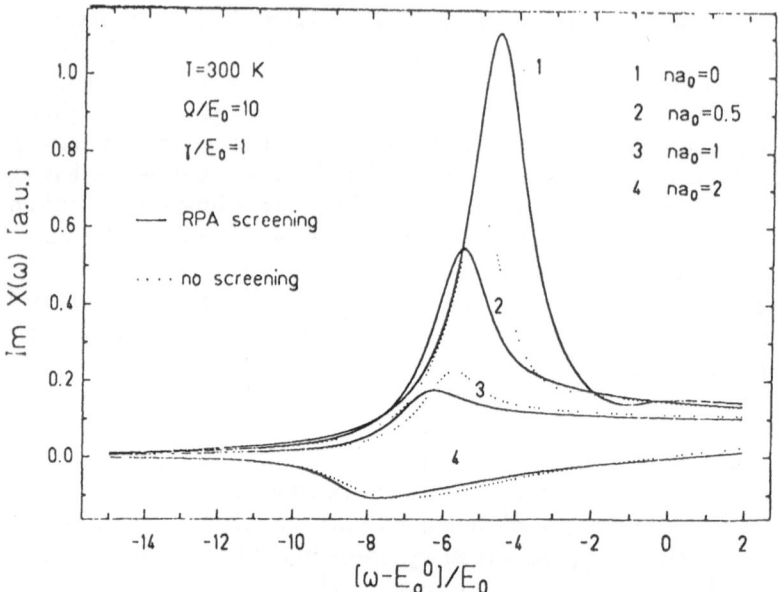

Fig. 1. Absorption spectrum given by the imaginary part of the optical susceptibility for various plasma densities n taking into account only one subband. RPA (full line), HFA (dotted line)

dotted lines give the results of the Hartree-Fock approximation (HFA) without any screening. The qualitative behaviour of the spectra is similar to those of two- or three-dimensional systems. The absorption spectrum for zero plasma density (curve 1) has a strong resonance at the exciton-ground state and is completely flat in the band edge region. With increasing plasma density the excitonic reso- nance is bleached connected with a weak red shift of the decreasing resonance. Because the RPA overestimates the screening below $na_0 < 1.5$ where most of the exciton bleaching is seen to occur, we calculated also the density-dependent spectra without any screening. It is clearly seen that in quantum wires the ex- citon ionization is mainly due to band filling and that the screening adds only small further contributions to the reduction of the oscillator strength and to the red-shift. This result shows again that screening becomes sucessively less impor- tant as the dimension is lowered by quantum confinement, because more field lines between two given charges lead through the embedding material. The exact spectra have to be in between the full and dotted lines of Fig. 1 for all densities. For a degenerated plasma (curve 4) one gets optical gain. In Fig. 2 we solved the polarization equation by considering the lowest three subbands. For small plasma densities one clearly sees the three excitonic resonances located below the corrseponding subband edges. By filling up the lowest subbands the lowest excitonic resonance bleaches more earlier than the higher excitonic resonances. We also expect that the accompanied red shift of the resonances should be more

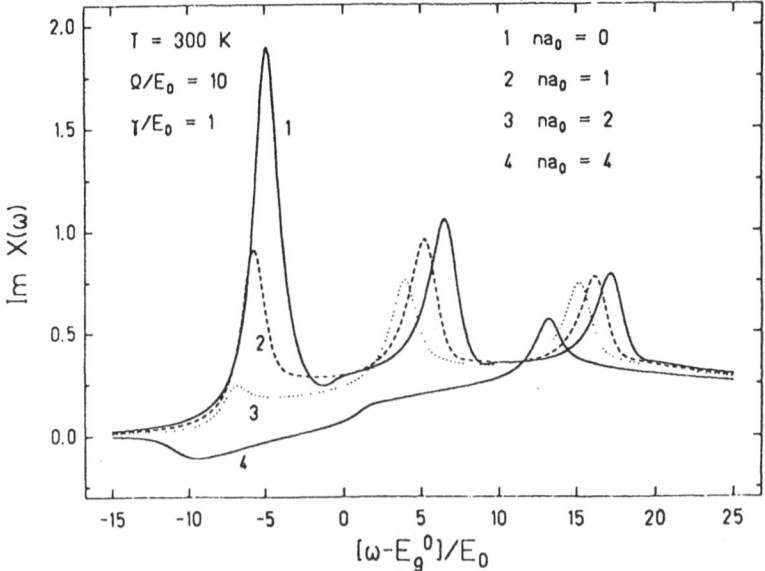

Fig. 2. Absorption spectrum given by the imaginary part of the optical susceptibility (calculated in HFA) for a QWW with three subbands

pronounced in the higher subbands, because these states are located in the continuum part of the lowest e-h state.

Acknowledgments

This work has been supported by the Deutsche Forschungsgemeinschaft and the Volkswagen Stiftung.

References

1. Sakaki, K., Kato, K., Yoshimura, H.: Optical absorption and carrier-induced bleaching effect in quantum wire and quantum box structures. Appl. Phys. Lett. **57** (1990) 2800.
2. Cibert, J., Petroff, P. M., Dolan, G. J., Pearton, S. J., Gossard A. C., English, J. H.: Optically detected carrier confinement to one and zero dimension in GaAs quantum well wires and boxes. Appl. Phys. Lett. **49** (1986) 1275.
3. Gershoni, D., Weiner, J. S., Chu, S. N. G., Baraff, G. A., Vandenberg, J. M., Pfeiffer, L. N., West, K., Logan, R. A., Tanbuk-Ek, T.: Optical transitions in quantum wires with strain induced lateral confinement. Phys. Rev. Lett. **65** (1990) 1631.
4. Kash, K.: Optical properties of III-V semiconductor quantum wires and dots. J. Luminesc. **46** (1990) 69.

5. Lehr, G., Bergmann, R., Rudeloff, R., Scholz, F., Schweizer, H.: Influence of carrier capture on the quantum efficiency of free standing and eptitaxially buried InGaAs-InP quantum wires. Appl. Phys. Lett., in press.

6. Haug, H., Koch, S. W.: Quantum theory of the optical and electronic properties of semiconductors. World Scientific, Singapore 1990.

7. Haug, H., Schmitt-Rink, S.: Electron theory of optical properties. Prog. Quant. Elect. 9 (1984) 3.

8. Ell, C., Blank, R., Benner. S., Haug, H.: Simplified calculations of the optical spectra of two- and three-dimensional laser excited semiconductors. J. Opt. Soc. Am. B 6 (1989) 2006.

9. Ell, C., Haug, H., Koch, S. W.: Many-body effects in gain and refractive-index spectra of bulk and quantum-well semiconductor lasers. Opt. Lett. 14 (1989) 356.

10. Hu, G. Y., O'Connel, R. F.: Eletron-electron interactions in quasi-one-dimensional electron systems. Phys. Rev. B 42 (1990) 1290.

11. Benner, S., Haug, H.: Linear response functions of a quasi-one-dimensional electron gas. Z. f. Physik B 84 (1991) 81.

12. Haug, H. (ed.): Optical nonlinearities and instabilities in semiconductors. Academic Press, Boston, p. 53 (1988)

13. Ogawa, T., Takagahara, T.: Optical absorption and Sommerfeld factors of one-dimensional semiconductors: An excat treatment of excitonic effects. Phys. Rev. B 43 (1991) 14325.

Nonlinear Optical Properties of II–VI Semiconductor Quantum Dots

A. Uhrig,[1] A. Wörner,[1] M. Saleh,[1] C. Klingshirn,[1]
N. Neuroth,[2] K. Remitz,[2] B. Speit[2]

[1] University of Kaiserslautern, Department of Physics,
   W–6750 Kaiserslautern, Germany
[2] Schott Glaswerke, W–6500 Mainz, Germany

We present results concerning optical properties of II–VI semiconductor quantum dots. In pump and probe beam experiments at room and helium temperature we found no holeburning but a strong bleaching of both maxima in the absorption spectrum up to 75% of the optical density under a pump intensity up to 60 MW/cm^2. This bleaching is nearly independent of the pump energy and we connect the resulting small energetic shift with the inhomogeneous broadening due to the size distribution of the crystallite. For both temperatures the halfwidth of the bleaching peak is comparable, which suggests a strong coupling of the excited states to the lattice. Corresponding results were obtained also from linear luminescence measurements. From both we calculated a Huang–Rhys factor from 1 to 2 for the system. Additionally we investigated the samples in a self diffraction experiment at room temperature. The efficiency spectrum is a very broad band corresponding to the broad structures in the absorption spectra of the sample with a halfwidth of about 20 nm. From these results we calculated the effective $\chi_{\text{eff}}^{(3)}$ of about 1×10^{-9} esu for the samples.

1 Introduction

For some years semiconductor quantum dots (QD) have been the aim of intensive investigation because of confinement effects for the excited quasi particles. This quantization is expected to increase the nonlinear effects which are a precondition for optical bistability and so are fundamental for optical logic switching devices. A second source of interest is the possibility of using the glass directly as an integrated optical device, with doping in the case of a nonlinear optical switch, or without doping as a waveguide.

Here we investigated $CdS_{1-x}Se_x$ quantum dots embedded in a glass matrix by various techniques of optical spectroscopy. A part of our samples were specially annealed OG570 glasses with radii of the dots between 2 and 6 nm. The other one was a commercial Schott glass OG550.

In the first part of our contribution we give a survey of linear measurements which we carried out and in the second one we introduce the nonlinear experiments and their interpretation.

2 Linear Optics

To characterize the samples we measured the linear absorption of the samples. In agreement with theory [1–4] we found the typical blue shift of the transition between the first electron and hole levels with decreasing radius R of the dots. The structures in the spectra are very broad due to size distribution, phonon coupling and eventually alloy disorder.

A special problem in the field of the QD is the assignment of the absorption maxima to the possible transitions. When we plotted the energetic shift of the first peak versus $(a_B/R)^2$ we found a linear relationship but not with the theoretically expected slope of C = 0.67 [1] but for different sample sets we found values for C between 0.3 and 10 [5]. This discrepancy can be caused by strain between the semiconductor and the glass matrix, by a deviation of the theoretically assumed spherical form or by errors in the determination of the radius by small angle X–ray scattering.

Another problem is the second peak in the absorption spectra seen e. g. in Fig. 1 at 2.7 eV. If this peak would correspond to the second quantized level in QDs, the energetic difference between the first and second peak as a function of the crystallite radius must be a line through the origin. The examination of different sample sets yields only for $a_B/R > 1$ a straight line with the expected slope. So we can attach the second peak for $R > a_B$ to the second quantized level in the dots. For $a_B/R < 1$ we get a horizontal line without a dependence on the dot radius. In [4,5] different possibilities for the assignments of the bands to this transition are discussed. In many samples the valence band split off by spin orbit coupling seems to be the origin for the second peak.

In luminescence measurements under low excitation [4,5]we found a large Stokes shift of the emission with respect to the absorption which can be attributed to a strong coupling of the excited one pair state to the lattice. The Coulomb interaction between electron and hole together with their different mass leads to a charge distribution [6] in the dot and so to a dipole moment which couples to the optical phonons as described in detail in [4,5].

3 Nonlinear Optics

We carried out pump and probe beam experiments with two synchronously pumped dye lasers. The laser for the pump beam produces pulses with ≈ 5 ns length which gives us quasi stationary excitation conditions in the experiments. As probe beam we used the spontaneous emission of an second small laser. The measurements were carried out under different excitation intensities and energies and for room and liquid helium temperature.

Figure 1 shows the optical density of a commercial OG 550 sample with and without the pump beam. The pump energy $\hbar\omega_p$ was resonant to the first absorption maxima and the pump intensity I_p was about 60 MW/cm^2. We can see a clear bleaching effect with a nearly complete disappearence of the first peak and a bleaching of the second peak although $\hbar\omega_p$ is too low to excite this

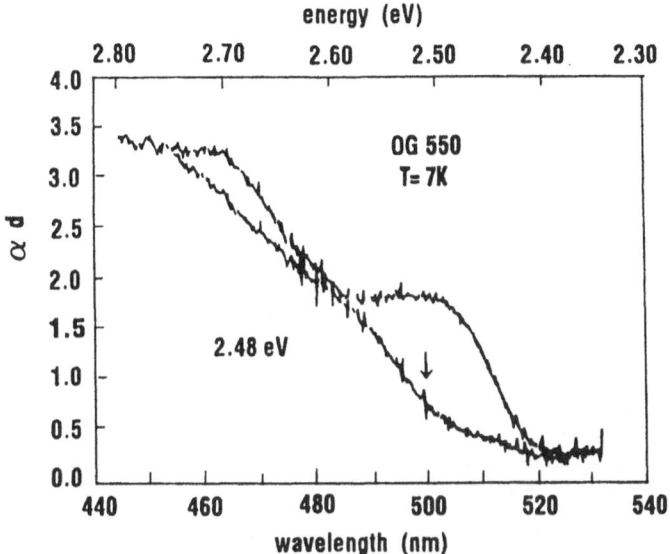

Fig. 1. Optical density of a sample OG 550 with and without excitation

second transition. This points to the same initial or final state being common to both peaks which is in good agreement with the discussion about the linear measurements above.

Figure 2 shows the differential transmission spectra (DTS) of an custom-made sample with an average crystallite radius of 3.6 nm where I_p is varied between 50 kW/cm^2 and 20 MW/cm^2. Here the peak grows continuously with I_p and shows the onset of saturation for higher I_p. The spectral width of the peak does not change with I_p. This is again a hint for a large homogeneous linewidth of the transition. An additionally interesting detail in the spectra is the negative value for $\Delta\alpha d$ because of induced absorption at the high energy side of the peak. This effect is the result of the excitation of a second electron hole pair in a dot in aggreement with [9], our calculation of the energy value in [4] and the luminescence measurements under high excitation [4]. For small I_p the probability of exciting a second electron hole pair increases increases as the number of dots which are excited with one electron hole pair increases, but for higher I_p, this transition also begins to saturate.

Figure 3 show an example of our measurement of the change of the optical density under different pump energies. Similar investigations were made on different samples, low and high pump intensities and all gave the same result: the the first peak shifts only slightly in the direction of $\hbar\omega_p$ essentially by changing the peak width on the high energy side, but the shift is not so strong as in similar measurements in [8]. The low energy wings of all spectra fall together. An explanation is the superposition of a very broad homogeneous linewidth with an

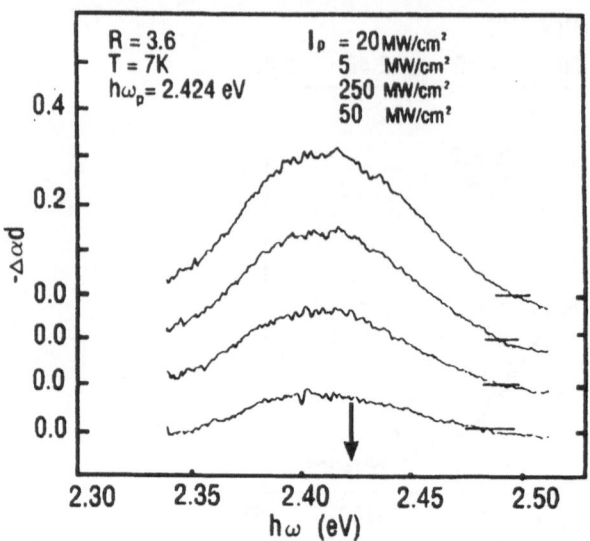

Fig. 2. DTS spectra of a sample with a dot radius of 3.6 nm under different pump intensities

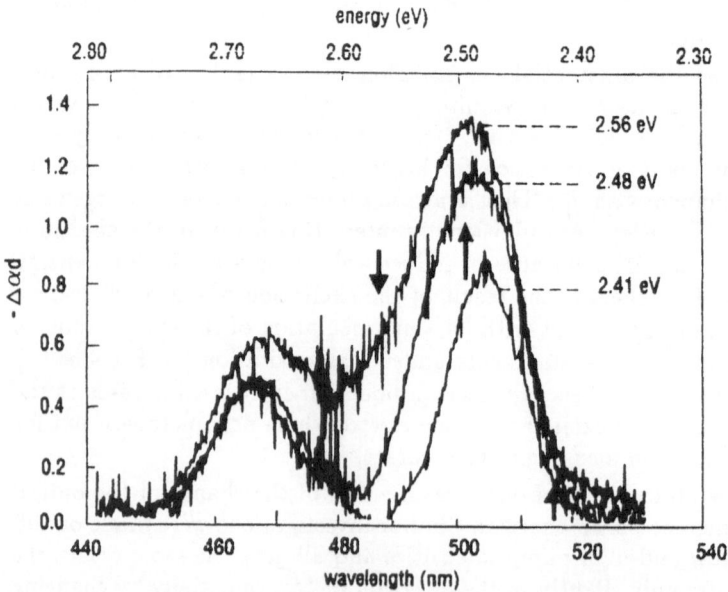

Fig. 3. DTS spectrum of the OG 550 sample under different pump energies $\hbar\omega_p$

inhomogeneous size distribution. If we pump at the low energy side only large dots are excited. So the DTS spectrum shows the homogeneous broadening of the sample. The more higher $\hbar\omega_p$ the better we reach smaller dots with higher transition energies. So the DTS spectrum shows the superposition of the homogeneous and inhomogeneous broadening. From the spectum with $\hbar\omega_p = 2.41$ eV we calculated the FWHM as 65 meV and from this the Huang–Rhys parameter $S \approx 0.9$ for the system. For more details see [5,10]

Finally we carried out a self diffraction experiment using ns laser induced gratings (LIG) at room temperature. The nonlinearities contributing to the LIG are of coherent and incoherent (e.g. bleaching) nature. The intensity of the diffracted order is given by [11,12]

$$\frac{I_d}{I_p} = \eta = \left(\frac{\pi \Delta n d}{\lambda_c}\right)^2 + \left(\frac{\Delta K d}{4}\right)^2 \tag{1}$$

where $\Delta K d$ is the absorption in the material and Δn the change of the index of refraction. In literature it is common to express both contributions together in an effective susceptibility $\chi_{\text{eff}}^{(3)}$.

$$\chi_{\text{eff}}^{(3)} = \frac{19 n_0^2 \lambda_c \sqrt{\eta}}{\pi I_p d} \tag{2}$$

Figure 4 shows the spectral dependence of the efficiency η for the sample OG550.

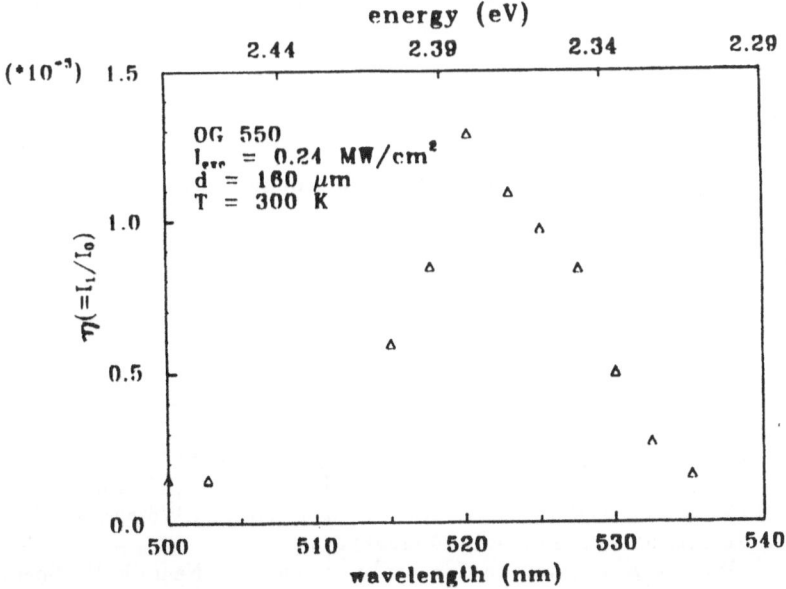

Fig. 4. Spectra dependence of the grating efficiency for the OG 550 sample at room temperature

The spectrum is a very broad band (≈ 20 nm) corresponding to the broad structures in the absorption spectra of the sample. The maximum of η for different I_p show a decrease of $\chi_{\text{eff}}^{(3)}$ with increasing I_p due to a strong bleaching of the absorption and the resulting saturation of the grating efficency. For the maximum of η we get $\chi_{\text{eff}}^{(3)} \approx 1 \times 10^{-9}$ esu which is much lower than in CdS bulk material where $\chi_{\text{eff}}^{(3)}(\text{CdS}) \approx 9 \times 10^{-3}$ esu [13]. This is a consequence of a filling factor of $\approx 10^{-4}$ of the dots in the glass matrix and of the the broadening of the resonances. The use of semiconductor doped glasses for phase–conjugation has been demonstrated already in [14].

4 Conclusion

In II–VI semiconductor quantum dot glasses we found a strong bleaching effect with a resulting change of the absorption up to 75%. The bleaching structures are very broad due to the large homogeneous and inhomogeneous broadening in the samples and they are nearly independent of the pump energy $\hbar\omega_p$. From the DTS spectra we determined a Huang–Rhys factor of $S \approx 0.9$. The change of the absorption under high laser intensities is a main contribution to the measured nonlinearity of the sample in a degenerate laser induced grating experiment. The resulting $\chi_{\text{eff}}^{(3)}$ is only around 1×10^{-9} esu due to the above mentioned causes. By using samples with higher filling factor and smaller size distribution the optical nonlinearities should increase. A possible material would be sol–gel glasses doped with semiconductor quantum dots in which it is possibles the reach filling factors around 20%. In principle this material shows the same physical behaviour and can also be directly used as device in integrated optics.

Acknowledgements

This work has been supported by the Deutsche Forschungsgemeinschaft and the Materialschwerpunkt of the University of Kaiserslautern. Stimulating discussions are acknowledged with Dr. U. Woggon (Berlin).

References

1. Efros, Al. L., Efros, A. L.: Soviet Phys. Semicond. **16** (1982) 772.
2. Brus, L.: IEEE J. Quantum Electon. QE–**22** (1986) 1909.
3. Ekimov, A. I., Efros, A. L.: Phys. Status Solidi (b) **150** (1988) 627.
4. Uhrig, A., Banyai, L., Hu, Y. Z., Koch, S. W., Klingshirn, C., Neuroth, N.: Z. Physik B **81** (1990) 385; Uhrig, A., Banyai, L., Gaponenko, S., Wörner, A., Neuroth, N., Klingshirn, C.: Z. Physik D **20** (1991) 345.
5. Uhrig, A., Wörner, A., Banyai, L., Gaponenko, S., Lacis, I., Neuroth, N., Speit, B., Remitz, K., Klingshirn, C., Proc. Int. Conf. II–VI, Japan (1991).
6. Hu, Y. Z., Lindberg, M., Koch, S. W.: Phys. Rev. B **42** (1990) 1713.
7. Banyai, L., Gilliot, P., Koch, S. W., Hu, Y. Z.: to be published.

8. Spiegelberg, C., Henneberger, F., Puls, J.: Superlattices Microstr. **9** (1991) 487; Henneberger, F., Puls, J., Spiegelberg, Ch., Schülzgen, A., Rossmann, H., Jungnickel, V., Ekimov, A. I.: In Proc. Advanced Research Workshop on Wide Gap II–VI Semiconductors, Montpellier, Jan. 1991, (Semicond. Sci. Technol., in press).
9. Peyghambarian, N., Fluegel, B., Hulin, D., Migus, A., Joffre, M., Antonetti, A., Koch, S. W., Lindberg, M.: IEEE J. Quantum Electonics QE–**25** (1989) 2516; Hu, Y. Z., Koch, S. W., Lindberg, M., Peyghambarian, N., Pollock, E. L., Abraham, F. F.: Phys. Rev. Lett. **64** (1989) 1805; Esch, V., Kang, K., Fluegel, B., Hu, Y. Z., Khitrova, G., Gibbs, H. M., Koch, S. W., Peyhambarian, N.: J. Nonlinear Opt., in press.
10. Wörner, A.: Diploma Thesis, Kaiserslautern, Germany, (1991)
11. Eichler, H. J.: Forced Light Scattering at Laser–Induced Gratings — a Method for Inverstigation of Optical Excited Solids in Advances in Solid State Physics XVIII, (Vieweg, Braunschweig, 1978)
12. Blau, P. Blau, W. Byrne H., Berglund, P.: Appl. Opt. **29** (1990) 31.
13. Kalt, H., Lyssenko, B. G., Renner, R., Klingshirn, C.: J. Opt. Soc. Am. B **2** (1985) 1188.
14. Oberhausen, D., Renner, R., Klingshirn, C.: Solid State Commun. **72** (1989) 913.

Part IV

Optical Computer Architectures

Optical Input and Output Functions for a Cellular Automaton on a Silicon Chip *

I. Seyd-Darwish,[1] P. Chavel,[1] J. Taboury,[1]
F. Devos,[2] T. Maurin,[2] and R. Reynaud[2]

[1] Institut d'Optique Théorique et Appliquée (CNRS), F-91403 Orsay, France
[2] Institut d'Electronique Fondamentale (AXIS Group, CNRS), F-91403 Orsay, France

1 Introduction

Parallel architectures for cellular automata are adapted to optical computing. Electronic implementations have the interconnection as the limiting factor. Our purpose is to couple the performance of electronics (which offers nonlinear components with large-scale integration and high speed) with the performance of optics (which allows the implementation of dense and rapid interconnections, in particular for the input and output), in a compact optoelectronic cellular automaton setup. The specific case that we selected for the purpose of illustration is the "Lattice-Gas" [1] automaton.

In this article we present the architecture of the cellular automaton implementation. After a brief description of the "Lattice-Gas" algorithm, we introduce the implementation of the different parts of this automaton.

2 Cellular Automaton Architecture

The cellular automaton contains a number of elementary processors (EPs) distributed on a regular lattice. These EPs perform nonlinear operations in parallel, the same for all the EPs. Each EP is connected to some of its neighbours.

It is therefore an SIMD machine (single instruction, multiple data). In our case (Fig. 1), The EPs are implemented electronically on a silicon chip (1). Each EP performs simple operations like memorizing, comparing and copying binary words. The inteconnections between neighbouring EPs are made electronically because they are short interconnections. The required instructions that are performed simultaneously by all EPs are projected optically using modulated light sources and an array illuminator (3). The initialization of the EPs is made optically by projection of an image of the data directly onto the EPs (2). The results are read out in parallel using another chip made of optoelectronic modulators formed of pixels connected to the EPs and project optically an image of their contents onto an external component (4) [2].

---

* With the contribution of: Thomson C.S.F. (L.C.R.) and LAAS (Toulouse)

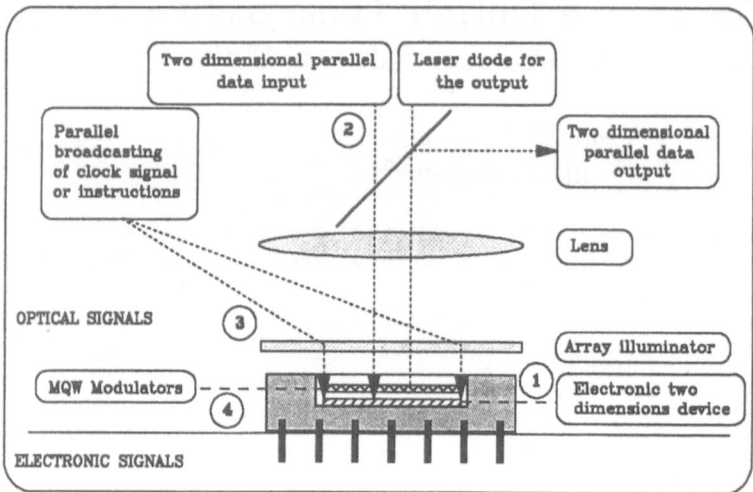

Fig. 1. Architecture of the cellular automaton

3 Lattice-Gas Algorithm

"Lattice-Gas" is an algorithm for the simulation of the Navier-Stokes equation in the case of aerodynamic flows using the statistical behaviour of particles in a gas [3]. Although three dimensional versions exist, we consider here a two-dimensional case.

In this model, the gas particles are restricted to move only on a hexagonal lattice. Their speed modulus can take only two values, zero and unity. Unity corresponds to a motion between two cells covered in the time of one iteration of the algorithm. Collisions between the particles occur at the lattice vertices . These collisions follow laws that respect the conservation of the particle number, the momentum and the energy of the gas. The laws are given in a table (lookup table, LUT) and the collision at each vertex is performed by symbolic substitution using that table. With the symbolic substitution the execution of the algorithm contains three steps at each vertex (Fig. 2):

Recognition step: A pattern of incoming particles is present at each vertex. The first step consists of recognizing this pattern by comparing it sequentially with all the possible patterns. As we use a hexagonal lattice, the maximum number of incoming particles at each vertex is 7 (including the one that may be present at the vertex), so we have $2^7 = 128$ comparisons to make.

Substitution step: When an incoming particle pattern is recognized at a vertex, we substitute it by the corresponding outgoing particle pattern resulting from the collision.

Propagation step: After 128 recognition/substitution steps, all vertices contain their outgoing particle patterns. We then propagate these particles to the neighbouring vertices and they will constitute the new incoming particles pattern for the next iteration.

Those three steps constitute one iteration. To have meaningful results of "Lattice-Gas," we need many thousands of iterations.

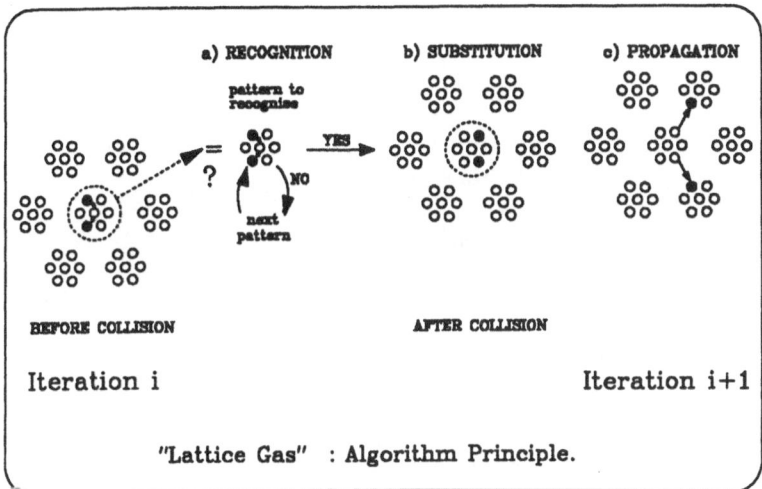

Fig. 2. Treatment steps

4 Implementation on Cellular Automaton

The electronic circuit contains EPs which contain the distribution of the particles and perform the collision. The LUT is memorized in an external memory named the "correspondence memory". This memory contains all the 128 possible incoming particle patterns, each pattern is followed by the appropriate outgoing particle patterns. The contents of this memory are projected optically and sequentially onto all the EPs of the electronic circuit.

4.1 Array Illuminator Based on Talbot Effect

The array illuminator (AI) broadcasts the same instruction to all the EPs on the electronic circuit. These instructions are words of 7 bits.

This AI illuminated by 7 sources images them onto each EP. As shown in Fig. 1, in addition to the instruction broadcast, two other optical signals have to access the chips through the AI. For this reason, a diffractive AI with wavelength or angular selectivity is necessary (Fig.3). We have implemented a new process to make a diffractive AI; it is based on the Talbot effect [4].

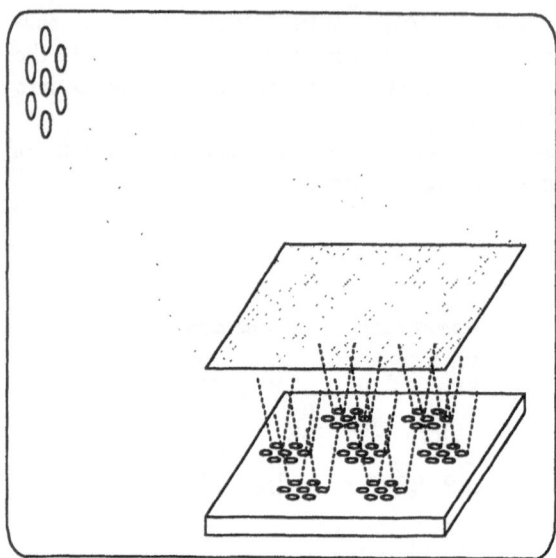

Fig. 3. Array illuminator

Array illuminator principle The Talbot effect [5] produces an image of regular gratings, illuminated by coherent light, at periodic distances a multiple of the Talbot distance $(2p^2/\lambda)$ (p is the grating period and λ is the wavelength). In addition, a shifted version of the grating image is reproduced at half the Talbot distance.

This effect could be used directly for an array illuminator. However the efficiency is then very poor: for example, in the case of a square grating of 300 μm with square holes of 50×50 μm^2, only 3% efficiency can be expected, just because of the duty cycle of the grating. These numerical values correspond to a typical test case that we developed in reference [4].

To improve this efficiency we record an hologram at some appropriate distance between the grating and its first Talbot image.

The position of this holographic plate is optimized according to the holographic techniques. To have the maximum gain of efficiency we have to use thick phase holograms. One important condition to have maximum efficiency is that the wave recorded at the plate shows a uniform intensity and all information must be encoded in phase modulation. This condition is hardly satisfied with a wave diffracted from a grating such as the one described above. We have studied the diffracted wave from such a grating in the Fresnel approximation to minimize the intensity variation at some diffraction plane where the holographic plate will be placed. Figure 5 shows the intensity variation at the optimal plane located, in our test case, at 3.1 cm from the half Talbot distance using a wavelength of 488 nm.

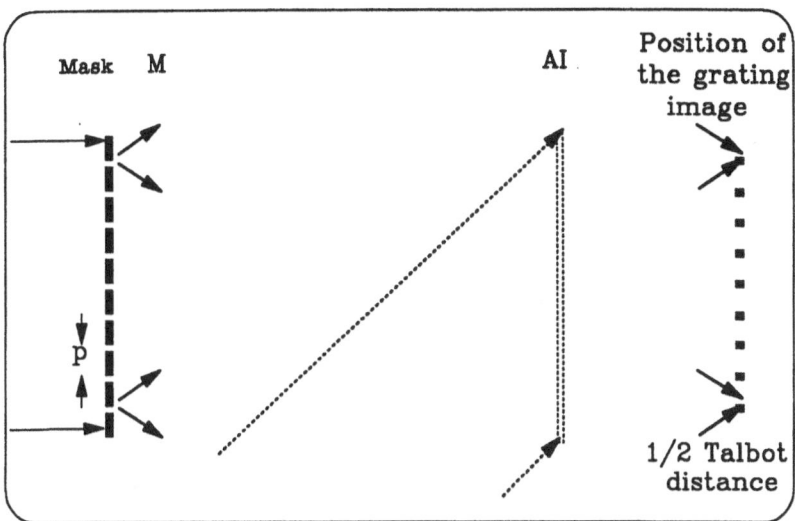

Fig. 4. Holographic array illuminator principle

We have recorded a thick phase hologram on dichromated gelatin using the grating described above with 90×90 holes. The recording wavelength was the 488 nm of Argon laser. The holographic plate was located at the optimal plane. The reconstruction of this hologram with the same wavelength reproduces the grating image with 70% efficiency. The study shows that the structure of the hologram is similar to a microlens hologram. Bragg condition for the diffracted wave offers transparency for the other wavelengths. The advantage of this technique is the simultaneous recording of a large number of microlenses.

Change of wavelength The recording wavelength is imposed by the DCG sensitivity. In the automaton setup we use a laser-diode for reasons of compactness. As the wavelength of these lasers is situated in the near infrared, the hologram will present chromatic aberrations.

We have conducted a study to minimize the influence of these aberrations by a simple change of the recording conditions, specially the incidence angle of the diffracted wave on the hologram plate. We take the energy concentration in the $50 \times 50 \ \mu m^2$ detector area as the correction criterion.

The simulation shows that for a given angle the influence of the aberration can be minimized. Figure 6 shows the simulation of this correction compared with the case of no correction. We observe that the energy concentration is practically doubled.

4.2 Electronic Circuit

The electronic demonstration circuit contains EPs distributed on an hexagonal lattice, each hexagon represents a collision vertex in the "Lattice-Gas" model.

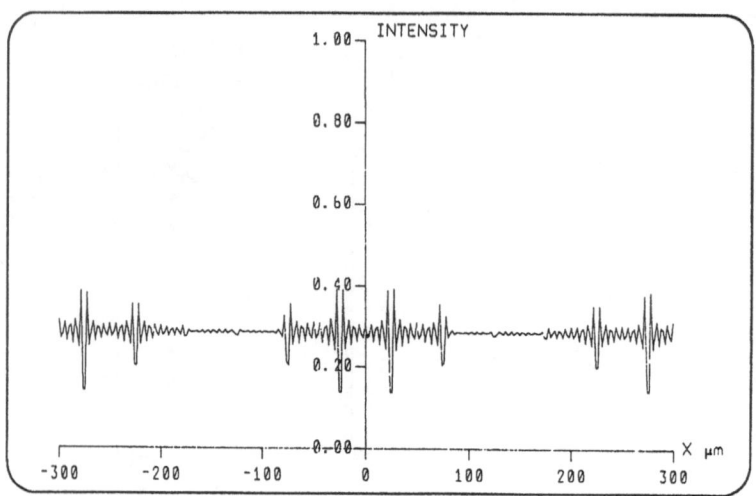

Fig. 5. Minimum intensity modulation plane

The function performed by each EP is reduced to the strict minimum i.e. memorizing of particles patterns (incoming particles and resulting particles from the collision), comparison and copying. In this way we can reduce the area occupied by each EP so as increase the number of these processors on one chip.

Figure 7 shows the schematic diagram of one elementary processor which is composed of 7 units (i.e. the maximum number of particles at each vertex). Each unit represents one direction of the hexagonal lattice. One unit is composed of one input memory of 1 bit, one ouput memory of 1 bit and one photodiode. In addition, the EP contains one recognition unit.

In the set of the 7 units, the incoming particle pattern is first memorized. In the recognition step, the recognition pattern is projected on the 7 photodiodes. The EP compares the input memory contents with the illumination state of the photodiodes. In the case of an exact match, the recognition unit is set to logical level "1". Otherwise this unit stays at logical level "0".

For the substitution step, the substitution pattern is projected just after the recognition step. If the recognition unit is at logical level "1" the illuminating state of the photodiode is transfered to the output memory, otherwise no copy occurs and the EP waits for the next pattern.

We repeat the projection of recognition and substitution patterns sequentially 128 times. At the end of this cycle the output memory contains the substitution pattern corresponding to the initial incoming pattern in the input memory. The propagation step is made by copying the contents of each ouput memory onto the input memory of the neighbouring EPs.

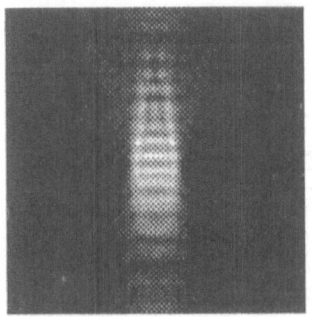

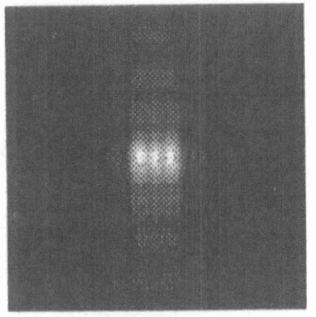

without correction

Energy concentration
on 50*50 µm² area
22.9 %

with correction

Energy concentration
on 50*50 µm² area
44.4 %

Recording wavelength : 488 nm
Reconstruction wavelength : 904 nm

Fig. 6. Aberration correction

4.3 Interface problem

Two interface problems are encountered with the electronic circuit:

Input interface: The response time of the integrated photodiodes must be adapted to the operation frequency of the Silicon components. We have modelled the sensitivity of the integrated photodiode knowing that the structure of the photodiode is a reverse biased P-N junction with an internal capacity. Absorption of photons generates electron-hole pairs which discharge this capacity and switch the voltage of the photodiode between two logical levels. The commutation time depends on the photodiode sensitivity as a function of illuminating wavelength and of the optical power. Imposing a commutation speed of 100 mV/ns, we have deduced the optical power appropriate for the photodiode illumination light (Fig. 8c).

Output interface: Optoelectronic modulators implemented on a GaAs circuit (see below) are used for parallel output. For a good contrast between logical levels "0" and "1", these modulators require a 10 V voltage change and a drive current of about 100 mA. The Silicon electronic circuit works with a logical level "1" of about 5 V and a current of a few mA only. We have implemented specific pads which can increase the logical level "1" from 5 to about 10 V using two transistors (T1,T2) (Fig. 9) and are protected with two inverters. The delivered current is controlled by image current transistor T3.

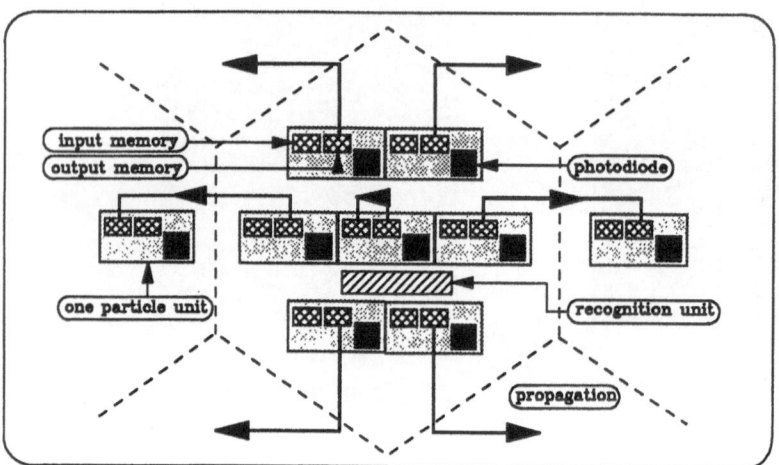

Fig. 7. Electronic circuit

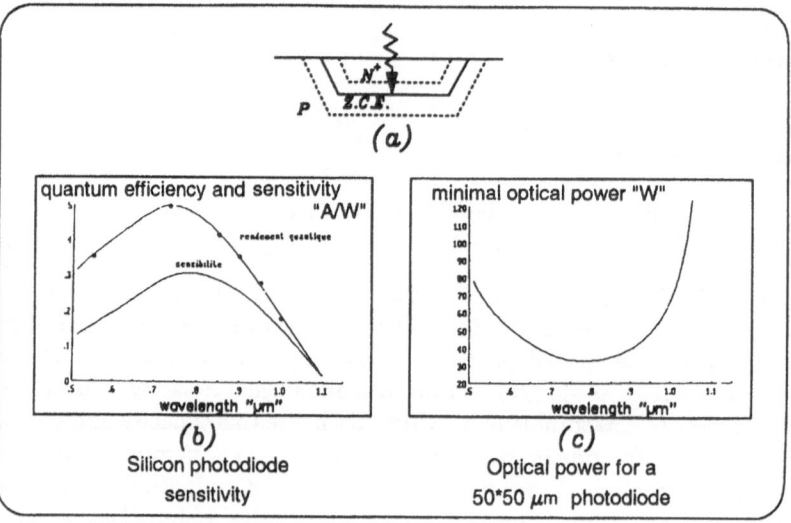

Fig. 8. Integrated photodiode sensitivity

4.4 Optoelectronic Modulators

Optics offers the possibility of parallel output. We use MQW optoelectronic modulators distrubuted on an hexagonal lattice, arranged in units of 7 pixels each.
One pixel is composed of Multiple Quantum Wells above Bragg reflectors. Applying an electric field modifies the reflection state of the pixel at one determined
wavelength.

Each pixel reads the contents of one ouput memory of one EP by one to one
connection between the pixels and the 1 bit output memories. The voltage of each

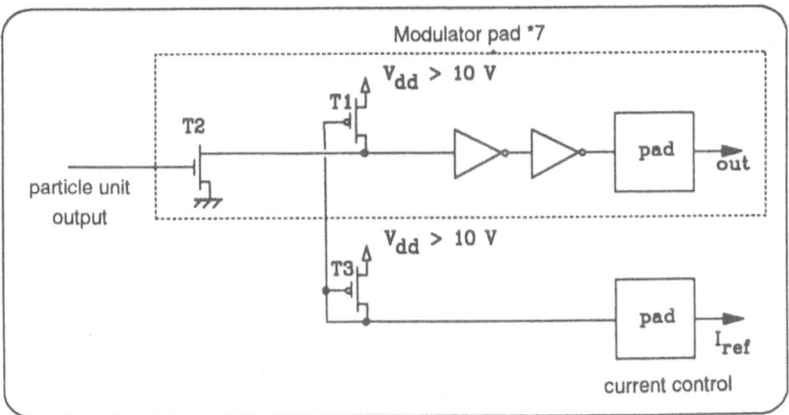

Fig. 9. Modulator output pads

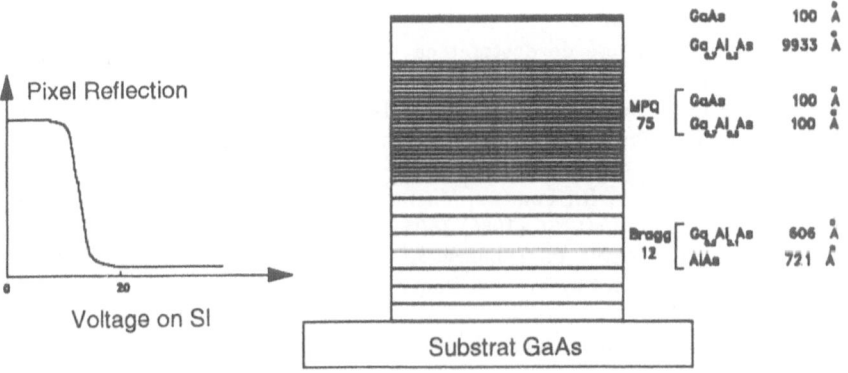

Fig. 10. MQW modulators

output memory drives the reflection state of the corresponding optoelectronic modulator pixel.

In the final setup of the automaton the GaAs circuit containing the modulators would be put on the Silicon circuit and bonded by spherical Indium balls.

5 Conclusion

We have studied the implementation of a parallel optoelectronic automaton for a specific algorithm, the "Lattice-Gas". We have studied and implemented three components for this automaton :

- a microelectronic circuit: for the treatment
- a Talbot effect hologram: for instructions broadcast.

– Optoelectronic modulators: for parallel readout.

This automaton exploits the performances of the electronics for nonlinear operation and the optics for the input and output interconnection.

A further step in our plane is to connect all these components in a complete setup. We have estimated the performances of this setup for 2 mm CMOS technology (french MCP process) :

– 15000 iteration per second
– 0.2 W power dissipation for 10^2 EPs
– High connectivity about 1.68×10^{10} bits/s/cm^2

Some generalization of the use of such an automaton for other algorithms using symbolic-substitution is possible. It can be made by simply changing of the contents of the lookup memory.

References

1. Seyd-Darwish, I.: Thèse de doctorat en sciences. Université Paris XI, Orsay, France (1991).
2. Seyd-Darwish, I., Chavel, P., Taboury, J., Devos, F., Reynaud, R., Maurin, T.: Opto-electronic automata for lattice-gas. In Optics in Complex Systems. Proc. SPIE **1319** (1990) 173–174.
3. Firsch, U., Hasslacher, B., Pomeau, Y.: Lattice-gas automata for Navier-Stokes equation. Phys. Rev. Lett. **56** (1986) 1505–1508.
4. Seyd-Darwish, I., Chavel, P., Taboury, J.: Optimization of an array illuminator hologram. Annales de Physique, Colloque n°1, Supplément au n°1 **16** (1991) 153–161.
5. Talbot, F. A.: Philos. Mag. **9** (1836) 401.

Demonstration of an Optical Pipeline Adder

W. Eckert and C. Passon

Angewandte Optik, Physikalisches Institut der Universität Erlangen-Nürnberg,
W-8520 Erlangen, FRG

We have demonstrated a fully functional optical adder based on systolic arrays and symbolic substitution (SSL). This application shows the feasibility of our optical design concepts based on incoherent data processing systems with opto-electronic threshold amplifier arrays. The optical setup applies four SSL rules in parallel to a 16^2 pixel data plane. Low hardware effort with integrable modules has been achieved.

1 Introduction

Numerical algorithms can be implemented optically by applying a set of SSL rules in parallel to a data plane [1,2]. The combination of SSL with systolic arrays (SAR) allows to simultaneously fulfil the optical hardware requirements and the processing requirements. Reduced hardware effort can be achieved by using dual-rail coded data. Together with a light efficient optical setup one gets systems which are also well suited for micro-integration. A macroscopic demonstration setup is shown here, implementing a full 8-bit pipeline adder.

2 Ripple Carry Pipeline Adder

To implement an adder optically the ripple carry method is very well suited. It is an iterative process, performing the same operations on the bits a_i, b_i of two numbers $\mathbf{A}$ and $\mathbf{B}$. The operation $a_i + b_i$ leads in the 1st step to the sum $s_i(t) := a_i(t)$ XOR $b_i(t)$ and a carry $c_i(t) := a_i(t)$ AND $b_i(t)$. These values define the inputs $a_i(t + 1) := c_{i-1}(t)$ and $b_i(t + 1) := s_i(t)$ for the next step of the iteration.

The advantage for the optical implementation compared with other methods [3] is, that the process can operate in parallel on all data bits during all iteration steps. Only two operations are required to calculate the sum and the carry. There are n iterations needed to add two n bit numbers. The throughput can be improved by pipelining the iteration steps. At each clock cycle a pair of numbers is fed into the pipeline; the result reaches the output n cycles later. In a parallel optical adder one dimension of a 2D data plane can be used for the pipeline operation.

3 Systolic Arrays Implemented with Symbolic Substitution

In a systolic array (SAR) approach the data plane is divided into an array of M functionally equal cells (Fig. 1). The processing operations modify the cells in the neighbourhood of each cell, depending on the data in this cell. Various iterative numerical SAR-algorithms have been developed for transputer based SIMD architectures [5].

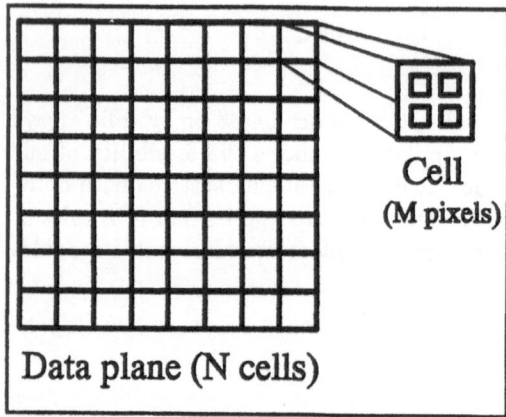

Fig. 1. Structure of a systolic array

An optical implementation is possible by applying a set of SSL rules in parallel to the cells of the SAR [1,4], i.e. the recognition patterns are searched in a data plane and replaced by the corresponding substitution patterns.

An optical ripple carry adder is achieved by coding a pair of numbers **A** and **B** in a row of SAR-cells. Each cell contains a pair of bits a_i and b_i. These bits are dual rail coded (Fig. 2) to reduce the hardware effort for the SSL operations. A set of four SSL rules (Fig. 3) is necessary to create s_i and c_i for the four possible combinations of a_i and b_i, thus implementing the ripple carry operation.

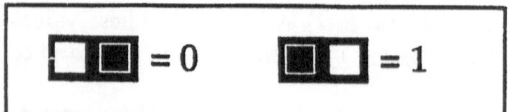

Fig. 2. Dual rail coded bits

For the pipeline operation the substitution patterns are generated one cell below the recognized patterns, thus causing a data flow from top to bottom of the data plane. The carry bit c is additionally shifted one cell to the left (Fig. 4). The input data are fed into the system at the top of the data plane (together

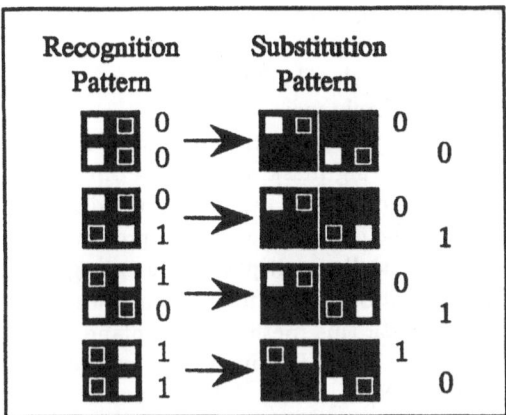

Fig. 3. The set of SSL rules for the adder

with hard-wired carry bits at the right border). The output at the bottom is reached after n iterations.

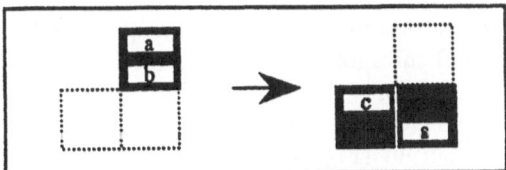

Fig. 4. Alignment of recognition and substitution patterns

4 Optical Implementation

The data flow through the system is shown in Fig. 5a). Data from the input are combined with the systems feedback and reach the SSL stages. The recognition module detects the recognition pattern of the four rules in parallel over the whole data plane. The recognitions are limited to cell aligned patterns by the mask module. The interlace- and deinterlace modules together with a mask, allow to process the four recognition results with only one NOR-device. The four resulting data planes contain bright pixels at all positions where the recognition patterns are found. For each of these pixels, the corresponding substitution pattern is generated by the substitution module. The four substitution planes are superposed. After a regeneration of the data by an OR device the system is fed back and the output leaves the cycle.

The details of the recognition- and the interlace operations are shown in Fig. 5b) for one of the four rules. Because of the redundancy in the dual-rail code only two of the four pixels in the recognition pattern have to be considerd. By searching only dark pixels, the hardware effort for the SSL stages is reduced. A pair of copies of the data plane is created for each rule. These copies are shifted according to the recognition patterns and recombined.

All pixels with no intensity in the resulting plane mark positions in the data plane where the pattern was found. The following mask allows only cell aligned recognitions to occur. Because there is only one pixel per cell required for the recognition, the remaining $M - 1$ pixels of the cell can be used to process other recognition patterns. This is achieved by spatially multiplexing the recognition patterns in the interlace module. The recognition results for all four rules in the optical adder are coded in one data plane with this method. This plane can be processed by one nonlinear device instead of four. The interlaced recognitions are threshold amplified and inverted by a NOR-device. The output contains bright pixels in the cells where the pattern was found. In the adder always one pixel in each cell is bright, marking one of the four possible input patterns. The deinterlace- and substitution stages perform the same operations but in reversed order.

5 Nonlinear Devices

The nonlinear devices we used are optoelectronic inverter arrays (OEIA) [6]. They act as an active (i.e. light emitting) threshold amplifier for the OR operation and as an inverting threshold amplifier for the NOR operation. Each device consists of a CCD-camera as its input, an electronic threshold amplification system and a 16^2 pixel LED-array at the output (Fig. 6). The optical part of the device module reduces the size of the LED-array to fit to the CCD-field. The size of the SAR is limited by the dimensions of the data plane to $N = 8^2$ cells (each $M = 2^2$ pixels in size). The array allows 8 bit numbers to be added. This device is only intended for demonstration purposes, but the sensitvity of the detector and the intensity of the LEDs lie in a typical range for active optoelectronic devices. Therefore the optical system would also work with other smart pixel devices.

6 Optical system

A schematic view of the optical setup is shown in Fig. 7b. The basic operations are performed by four identical modules. They implement the recognition-, interlace-, deinterlace- and substitution operations. The data plane at the modules input is split into several copies by aperture division in the pupil plane by a mirror array. Light efficiency is achieved by using large aperture lenses. Utilizing only a part of the aperture for each copy results in reduced aberrations. Shifts of the copies are performed by tilted mirrors in the pupil plane. For each module the superposed copies are imaged onto its output. This output is either directly processed by an OEIA or it enters a second module after passing a mask and a field lens. The field lens images the mirrors of the first module to the mirrors

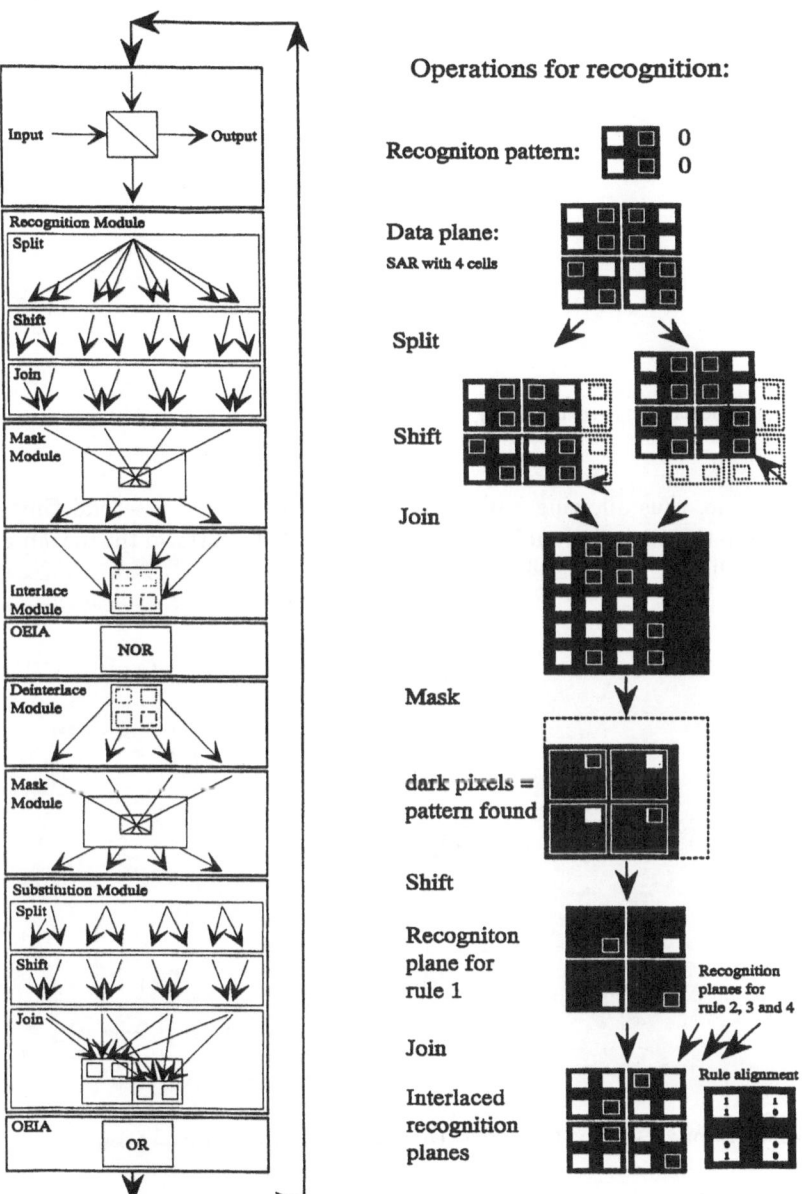

Fig. 5. (a) Data flow through the system. (b) Operations required for the recognitions

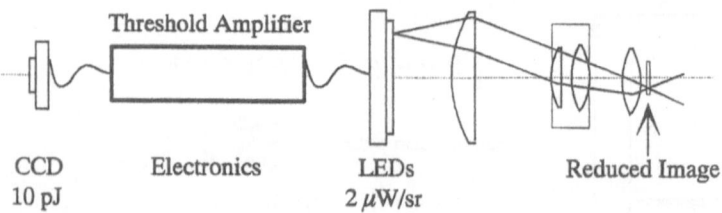

Fig. 6. The optoelectronic inverter array (OEIA)

of the second one, thus allowing individual copies to be shifted a second time. Note that due to this direction multiplexing it is possible to perform the masking operations for all four channels simultaneously.

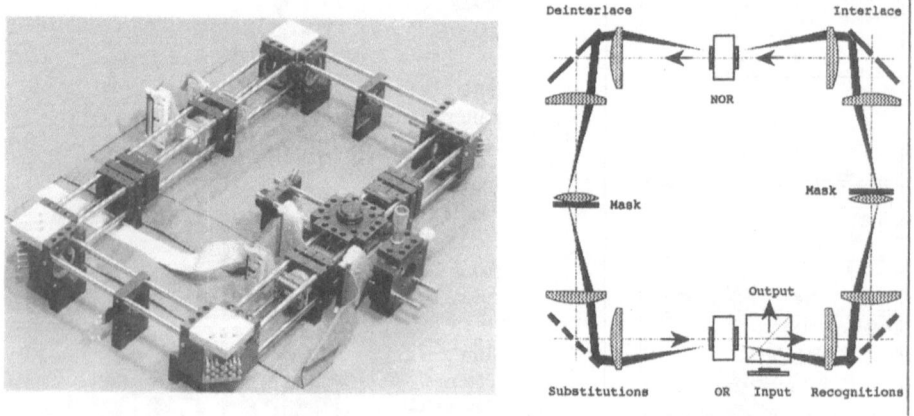

Fig. 7. (a) Laboratory setup for the adder. (b) Schematic of the optical setup

Data from the input first enter the recognition module. The four recognition planes are imaged onto the first mask. The masked data plane enters the interlace module where the four recognition planes (consisting of two superposed copies) are spatially multiplexed by additional shift operations (see also Fig. 5b). The resulting plane is processed by a OEIA acting as a NOR device. The output of the device passes the deinterlace module and a mask providing the four separated recognition results. The substitution module generates the corresponding substitution patterns. The resulting superposed image of the substitution results is then regenerated by the OR device.

In the demonstration system, shown in the Fig. 8, the number of applicable rules is limited by the fan-out of the nonlinear devices required for SSL. With two OEIAs it was possible to implement the four rules (required for the adder) with two pixels to be recognized and two pixels to be substituted. A light efficient optical setup is needed to maximize the number of implementable rules with minimum optical power regeneration. For sources that emit into a large solid angle, such as incoherently radiating or very small data pixels, directional multiplexing is the optimum mode of operation for several reasons. First the splitting into several channels can be done without loss of light, thus allowing to make use of most of the emitted light and resulting in a very light efficient setup. Second, because for each channel only a fraction of the aperture is used, the aberations resulting from lens errors are reduced.

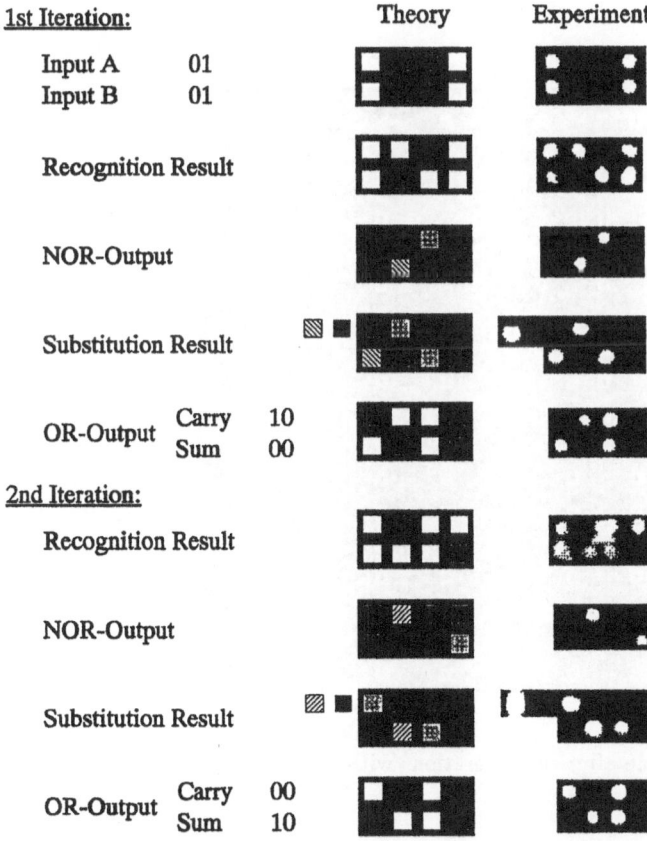

Fig. 8. Example with two two-bit numbers

7 Experimental results

If the pair of input numbers is set constant, a steady state is reached after 8 iterations (Fig. 9). The figure demonstrates the data flow from the top, where the dual rail coded input numbers 53 and 77 appear, to the bottom, where the result 130 is coupled out. During the iteration process the arithmetic sum of each row of cells (containing a pair of numbers) remains constant, while the upper number in the rows is reduced from step to step, until zero.

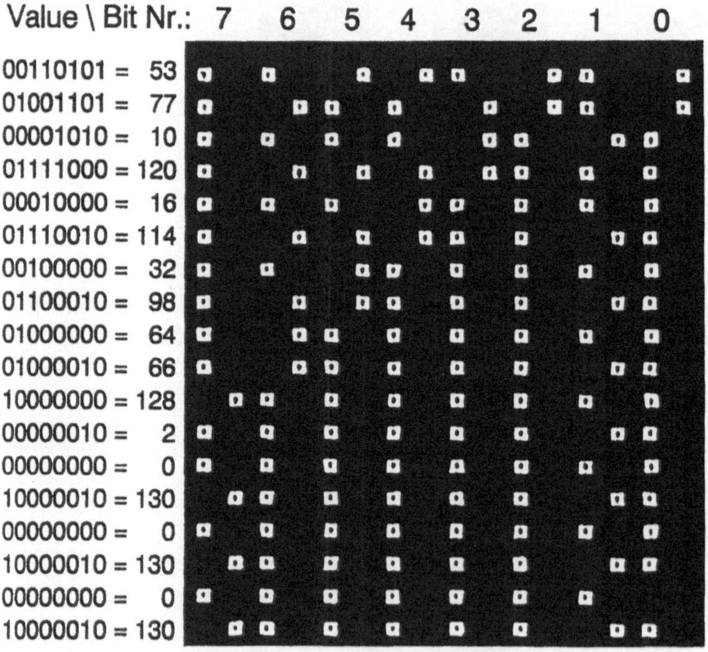

Value \ Bit Nr.:	7	6	5	4	3	2	1	0
00110101 = 53								
01001101 = 77								
00001010 = 10								
01111000 = 120								
00010000 = 16								
01110010 = 114								
00100000 = 32								
01100010 = 98								
01000000 = 64								
01000010 = 66								
10000000 = 128								
00000010 = 2								
00000000 = 0								
10000010 = 130								
00000000 = 0								
10000010 = 130								
00000000 = 0								
10000010 = 130								

Fig. 9. A whole data plane after eight iterations with constant input

In the right upper 4 by 4 pixel area the binary numbers **A** = 01 and **B** = 01 are summed up during the first two iterations. We discuss the applied operations for this part of the data plane in detail (Fig. 8). The recognition operations for these numbers are shown in Fig. 5 for one rule. In Fig. 8 the recognition result marks the lower right corner of the left cell as dark for the 0/0-case and the upper left of the right cell as dark for the 1/1-case. After the NOR operation the substitution module generates the substitution patterns corresponding to the different locations within the 2×2 cell. The result at the OR array is the superposition of the substitution patterns of all the rules. A second iteration process is needed to process the remaining carry bit.

8 Conclusion

The optical system of the pipeline adder demonstrated, implements a SAR in a modular way with relatively low hardware expense. SARs for different applications can be constructed by using the design concept based on aperture division. The adder is presently constructed with standard optical components, resulting in a size of approx. 30 cm × 50 cm. Due to its modular design the size can be reduced considerably by micro-integration using micro lenses and microprisms-arrays.

References

1. Eckert, W.: Entwurf und Realisierung kaskadierbarer Mustersubstitutionssysteme. Diplomarbeit 1990 (unpublished).
2. Brenner, K.-H., Huang, A., Streibl, N.: Digital Optical Computing with Symbolic Substitution. Appl. Opt. **25** (1986) 3054.
3. Barua, S.: Single stage optical adder/subtractor. Opt. Eng. **30** (1991) 265-270.
4. Fey, D.: Digital optical arithmetic based on Systolic Arrays and Symbolic Substitution Logic. Opt. Comp. **1** (1990) 153–167.
5. Kung, H. T.: Systolic algorithms for the CMU warp processor. In Proc. 7th International Conference on Pattern Recognition. Montreal (1984) 570–577.
6. Zürl, K.: Opto-Electronic Inverter Array. Annual Report "Angewandte Optik, Universität Erlangen 1988". page 21.

Performance and Hardware Requirements of Parallel Addition Algorithms for Optical Implementation Using SEEDs

D. Rhein

Alcatel Austria – Elin Research Centre, Ruthnergasse 1-7, A-1210 Vienna, Austria

The aim of the present work is to give an overview of existing algorithms for parallel addition and to estimate the performance of optical implementations based on self-electro-optic effect devices (SEEDs). The main conclusion is that more processing capabilities in the active elements ("smart pixels") or new design principles are required in order to reduce the complexity of digital optical arithmetic units.

1 Introduction

The interest in optical computing has been growing very rapidly in the last few years. To exploit the advantages of optics for digital, optical parallel computing systems, several parallel arithmetic algorithms have recently been proposed and investigated. In this paper, these new methods will be reviewed and compared with conventional addition algorithms and their optical implementations.

For the addition of two binary numbers $A = (a_n, ..., a_2, a_1)$ and $B = (b_n, ..., b_2, b_1)$ there exist several algorithms, which can be classified into one of two different methods [1]: the bit-serial and bit-parallel methods, as shown in Fig. 1. In the first method all digits s_i $(i = 1, ..., N + 1)$ of the sum

$$S = A + B \tag{1}$$

are calculated separately from the corresponding digits of the two operands and the carry c_i of the foregoing digit operation.

In the first algorithm of this method the addition is executed bit-serial, but it can be performed in an optical computer on a large number of pairs of words (or operands) in parallel. It requires to arrange one bit of every operand to be processed in one data plane together with the carry bit, which is the result of the foregoing bit operation. This data plane is a "bit slice", and the process is called bit slice addition. The obtained sum of every digit operation is the final result, but the obtained carry bit must be included into the data plane of the next digit operation for each operand pair. The addition of N-bit words requires $N + 1$ operations to give the $(N + 1)$-bit result.

In the second group of method all digits of each operand are processed simultaneously. However, the result after one operation is not the final result since

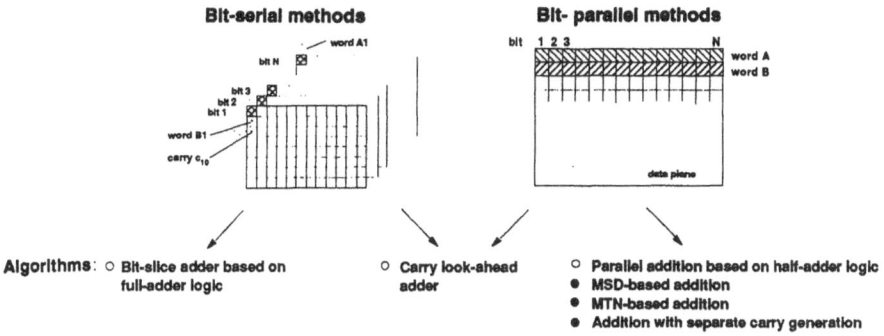

Fig. 1. Classification of addition algorithms

the carry must propagate through all the digits. Hence, the operations must be performed at least N times, or special carry handling procedures for "carry-free" addition must be implemented.

The following studies include the conventional bit-slice addition, look-ahead carry addition, and parallel addition using half-adder logic [2] as well as the new "carry-free" parallel algorithms based on modified signed-digit numbers [3,4], on modified trinary numbers [5], and the parallel addition with separate carry generation [1]. The considerations are restricted to the addition operation since other arithmetic operations such as subtraction, multiplication and division can be traced back to the addition operation.

For optical computing hardware design one of the possible optical components is the symmetric self-electro-optic effect device (S-SEED, see Fig. 2). These devices are well developed for mass fabrication and meet most of the requirements of digital information technology such as cascadability, signal regeneration properties, fan-in and fan-out. There are two possibilities for implementing general logic functions with SEEDs. The first is the Programmable Logic Array design approach to execute any logic function using arrays of equal logic gates (to perform SIMD operations) with fan-in and fan-out of two, and regular optical interconnects [6,7]. In this approach, to perform any logic function of m variables, the 2^m unminimized minterms are generated in $m + 1$ stages and are then combined in further $m + 1$ stages to implement the desired logic functions. In the dual-rail representation of variables, the network in the required $2(m+1)$ stages will be 2^{m+1} gates wide. Examples of network design using AND/OR logic and butterfly interconnections or using OR/NOR logic with crossover interconnections have been described by Murdocca [7].

With SEED devices, logic functions can also be realized using symbolic substitution logic [8]. Different architectures for bit-serial addition units based on this approach have been investigated and compared [9]. However, the comparison of parallel addition units in this paper will be restricted to the first approach.

For estimation of the addition time, a clock period T_c will be introduced,

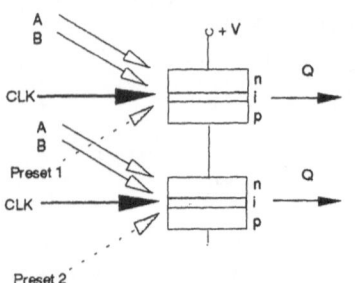

Fig. 2. Structure of S-SEED and its main properties

which is the time taken to propagate the information from one SEED array to the next (this period usually must contain two clock signals CLK, $\overline{\text{CLK}}$ and also a preset pulse to the SEEDs, i.e. at least three switching times of the SEED). For all addition units in this chapter it is assumed that they can add M binary numbers with N bits each.

In the second chapter of this paper the different addition algorithms will be described and their optical implementation using SEEDs will be discussed. For optical system implementation only regular interconnects and the application of the universal (i.e. redundant) PLA design will be assumed. In the third chapter the addition units based on these algorithms will be compared for their performance and their hardware requirements and finally some conclusions will be drawn.

2 Addition Algorithms and their Optical Implementation Using SEEDs

2.1 Conventional Addition Algorithms

Bit-slice addition based on full-adder logic. In this method, in every operation the digits a_i, b_i, and the carry c_i of the foregoing operation are logically processed to give the final sum digit and the new carry

$$s_i = \bar{a}_i \cdot \bar{b}_i \cdot c_i + \bar{a}_i \cdot b_i \cdot \bar{c}_i + a_i \cdot \bar{b}_i \cdot \bar{c}_i + a_i \cdot b_i \cdot c_i = a_i \otimes b_i \otimes c_i \ , \qquad (2)$$

$$c_{i+1} = a_i \cdot b_i + a_i \cdot c_i + b_i \cdot c_i \ , \qquad (3)$$

where ".", "+", and "$\otimes$" denote the AND, OR, and XOR (exclusive OR) logic functions, respectively. Equations (2) and (3) describe the logic functions of a full-adder and the result of (2) is the final ith bit of the sum. If these operations are performed on all groups of operands and carry bits of a data plane in parallel (SIMD) the parallel processing capabilities of optics can be utilized.

Another expression of the full-adder logic can be obtained using symbolic substitution logic [8,10]. Since the number of input variables is three, the full-adder can be realized by using 8 different substitution rules.

The bit slice adder can be modified to the pipelined ripple carry adder [2], where additions of the bits of different word pairs are executed successively, i.e. delayed by one operation for bits of successive words. This is especially useful if the result of an addition is an operand in a following operation, which then can be started with a delay of just one bit operation.

An optical implementation of a serial adder with banyan interconnected AND/OR logic using SEEDs or with cross-over interconnected OR/NOR logic has been designed by Murdocca [7]. It needs for each digit a full-adder corresponding to (2) and (3) with three input variables, leading to a $2(n + 1) = 8$ stages deep and $2^{n+1} = 16$ gates wide network. Assuming that at least two additional stages are required for separation of sum and carry from the result and for combining the carry with the next bit input information, the whole addition time can be estimated to be

$$t_{\mathrm{add}} = 10(N + 1)T_c \ .$$ (4)

If M pairs of words are to be added in parallel the whole gate count will be

$$z = 160M$$ (5)

in 10 stages (modules). Here also besides the 8×16 gates for one full-adder two additional stages with 2×16 gates for result and carry separation are taken into account.

Look-ahead carry addition. This method can be considered as a hybrid variant between the bit-serial method and the algorithms which process all bits of the operands in parallel. According to this method the carries for groups of digits (instead of all digits of the numbers) are calculated in advance by calculating first two intermediate vectors from the bits of the operands (see [2]). These are the *generator vector* g_i and the *propagation vector* p_i with the components

$$g_i = a_i \cdot b_i \ ,$$ (6)

$$p_i = a_i \otimes b_i \ .$$ (7)

With these values the equations (2) and (3) for the sum and the carry of i-th digit can be written as

$$s_i = a_i \otimes b_i \otimes c_i = p_i \otimes c_i \ ,$$ (8)

$$c_{i+1} = a_i \cdot b_i + a_i \cdot c_i + b_i \cdot c_i = a_i \cdot b_i + c_i \cdot (a_i + b_i) \equiv a_i \cdot b_i + c_i \cdot (b_i \otimes a_i) = g_i + c_i \cdot p_i \ .$$ (9)

In the look-ahead carry adder, the carries for all bits (or a group of bits) are computed in parallel by means of additional logic gates from the following equations, resulting from (9):

$$c_2 = g_1 + p_1 \cdot c_1$$
$$c_3 = g_2 + p_2 \cdot c_2 = g_2 + p_2 \cdot g_1 + p_2 \cdot p_1 \cdot c_1$$ (10)
$$c4 = g_3 + p_3 \cdot c_3 = g_3 + p_3 \cdot g_2 + p_3 \cdot p_2 \cdot c_1 + p_3 \cdot p_2 \cdot p_1 \cdot c_1$$

Note, that for the least significant bit (LSB) the carry c_1 equals zero, while for higher groups of digits, c_1 for these groups equals the carry of MSB of the foregoing group. After computing all carries (of the actual group) simultaneously according to (10), the final result of sum digits can be computed from (8) in one step. In this way the time for addition is shorter than with carry propagation, but additional hardware is required.

A scheme for optical implementation of a look-ahead carry adder based on SEEDs has been described in [2]. However, in that scheme unusual one-dimensional SEED arrays are used, and unlimited fan-in and fan-out are assumed to be possible. Instead we will consider the usual Programmable Logic Array design approach with regular interconnects as in the case before.

With 4-bit groups to be added in one step, 9 input signals (8 number digits and 1 carry of foregoing group) are to be combined logically. This requires a $2(9+1) = 20$ stages deep and $2^{9+1} = 1024$ gates wide SEED network. In the general case of n-bit groups, $2n+1$ input signals are to be combined to get the n final bits of the result and the carry to the next group. If taking into account two additional stages for combining the carries of each group with the next group input signals, $2(2n+2)+2$ stages are necessary with 2^{2n+2} gates per stage. The addition time for serial processing of N/n groups then will be

$$t_{\text{add}} = (4n + 6)\frac{N}{n}T_c \ , \tag{11}$$

and the number of gates required is

$$z = (4n + 6)2^{2n+1}M \ . \tag{12}$$

Note that from these equations the corresponding formulas of the bit-slice adder can be obtained as a special case for $n = 1$.

From (11) and (12) it can be seen that the addition time is smaller than in the previous case, but the advantage is not large if the expenditure of the exponentially increased gate count is considered. The addition time and the gate count depend on the number of bits executed in parallel. The smallest addition time with a large n is about $4NT_c$, while the number of gates required for parallel operation increases exponentially.

Bit-parallel addition using half-adder logic. If the different bits of the two binary numbers are simply added in parallel, the half adder operation results leading to preliminary sum bits (ignoring carries)

$$s_i = a_i \cdot b_i + a_i \cdot b_i = a_i \otimes b_i \tag{13}$$

and the carry to the next digit

$$c_{i+1} = a_i \cdot b_i \tag{14}$$

After doing the half-adder operation, the same operation must be repeated with s_i and c_i instead of a_i and b_i as the operands, until all carry signals become zero. This usually requires $N + 1$ operations for the addition of two N-bit words.

For addition of two input numbers with two bits to be combined in every digit according to logic equations (13) and (14), $2(2 + 1) = 6$ stages of SEEDs $2^{2+1} = 8$ gates wide are necessary. If the results can be arranged directly as required for the next half-adding process, additional stages may be not required. In this case the resulting addition time and gate count for parallel operation at M pairs of numbers to be added will be

$$t_{\text{add}} = 6(N + 1)T_c \; , \qquad (15)$$

$$z = 48(N + 1)M \qquad (16)$$

in 6 stages, respectively.

2.2 New parallel addition algorithms and their optical implementation

Besides, the conventional parallel addition approach based on half-adder logic, recently several "carry-free" addition algorithms have been proposed for optical computing systems. In redundant number systems as in the *residue number system* [11] arithmetic operations can be performed in only one step without occurrence of carry propagation. However in this system division and the necessary number conversions cannot be implemented easily.

Another redundant number system, the *modified signed-digit system* (MSD) [12], with proper implementation, can be used to add binary numbers in parallel in only a few steps without remarkable carry propagation as will be shown in the next paragraphs together with another similar "carry-free" method developed by Brenner et al. [1].

Parallel addition of binary numbers based on the MSD number system. In the MSD number system a number A can be represented by an expression

$$A = \sum_{i=1}^{N} a_i 2^{i-1} \; , \qquad (17)$$

where each digit a_i is an element of the set $[1, 0, \bar{1}]$, and $\bar{1} = -1$. The representation of (17) is ambiguous, e.g. the number 7 corresponds to either 111 or $10\bar{1}$. In the MSD number system, the carry propagation is limited to two positions to the left. Hence, the addition can be carried out in three stages (during addition transfer, weight digits $[w_i, t_i]$ are generated in parallel which correspond to the sum and carry digits $[s_i, c_i]$ of half-adder without carry propagation).

If the resulting $[t_i, w_i]$ are chosen accordingly (see [3,4]), the carry will propagate only one position to the left, i.e. the final result depends only on two digits. This can be done as follows:

1. If $[a_i, b_i] = [1, 1], [1, \bar{1}], [0, 0], [\bar{1}, 1]$, and $[\bar{1}, \bar{1}]$ there is only one possible value for $[t_i, w_i]$, i.e. $[t_i, w_i] = [1, 0], [0, 0], [0, 0], [0, 0], [\bar{1}, 0]$.

2. For $[a_i, b_i] = [1, 0]$ or $[0, 1]$ two possibilities exist, i.e. $[t_i, w_i] = [0, 1]$ or $[1, \bar{1}]$. In this case chose $[t_i, w_i]$ in dependence of the next lower digits of operands: as $[1, \bar{1}]$ if both a_{i-1} and b_{i-1} are non-negative, and as $[0, 1]$ in all other cases.

3. If $[a_i, b_i] = [\bar{1}, 0]$ or $[0, \bar{1}]$, from the two possibilities chose $[t_i, w_i] = [0, \bar{1}]$ if both a_{i-1} and b_{i-1} are non-negative, and $[t_i, w_i] = [\bar{1}, 1]$ in all remaining cases.

With these definitions and with the structure of the data as indicated in the box, the addition rules for the remaining two stages are as given in Fig. 3. As can be seen from the figure the final result for any sum digit s_i depends only on two of the operand's digits. These functions can be easily expressed as logical functions of four input bits or as 16 SSL rules for symbolic substitution logic (the latter can be found in [3]. After execution of the addition by application of the given rules to the digit's of the two binary operands, the result of addition is obtained as a MSD number. It must be transformed to a binary number (BN) for the next arithmetic or I/O operations. This can be performed based on the identity $0\bar{1} = \bar{1}1$. This procedure will be explained with trinary (MSD) number representation as given in the middle of Fig. 4(a), which can be implemented with bistable devices like the SEEDs. In this representation it is assumed that a "1"' of a MSD number is represented with a "1" in the corresponding digit of a lower row (A) and a "$\bar{1}$" by a "1" in the upper row (B) of a two-row field.

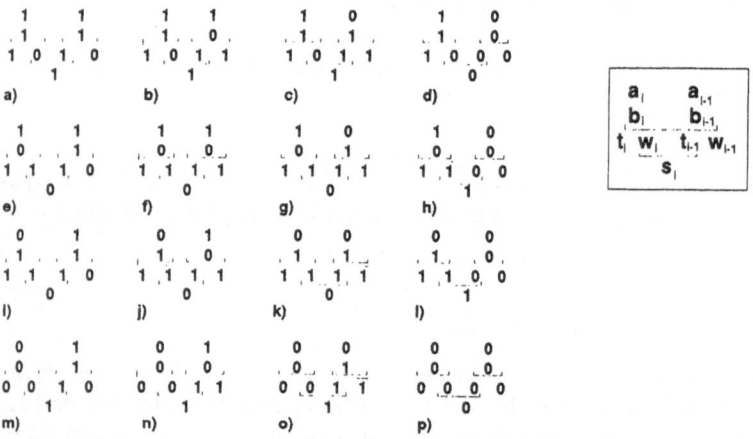

Fig. 3. Addition rules based on MSD number system (after [10])

The MSD to BN conversion then requires successive replacement of a "1" in the upper row with a "1" in the lower row and a "1" in upper row in the left higher digit (as shown by the first substitution rule in Fig. 4(b)). The process must be repeated until the "1" in the upper row will be compensated by a "1" in the lower row to give a "0" (second substitution rule in Fig. 4(b)). The

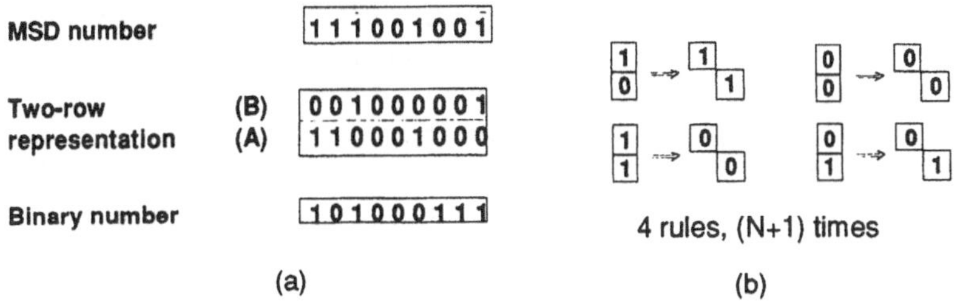

MSD number `1 1 1 0 0 1 0 0 1`

Two-row representation (B) `0 0 1 0 0 0 0 0 1`
(A) `1 1 0 0 0 1 0 0 0`

Binary number `1 0 1 0 0 0 1 1 1`

(a)

4 rules, (N+1) times

(b)

Fig. 4. Conversion of MSD numbers into binary numbers. (a) Number representation. (b) Basic substitution rules for successive conversion

conversion by successive application of the two rules in Fig. 4(b) has finished if no "1" remains in the upper row. The example of Fig. 4(a) shows the resulting binary number at the bottom.

Logically, the following functions have to be executed in successive steps, i.e. for $\nu = 1...(N+1)$:

$$a_i^{(n+1)} = a_i^{(n)} \otimes b_i^{(n)}, \quad b_i^{(n+1)} = b_{i-1}^{(n)} \cdot a_{i-1}^{(n)} . \tag{18}$$

The adder based on the MSD number system according to the algorithm described performs addition in two procedures. First the two binary numbers are added to an MSD number. Each final sum digit in this operation depends on four binary input variables and can be found from $2^4 = 16$ minterms in a PLA network consisting of 10 stages each 32 bits wide. The resulting MSD number can be represented in two binary SEED-words as illustrated in Fig. 4(a).

The conversion of the MSD number into a binary number takes place by successive 2-bit additions according to the rules in Fig. 4(b) until no "1" occurs in the "$\bar{1}$"-word (generally after $N + 1$ operations). This requires 6 stages more of a $2^3 = 8$ gates wide SEED network.

Hence, the resulting addition time of this type of adder is

$$t_{\text{add}} = 10T_c + 6(N + 1)T_c . \tag{19}$$

Note that the first term in this equation is independent of word length while the MSD to BN conversion (second term) is proportional to word length. The whole gate count for the adder is

$$z = (10 \cdot 32 + 6 \cdot 8)NM = 368NM \tag{20}$$

gates in 16 stages for addition of M pairs of numbers in parallel. Since the MSD into BN conversion has the same gate and time requirements as the parallel half adder, this addition variant cannot be competitive with that in Sect. 2.2 if the conversion is executed after every operation.

Addition based on "modified trinary number" (MTN) system. Another similar addition algorithm was developed based on a modified trinary
number representation [5]. For parallel addition, the basic idea of this method
is to use one operand in the usual binary form (binary number, BN) with only
0s and 1s as digits, and the other operand in a modified trinary number (MTN)
system introduced in such a way, that this representation contains only 0s and
$\bar{1}$s except in the most significant bit (MSB). The binary number and the MTN
are then added following the simple rules

$$
\begin{aligned}
1 + 0 &= 0 + 1 = 1 \\
0 + 0 &= 1 + \bar{1} = 0 \\
\bar{1} + 0 &= 0 + \bar{1} = \bar{1}
\end{aligned}
\tag{21}
$$

without any carry being involved. The result of addition, however, is a trinary
number (TN) containing 0s, 1s, and $\bar{1}$s and must be converted into either a
binary number for output or into a MTN for successive additions.

The MTN representation of a number is defined as follows. A binary number
(BN)

$$
(A)_{\text{BN}} = \sum_{i=1}^{N} C_i 2^{i-1} , \tag{22}
$$

with C_i taking the values 0 or 1, can be converted to its two's complement

$$
\sum_{m=1}^{N} C_m 2^{m-1} = 2^N - \sum_{i=1}^{N} C_i 2^{i-1} , \tag{23}
$$

which is always a positive number. The MTN corresponding to A is obtained
by replacing all 1s by $\bar{1}$ in the left side (giving the negative value of the two's
complement) and adding 2^N:

$$
(A)_{\text{MTN}} = 2^N + \sum_{i=1}^{N} C'_m 2^{m-1} , \tag{24}
$$

where C'_m may take on the value 0 or $\bar{1}$. From (24) results that a MTN has a 1
as its most-significant bit, whereas all other digits are either 0 or $\bar{1}$.

The conversion of a BN to a MTN is done by the following steps [5]:

1. If the BN has 0's for its least-significant bits, they are deleted. If the BN has
   a 1 as its LSB, this processing step is by-passed.
2. Complement all binary digits that are obtained from step 1.
3. Replace all 1s in the result of step 2 by $\bar{1}$ and add a 1 to the left of the MSB.
4. Add a $\bar{1}$ to the result of step 3 (replace the zero in the right-most position
   by a $\bar{1}$).
5. Add the zeros that have been deleted in step 1 to the right-hand side to get
   the converted result as a MTN.

The following example illustrated the process:

Binary number 10011010
       step 1 1001101
       step 2 0110010
       step 3 10$\bar{1}\bar{1}$00$\bar{1}$0
       step 4 10$\bar{1}\bar{1}$00$\bar{1}\bar{1}$
MTN (step 5) 10$\bar{1}\bar{1}$00$\bar{1}\bar{1}$0

Since the result of addition is a TN (or MSD number), its conversion into a BN or into a MTN is essential. The first task has already been described in the foregoing section (see Fig. 4). For conversion of a TN into a MTN, corresponding to the identity $01 = 1\bar{1}$, all 1s in the TN have to be replaced successively by $1\bar{1}$ until no 1s occur in the number, except in its most-significant digit. The requirements of these conversions are identical to those in the conversion of a TN (or MSD number) into a binary number, as described in Fig. 4.

The MTN adder according to the algorithm must first perform the conversion of one of the operands from binary into MTN representation. Although the process described above is not well adapted to SEED implementation, it can be easily executed. The first step of masking out the zeros at the end may take in the worst case N operations, while step 2 requires only one cycle. Step 3 does not need any time if the 1s are simply defined as $\bar{1}$s, and the addition of a 1 to the position left of the MSB can be performed to the other operand which contains only 1s and 0s (this also simplifies the addition phase itself). Steps 4 and 5 then can be performed in one cycle each, leading to a conversion time of $(N + 4)T_c$.

The addition process according to (21) itself is very simple. The logic functions of the two input variables per digit can be executed in 6 stages of 8 SEED gates per digit. The result must be represented as two binary numbers as in the case of the MSD adder. The conversion of the trinary result into a binary number has already been described and requires 6 stages of SEEDs, 8 gates wide which have to be executed successively $(N + 1)$ times.

Summarizing, the MTN adder has an estimated addition time of

$$t_{\text{add}} = [(N + 4) + 6 + 6(N + 1)]T_c = (7N + 16)T_c \ . \tag{25}$$

If we realize the BN into MTN conversion with 4 stages of SEEDs, the whole gate count of the adder can be estimated to be

$$z = (4 + 6 + 6)8NM = 128NM \ . \tag{26}$$

Parallel addition with separate carry generation. Brenner et al. [1] have developed a method which first calculates an intermediate sum S' and a carry vector C in parallel like the parallel adder with half-adder logic. The bits of the intermediate sum vector are then changed or not, depending on the digits of an inverting vector I, which is derived from the carry vector and an additional auxiliary vector N. The elements of these vectors for the i-th digit are defined as

$$\text{intermediate sum:} \quad s'_i = a_i \otimes b_i$$
$$\text{carry vector:} \quad c_i = a_i \cdot b_i$$
$$\text{auxiliary vector:} \quad n_i = \overline{a_i + b_i} \tag{27}$$

The carry vector indicates carry bits, while the vector N indicates the end of a carry condition, i.e. n_i is a "one" only in the case if $a_i = 0$ and $b_i = 0$. To build up the inverting vector the following procedure must be executed:

1. Shift C and N by one digit to the left (since the carry acts on the next higher digit, and with $n_i = 1$ ($a_i = b_i = 0$) there is no carry in the next higher digit). To each word a zero is added to their right side.
2. Set $I_i = c'_i$ (after shifting of C and N), and for $I_i = 1$ (carry condition) fill

$$I_{i+1}...I_{i+k} = 1 \text{ for } I_i = 1 \text{ and } n'_i + m = 0 \ (m = 1...k) \ ,$$

i.e. fill in ones into the inverting vector until the digit before the first "one" in the N vector indicates the end of carry propagation. For $n'_i = 1$ the corresponding digit of inverting vector must be zero ($I_i = 0$). The latter condition is fulfilled automatically since from (27) c_i and n_i cannot be simultaneously "1".

To perform the generation of the vector I after shifting, the operations must be executed which correspond to the SSL rules in Fig. 5. The two rows herein correspond to the digits of C and N, where C is transformed after execution into I. It is of interest to notice, that the operations are very similar to those in MSD to BN number conversion, as described in Fig. 4(c). As in that case, either the first rule from Fig. 5 can be executed successively N times, or all rules of the figure can be executed in parallel.

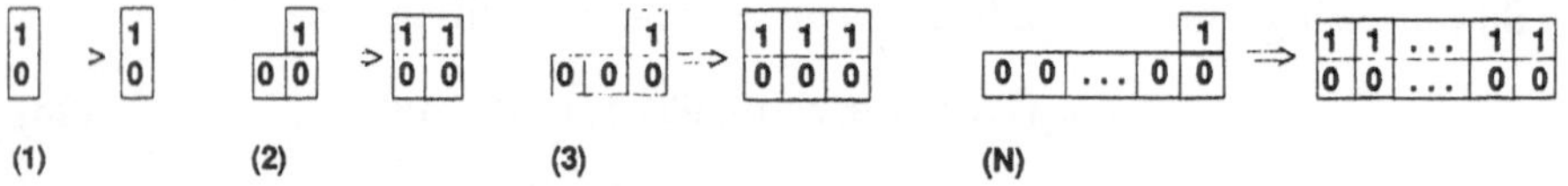

Fig. 5. SSL rules for inverting vector generation from C' and N'

After generation of the inverting vector in the way described, the final sum of the two operands is obtained in one single step by parallel execution of the logic function

$$s_i = s'_i \otimes I_i \ . \tag{28}$$

An example is given below to illustrate the algorithm:

$$
\begin{array}{ll}
A & = 1\ 0\ 1\ 0\ 0\ 1\ 1\ 0 = 166 \\
B & = 0\ 0\ 1\ 0\ 0\ 0\ 1\ 1 = \ \ 35 \\
\hline
S' & = 1\ 0\ 0\ 0\ 0\ 1\ 0\ 1 \\
C & = 0\ 0\ 1\ 0\ 0\ 0\ 1\ 0 \\
N & = 0\ 1\ 0\ 1\ 1\ 0\ 0\ 0 \\
C' & = 0\ 1\ 0\ 0\ 0\ 1\ 0\ 0 \\
N' & = 1\ 0\ 1\ 1\ 0\ 0\ 0\ 0 \\
I & = 0\ 1\ 0\ 0\ 1\ 1\ 0\ 0 \\
\hline
S & = 1\ 1\ 0\ 0\ 1\ 0\ 0\ 1 = 201
\end{array}
$$

In this adder, first the vectors S', C, and N must be calculated by logical combination of the two input bits per digit which can be performed in 6 stages each 8 gates wide. The shift operation requires one step and the generation of the inverting matrix can be done by logically combining the two elements of vectors C and N in 6 stages 8 gates wide. Similar to the MSD into BN conversion, this operation (according to the rules in Fig. 5) needs to be processed successively $(N + 1)$ times. Then with the single parallel logic operation according to (28), the result is obtained in one additional step (using a special structure of S-SEED to perform XOR operation)

Hence, the resulting addition time will be

$$t_{\text{add}} = 8T_c + 6(N + 1)T_c = (6N + 9)T_c \tag{29}$$

and for execution, 14 stages with 8 gates per digit are necessary, i.e. the gate count will be

$$z = 112NM \ . \tag{30}$$

3 Comparison of Various Parallel Adder Types with SEEDs

The results of the estimations of addition time and gate count of different adder types as investigated in section 2 are summarized in Table 1.

The comparison in the table and especially the data for a concrete adder for 32-bit numbers in the last two columns show that the velocity advantage of the bit-parallel methods is – at least for the SEED implementations with the assumptions made – not larger than about 30%, and it must be paid for with a gate count increased by more of one order of magnitude. Among the parallel addition methods, the differences are considerably small and the conventional parallel half-adder gives better results than the recently published parallel addition algorithms.

4 Conclusions

The results of the comparison of different addition algorithms for parallel digital optical computing given in this paper show that the design of high-perfomance

Table 1. Addition time and gate count of different parallel adder types with SEEDs

Adder type	t_{add}	z	t_{add} for N=32	z for N=32
Bit-slice adder	$10(N+1)T_C$	160M	$330T_C$	160M
Look-ahead adder (n=4)	$5.5NT_C$	22528M	$176T_C$	22528M
-------- " -------------- (n=2)	$7NT_C$	896M	$224T_C$	896M
Parallel half-adder	$6(N+1)T_C$	48(N+1)M	$198T_C$	1584M
MSD adder	$(6N+11)T_C$	368NM	$203T_C$	11776M
MTN adder	$(7N+16)T_C$	128NM	$240T_C$	4096M
Parallel adder with separate carry generation	$(6N+9)T_C$	112NM	$201T_C$	3584M

parallel arithmetic units with SEEDs based on the known design methods is not primarily a question of having the right algorithm. It appears that for efficient optical arithmetic units the hardware side must be developed considerably. The assumptions taken for SEEDs, e.g. limited fan-in and fan-out of two, application of regular interconnects only and the PLA design method based on universal AND stages for generation of all possible minterms, seriously limit the performance of the computing systems, independently of the used algorithm.

Hence the investigation of other design methods for SEED modules, the application of irregular (holographic for instance) interconnections, and/or the development of more powerful components like "smart pixels" seem to be necessary in order to benefit from the advantages of digital optical parallel processing. Furthermore, the application of symbolic substitution in parallel adder design using SEEDs should be taken into account.

Acknowledgments

Several useful discussions with J. Doppelbauer, T. Cloonan, K.-H. Brenner and A. Wachlowski are gratefully acknowledged.

References

1. Brenner, K.-H., Kufner, M., Kufner, S.: Highly parallel arithmetic algorithms for a digital optical processor using symbolic substitution logic. Appl. Opt. **29** (1990) 1610–1618.
2. McAulay, A.D.: Optical computer architectures. Wiley, New York, 1991.
3. Barua, S.: Design of a high-speed optical arithmetic unit. Proc. SPIE **1347** Optical Information-Processing Systems and Architectures II, (1990) 526–535.
4. Barua, S.: Carry-free optical binary adders. Proc. SPIE **1347** Optical Information-Processing Systems and Architectures II, (1990) 573–579.

5. Datta, A.K., Basuray, A., Mukhopadhyay, S.: Arithmetic operations in optical computations using a modified trinary number system. Opt. Lett. **14** (1989) 426–428.
6. Murdocca, M.J., Huang, A., Jahns, J., Streibl, N.: Optical design of programmable logic arrays. Appl. Opt. **27** (1988) 1651–1660.
7. Murdocca, M.J.: A design methodology for optical computing. The MIT Press, Cambridge, Mass., 1990.
8. Brenner, K.-H., Huang, A., Streibl, N.: Digital optical computing with symbolic substitution. Appl. Opt. **25** (1986) 3054–3060.
9. Cloonan, T.: Performance analysis of optical symbolic substitution. Appl. Opt. **27** (1988) 1701–1707.
10. Huang, A.: Parallel algorithms for optical digital computers. IEEE 10th International Optical Computing Conference, H.J. Caulfield, ed., Cambridge, MA, 1983.
11. Huang, A., Tsunoda, Y., Goodman, J.W., Ishihara, S.: Optical computing using residue arithmetic. Appl. Opt. **18** (1979) 149–162.
12. Bocker, R.P., Drake, B.L., Lasher, M.E., Henderson, T.B.: Modified signed-digit addition and subtraction using optical symbolic substitution. Appl. Opt. **25** (1986) 2456–2457.
13. Louri, A.: Three-dimensional optical architecture and data-parallel algorithms for massively parallel computing. IEEE Micro pp. 24–27 & 65–81 (April 1991).

2-D Parallel Optoelectronic Interconnect Using a Highly Light Sensitive Monolithic Receiver Array

K. Zürl

Angewandte Optik, Universität Erlangen-Nürnberg, Staudtstr. 7, D-8520 Erlangen, F.R.G.

1 Introduction

Optical interconnects are already well established in the field of long distance communication. In this paper we present a board-to-board interconnect system. The interconnect distance is in the range of some centimeters to some meters. We present an optoelectronic interconnect comprising a 2-D emitter array, a fibre bundle and a monolithic custom-designed receiver array (opto-ASIC) for application as a link between CPUs and shared memories within a multiprocessor system. Each channel has a data rate of 10 MBit/s (similar to the system clock rate), which yields with, 128 parallel channels, an overall data throughput of more than 1 GBit/s.

Even today, for high-speed applications in electronics separate point-to-point transmission lines are used for proper impedance control instead of bus systems. An optical point-to-point interconnect is demonstrated which is parallel in two dimensions and which is expandable to a one-to-many architecture.

The advantages of optical vs. electrical interconnects are the galvanic decoupling of the boards, higher spatial density of the data channels and a higher temporal bandwidth [1,2,3,4].

Another important aspect, especially of short distance interconnects with a very high overall data throughput (like in high performance computers), is the energy required to transmit a bit [5]. All energy necessary to transmit a bit (starting from the electrical signal at the system clock rate on one board and ending with a similar signal on another board) is to be taken into account. High-speed multiplexing in particular takes a lot of energy. For example, a 8:1 ECL-multiplexer for an output data rate of 1 GHz consumes 1.5 nJ/Bit [6]. Multiplexing up from the bus clock rate to the inputs of this 8:1 multiplexer requires still more "slower" multiplexers consuming some more energy. Similar amounts of energy are required for the demultiplexing on the other side. All this energy is turned into heat that has to be removed by the cooling system and that ultimately prevents high packing densities. With our optical data link which works at the bus clock rate no unnecessary multiplexing is required.

2 The System Concept

For optoelectronic point-to-point interconnects two different basic concepts are possible:

a) A *few* optical channels, each with a high data rate: This concept is successful in long-distance communications where the cost of the optical channel (e.g. the fibres) is important. In such systems, the energy to produce high-speed data streams, e.g. for the multiplexing, is not important. Systems such as telephone switches which are based on high-speed data, are other applications for this approach.

b) Highly *parallel* optical channels with moderate channel data rate, that is parallel transmission of data at the system clock rate. A typical application would be a multiprocessor system. For short distances the energy required to create the optical signal and to transmit one bit is important. On the other hand, the the cost of a few meters of fibre is not important, as long as a connector system for arrays is affordable.

We developed a demonstration for a multiprocessor system working at a bus clock rate of 10 MHz (worst case). This is not at all a fundamental limit, but many systems have in practice such a clock rate because the access time of memories is in this order of magnitude. To save energy, a highly sensitive receiver was developed as an ASIC in CMOS technology. To achieve a good BER, the data transmission is performed in a differential way. Figure 1 shows a block diagram of the system.

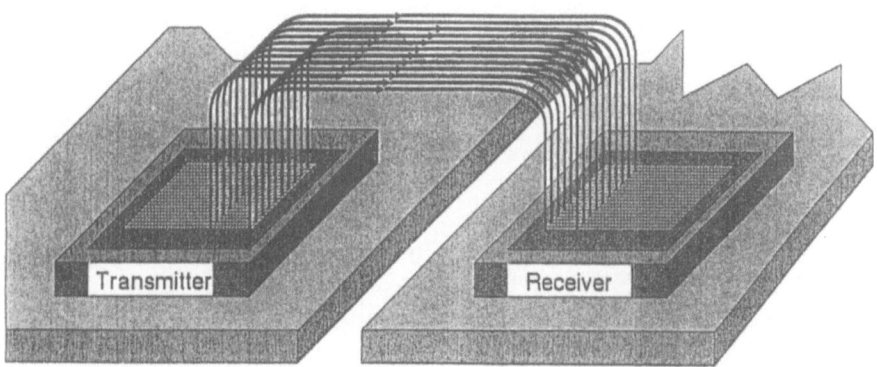

Fig. 1. Optoelectronic interconnect system

For the data conversion from electrically to optically coded data both modulators and active emitters (LEDs or laserdiodes) are possible. To avoid an external optical power source, active emitters are used. Since 2-D, parallel adressable monolithic laserdiode-arrays are not yet commercially available (but under development [7,8]), a two-dimensional hybrid array of laserdiodes was used for the first setup. Due to the high sensitivity of the receiver, the system concept also allows the use of 2-D LED-arrays.

For the optical system 2-D arrays of multimode fibres are used (Fig. 2). The ends are held by a special fixing. Such arrays can be produced by a drilling technique similar to the way SMA-fibre-connectors are produced. To avoid problems with critical mechanical adjustment, we used multimode fibres with 100 μm diameter and a pitch of 420 μm.

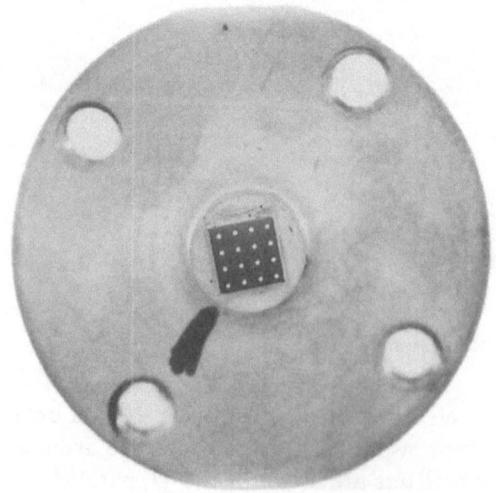

Fig. 2. Prototype of a 2D fiber array

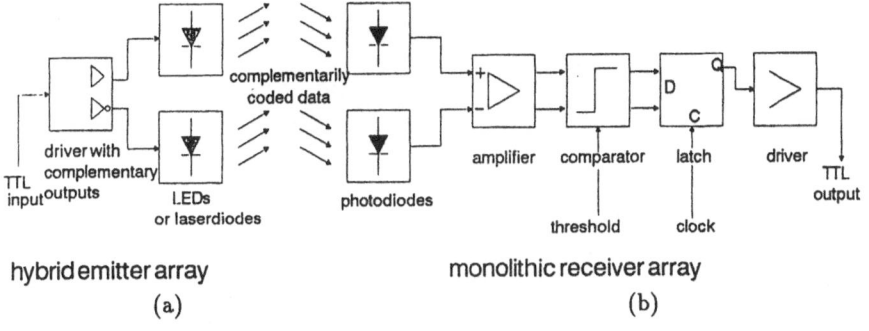

Fig. 3. (a) Hybrid emitter array. (b) Monolithic receiver array

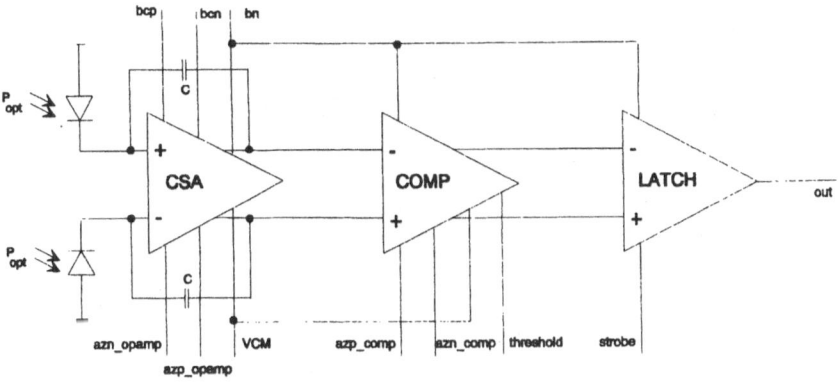

Fig. 4. Block diagram of the front end of the opto-ASIC

3 The Optoelectronic Receiver IC (opto-ASIC)

Figure 3b shows the functional diagram of one logical channel of the receiver chip. On the final chip there will be an array of 8×16 logical channels each of them including 2 photodiodes. Our first experimental sample is a 2×4 receiver array. At a data rate of 10 MBit/s per channel (worst case), we project an overall data throughput of about 1 GBit/s.

To provide high signal-to-noise margins, a differential concept was used both in the optical part and in the receiver (Fig. 3). Stray light impinging with the same intensity on both receiver diodes does in principle not effect the performance of the system. In addition to that, drift effects due to thermal shifts can widely be compensated in the receiver electronics.

The complete receiver front end is fully custom designed. So it was possible to achieve a very high sensitivity. In worst-case simulations an optical energy of

Fig. 5. Photograph of the layout of the prototype IC

10 fJ/bit was sufficient for a signal to noise ratio (SNR) of 23 dB. This SNR correponds theoretically to a bit error rate of better then 10^{-11} [9]. The received energy per bit corresponds to about 30.000 photons/bit.

The very high sensitivity was only made possible by the use of a completely symmetric, differential input stage (Fig. 4) and by providing an auto-zero adjustment at every clock cycle. Fig. 5 shows the geometrical layout of two logical channels. The four large grey areas are the photodiodes (150 μm $\times$ 150 μm, pitch 420 μm).

4 The Data Link, Measurements

In our first experimental setup, commercially available laserdiodes were used which are focussed with an objective directly onto a prototype of the receiver ASIC. Figure 6 shows the setup for BER measurements. The data of the word generator, which is fed to the laserdiodes, and the received data of the ASIC are compared and any errors are counted. As the data throughput is known, the BER can be calculated. With a light pulse duration of 50ns and a peak power of 0.21 μW we optimized some timing parameters of the receiver and achieved a BER better than 10-12. The system is also running at higher frequencies, but our setup limited the maximum data rate to approximately 20 Mbit/s/channel.

For the final setup, we calculated the tolerances within which a fiber array has to be placed above the receiver photodiodes on the ASIC. Figure 7 shows some results. The geometry is tolerant to some misalignment even for 100% coupling efficiency. Therefore, a reasonably priced fibre array connector with moderate tolerances can be designed.

5 Limits and Perspectives

Such a 2-D parallel system is limited in its overall data throughput by two factors:

a) The receiver sensitivity, that is the number of photons per bit that are required to obtain a good BER. With our system, we are currently near to the physical (thermal noise) limit.

b) The light which can be emitted from a source array is limited by the thermal power which can be removed from a given area. LEDs have a very poor efficiency (on the order of 1%) and emit into a very wide spatial angle. Therefore only a very small part of the light can be fed into a fibre. With 2-D laserdiode arrays [7,8] the system could be upgraded by 2 orders of magnitude. This would allow both a higher data throughput and a splitting onto several receivers, i.e. fan-out, for example using free space modules.

(a)

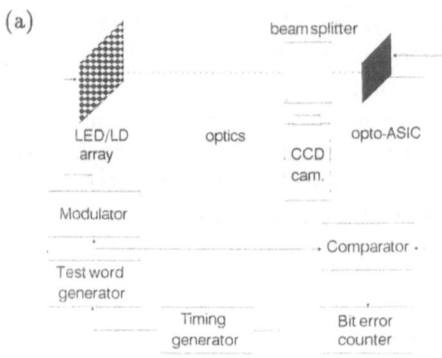

(b)

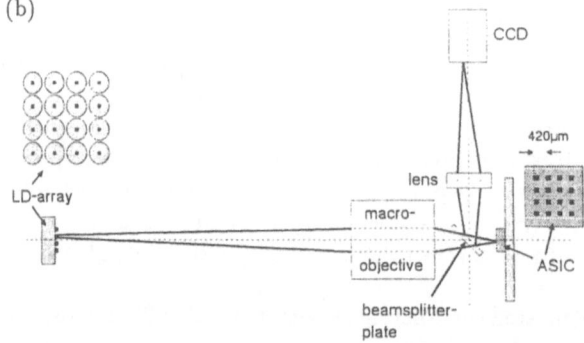

(c)

Fig. 6. BER measurement. (a) schematic of the BER measurement. (b) scheme of the optical setup. (c) photo of the optical setup

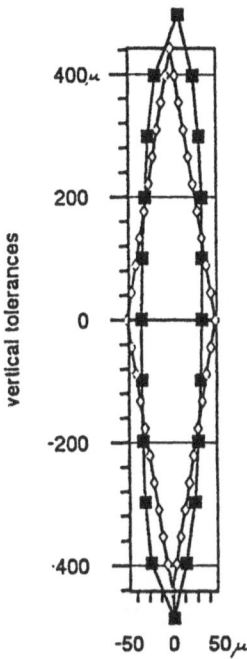

vertical tolerances

400μ

200

0

-200

·400

-50 0 50μ

horizontal tolerances

Fig. 7. Vertical and horizontal tolerances for 100% coupling of a 100 μm fiber onto a 150 μm × 150 μm photodiode

References

1. Goodman, J. W., Leonberger, F. I., Kung, S.-Y., Athale, R. A.: Optical Interconnections for VLSI Systems. Proc. IEEE **72** (1986) 850.
2. Feldman, M. R., Esener, S. C., Guest, C. C., Lee, S. H.: Critical issues in free space intrachip optical interconnect technology. Proc. SPIE **836** (1987) 336.
3. Streibl, N., Brenner, K.-H., Huang, A., Jahns, J., Jewell, J., Lohmann, A. W., Miller, D. A. B., Murdocca, M., Prise, M. E., Sizer, T.: Digital Optics. Proc. IEEE **77** (1989) 1954.
4. Kostuk, R. K., Goodman, J. W., Hesselink, L.: Optical imaging applied to microelectronic chip-to-chip interconnections. Appl. Opt. **24** (1985) 2851.
5. Miller, D. A. B.: Optics for low-emergy communication inside digital processors: quantum detectors, sources, and modulators as efficient impedance converters. Opt. Lett. **14** (1989) 146.
6. Giga Bit Logic: GaAs IC Data Book & Designer's Guide. Newbury Park, CA, 1989.
7. Sakaguchi, T., Koyama, F., Iga, K.: Vertical cavity surface-emitting laser with AlGaAs/AlAs Bragg reflector. Electr. Lett. **24** (1988) 928.
8. Jewell, J., Scherer, A., McCall, S. L., Lee, Y. H., Walker, S., Harbison, J. B., Florez, L. B.: Low threshold electrically pumped vertical-cavity surface-emitting microlasers. Electr. Lett. **25** (1989) 1123.
9. Yariv, A.: Introduction to optical electronics. Wiley, New York, 1976.

O-CLIP – A Demonstrator All-Optical Processor

B. S. Wherrett

Department of Physics, Heriot-Watt University, Edinburgh EH14 4AS, U.K.

This paper reviews briefly the Heriot-Watt activity over the past decade on the design and construction of all-optical computational circuits, leading to the present demonstrator Optical Cellular Logic Image Processors (O-CLIP).

1 Introduction

Initial pioneering experiments on nonlinear refraction and optical bistability in InSb and ZnSe led to the first iterative synchronised optical circuit, constructed during the European Joint Optical Bistability project, in 1986. The incorporation of a programmable logic unit then allowed the implementation of a series of algorithms on a single channel of information input. Loop circuits of 16×8 and 16×16 parallelism have now been constructed, with added functionality attained using nearest-neighbour interconnect modules. Theoretical studies have spanned the microscopic physics of nonlinearities, the optimisation and tolerancing of devices within processing circuits, algorithmic code writing, simulations, and benchmarking of non-local interconnect schemes.

The philosophy of the Heriot-Watt programme has been to use the construction of research demonstrators in order both to drive the component development and to provide a focus for practical processor design. This paper will indicate how technology advances over the decade have influenced the circuit designs, raising issues of optical, mechanical, thermal and interface nature. These issues we see as equally important generically as the more obvious target of a processor, characterised by

$$(\text{performance}) = (\text{device speed}) \times (\text{optical parallelism})$$
$$\times (\text{interconnect bonus factor}) \ , \quad (1)$$

that could complement the performance of electronics, in a specialised or hybrid system.

2 First-Generation

Following a history of activity in the nonlinear optics of semiconductors the Heriot-Watt group discovered in 1978 [1] the giant nonlinear refraction achieved at frequencies just below the band edge of passive semiconductor samples. Values as high as 10^{-3} per W cm^{-2} (1 esu) were measured, some nine orders of magnitude higher than reported previously for mid-gap frequencies. In hindsight this was not a new phenomenon, however, corresponding in the absorption saturation regime to the refraction changes accompanying gain saturation that lead to mode-pulling of semiconductor lasers. Optical responses can be far more sensitive to small refractive index changes than to the small absorption reduction. This was shown dramatically in 1979 [2] by the first semiconductor cw optical bistability observations, achieved for index changes of order 10^{-3} in a plane-parallel sided InSb cavity (a Fabry-Perot etalon of low finesse).

The InSb band-gap corresponds to a radiation wavelength of 7 μm at room temperature, 5.5 μm at liquid nitrogen temperatures. The most suitable available cw system for InSb studies was therefore the CO laser (operating near 5 μm) and the use of 77 K cryostats. InSb has proven to be an excellent material for fundamental studies of optical bistability – temporal studies, spatial aspects, crosstalk between adjacent pixels, the influence of noise, competition between electronic and thermal effects. However, from the circuit viewpoint the need for operation within cryostats, the bulkiness of the optical power sources, and the impracticality of producing highly uniform plane-parallel samples of InSb are serious limitations. The most sophisticated optical logic demonstration using InSb, achieved in 1983, was the coupling of two areas of a single sample to produce an XNOR response [3]. Designs for an R-S flip-flop and an optical clock [4] – based on the coupling of two and three optical NOR (reflective) gates respectively, by analogy with electronics – were not to be implemented in this first generation technology.

3 Second-Generation

The European Joint Optical Bistability project (EJOB) began in 1984 [5], involving eight Universities from throughout Europe as the major partners, and a number of associates. As well as fundamental and applied studies of bistability, one target that the community set itself was to construct the core of an optical computational device. The functionalities of an optical computer had been proposed earlier – a loop circuit with processing stages and gain stages. How the functionalities were to be implemented would be formulated during EJOB. The difficulties of the InSb-system were overcome early in EJOB by use of ZnSe in a multilayer dielectric filter configuration. Bistability was achieved at room temperature, using argon-ion laser radiation at 514 nm, in this visible band-gap material [6]. The filters themselves were produced commercially by well established thermal-vapour-deposition techniques, considered at the time to give excellent spatial uniformity and reproducibility. Earlier work in Minsk [7] prompted our attention to ZnSe filters and although the bistability mechanism proved to be of thermal origin, and therefore relatively slow, the filters provided a source of available devices that worked sufficiently well to allow a series of second and third-generation circuits to be put together.

In particular the architecture that is now called lock-and-clock was devised in 1984 [8], for implementation under EJOB. Lock-and-clock now forms the basis for all of our iterative loop circuits; its function is to synchronise the flow of information, in the form of binary images, around a loop in which there is a processing (arithmetic or logic) unit. A series of three, bistable plates is adequate for this purpose, such that two can hold (different) images at a given time and the third acts as a shutter or buffer between them. The electronic equivalent is a shift register for 2-D parallel information channels, in a wiring harness. By appropriate phasing of power (holding) beams on the plates the images may be shunted around the loop; control of this phasing provides a time delay to prevent any processed image circulating at the speed of light and thereby contaminating (potentially) the process stage itself. Circuit gain can automatically be provided by bistable operation and the lock-and-clock also produces the logic-level restoration that is essential to prevent error build up over multiple cycles.

By the end of EJOB the synchronous circuit had been implemented [9], operating with only a single channel of information initially and using argon powered ZnSe nonlinear etalons with independent optical mounts for each circuit component. With this implementation the principle of indefinitely extensible restoring optical logic had been established [10].

Whilst no logic processing was incorporated in the circuit, its invention had come about from considerations of an Ising-model binary processor in which a limited number of logic functions were required, differing on successive cycles; a scheme for such a programmable processor had been devised [8]. By coupling just two nonlinear etalons together a component with eight-function programmability was designed, the required function (OR, AND ... XNOR) being selected by the power-level settings of two control laser beams. In 2-D parallel (image) format the fanning out of these two control beams each to a 2-D beamlet array would therefore enable the same function to be carried out across the image plane. In like manner the fanning out of the power beams to the bistable plates would produce lockstep synchronisation of the data flow around the circuit. The picture of a single-instruction-multiple-datastream (SIMD) computer architecture was thus evolving, with the control of dataflow via just three power beams and the determination of logic functions on each cycle by two control beams. Scaling to massive parallelism is then trivial (in principle) by the insertion of linear optical components that produce one-to-many fan-out. The development of holographic and binary phase grating components would enable precisely such scaling [11].

Four-channel versions of the ZnSe-etalon lock-and-clock circuits and of the RS flip-flop were constructed [12] and just one loop circuit in which processing was achieved [13]. The latter was an implementation of the optical full-adder, designed in 1984 [14]. The adder takes advantage of the simultaneous availability of reflected and transmitted beams from a nonlinear etalon which, under suitable thresholding, satisfy the logic truth table for the sum and carry respectively of full addition. Bit-slice addition on multiple independent data channels would be the natural use of this circuit.

The optical full-adder pushed the second generation technology to its limits. Whilst providing full adjustability the use of independent mounts led to complex alignment routines and to mechanical instability. The nonlinear filters themselves proved insufficiently stable under the 514 nm radiation from the, remarkably inefficient (0.1% electrical to optical), argon lasers.

4 Third-Generation

A series of arithmetic processor designs based on the parallel channel full-adder (multiplier, fast Fourier transformer [15]) were well beyond the capacities of existent technology. However, a design for a cellular logic image processor has now been successfully implemented. Based on the electronic CLIP machines the optical-CLIP [16] is topologically very similar to several other designs that preceded it. The DOCIP (Digital Optical Cellular Image Processor), one cell of which has since been constructed by Sawchuk's group at USC [17]; the symbolic substitution architecture of Alan Huang at AT&T Bell Labs. [18]; and the Optical Parallel Array Logic of Ichioka at Osaka [19] are contemporary examples. By reducing the CLIP functional cell to a minimum complexity, however, all of the functional modules of O-CLIP could be constructed all-optically and a set of algorithms could be run on the machine.

Figure 1 shows schematically the O-CLIP architecture. Input, from a spatial light modulator, is a sequence of binary images. At a given cycle one such image is merged with that processed during the previous cycle and SIMD logic is undertaken in one of the two programmable logic units (F_N or F_A). Each information bit emerging from F_N at a given pixel is then redirected to one or more pixels on a further logic (threshold) plane. This function is carried out simultaneously for all pixels, using a space-invariant or space-variant 2-D optical interconnect. The resulting thresholded image is then clocked back around the loop circuit to be merged with the next input, or to be output to a CCD detector array. If the interconnect is chosen to be a regular nearest-neighbour (North, South, East, West) connection then the O-CLIP is able to carry out any of the binary image algebra tasks that do not distinguish between the four directions; these include local (single-cycle) operations such as noise removal, global operations such as labelling, and dynamic tasks such as target tracking. A range of such tasks have been simulated on our AMT DAP (Distributed Array Processor) during the period of O-CLIP construction. In combination with the tolerancing of active devices within loop circuits [20], and the generation of codes based on different instruction sets, it has been possible to determine the trade-offs, for example, between hardware complexity and optical tolerances, or between numbers of cycles per operation and hardware tolerance.

An eight-function programmable logic unit was operated in 1988, the last ZnSe-filter-argon laser demonstration [21]. It employed filters in the so-called BEAT structure [22], in which heating produced by absorption in an external layer diffuses to the Fabry-Perot spacer layer, where the induced refractive index change controls the nonlinear optical response. The requirement of absorption within the large band-gap ZnSe is thereby relaxed and operation at longer wavelengths is achievable. As a result the third-generation circuitry from which the

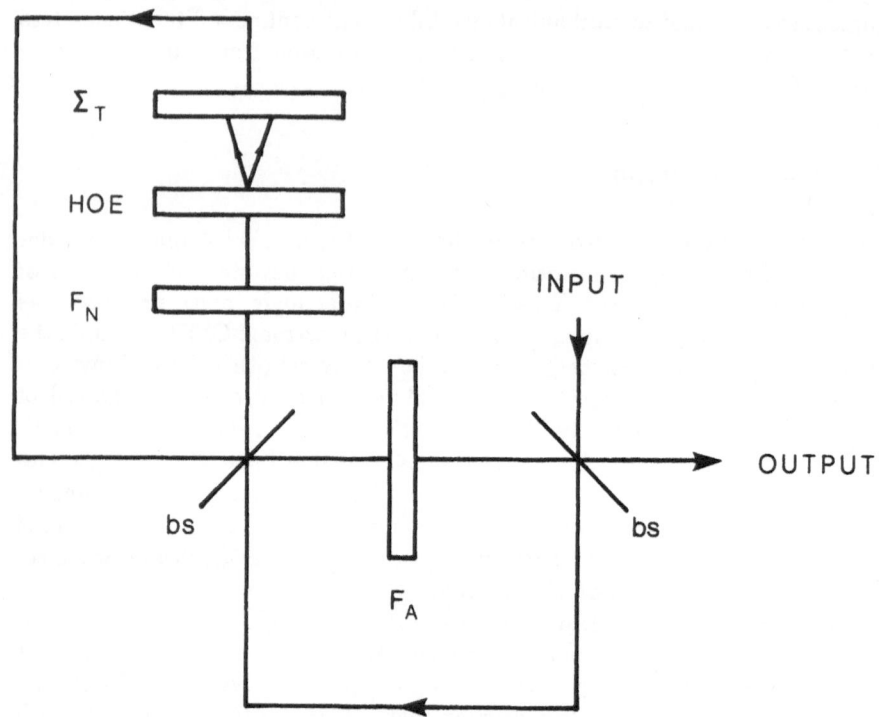

Fig. 1. Schematic of the optical-cellular-logic-image-processor (O-CLIP)

first O-CLIP system was built consisted of ZnSe-BEAT filters addressed by efficient and compact GaAs diode lasers operating at 830 nm. Optomechanical
rigidity was also improved by use of Spindler-Hoyer rail mounting. O-CLIP I,
built in 1989 [23], is a single-channel circuit with (i) laser diode input, (ii) an
eight-function logic unit – two etalons controlled by two diodes and powered by
two further diodes, (iii) a dummy threshold unit with one power diode (there is
no interconnect for a single-channel), (iv) one etalon plus diode to complete the
lock-and-clock synchronisation (bistable logic and threshold etalons avoid the
need for a separate three-element lock-and-clock). A host electronic computer
controls the output power levels of the seven diode lasers in order to (i) produce
a bit-stream of data input, (ii) clock the data flow, and (iii) control the logic
functionality at each cycle.

Even such a simple circuit is able to run a number of algorithms on an
arbitrary input bit-stream – bit-stream recognition, comparison to, addition to
or substraction from a pre-known bit stream [24]. Figure 2 shows the measured
optical power levels for a recognition task. The number 55 is represented as the
8-bit input sequence 0110111 (with most significant figure introduced first). In
order to recognise such a sequence the host computer sets up a logic algorithm
such that the next correct digit is output from the logic unit if, and only if, the
present digit is correct. The sequence OR, NOR, AND, NAND, NOR, AND,

AND, AND accomplishes the task and outputs a 1-bit on the ninth cycle only when presented with the number 55, as indicated in Fig. 2. The figure also indicates the format of the optical signals and the fact that whilst output contrast levels are of order 2:1 the results are quite unambiguous, temporal transients exist but their influence on the information output is avoided by the correct choice of clock timing.

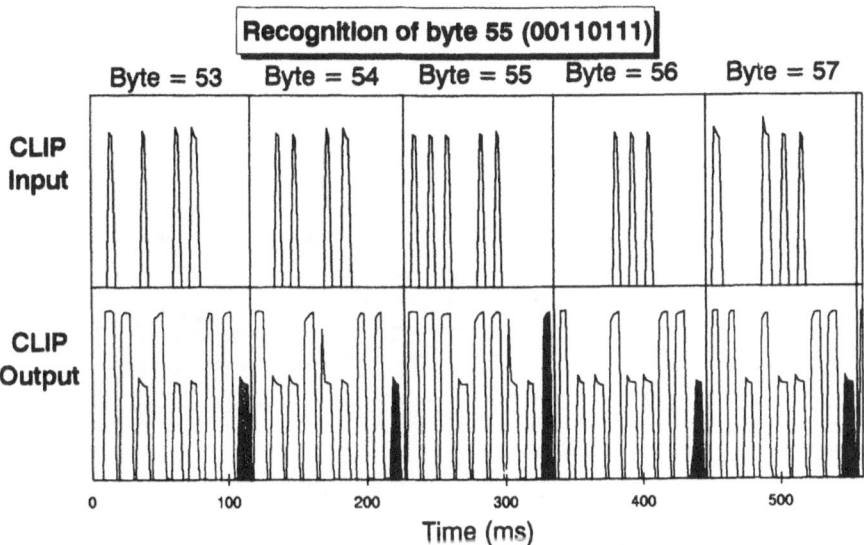

Fig. 2. Measured optical power levels for a recognition algorithm operated on O-CLIP I

A parallel version of the single-channel O-CLIP would carry out for example multiple recognition or additions simultaneously. We are not, however, concerned to employ independent (non-interconnected) processing channels. The second version of O-CLIP, built during 1990/1991, has 16×16 parallelism and a nearest-neighbour (1 to 4) interconnect achieved using either a segmented mirror or a binary phase grating. A liquid-crystal, electro-optic transmission-SLM provides the electronically-addressed image input, and binary-phase-grating elements are used to produce arrays of power beams uniform to within a few percent [25]. Figure 3 shows one of the local binary image processing algorithms implementable on O-CLIP II. To reduce the hardware complexity just one etalon is used in the programmable logic unit, reducing the instruction set to NOR or NAND. The power required for 256 channels could be obtained using diode-array pumped YAG lasers but for expedience was initially provided by a pair of flashlamp pumped YAG systems. Considerable linear optical design is required in order to achieve the beam delivery and image registration around the circuit to within the tolerances set by the etalon bistable responses. No sample physical pixellation is used, however ambient temperature control proves to be critical both to offset the variation brought about by the clocking procedures and the non-uniformity produced by binary images, and to bring the nonlinear etalons to optimum detuning. The use of active devices that are highly wavelength-sensitive along with fixed wavelength laser sources has proven to give a very low yield of acceptable etalons, even with the adjustments achievable by temperature tuning.

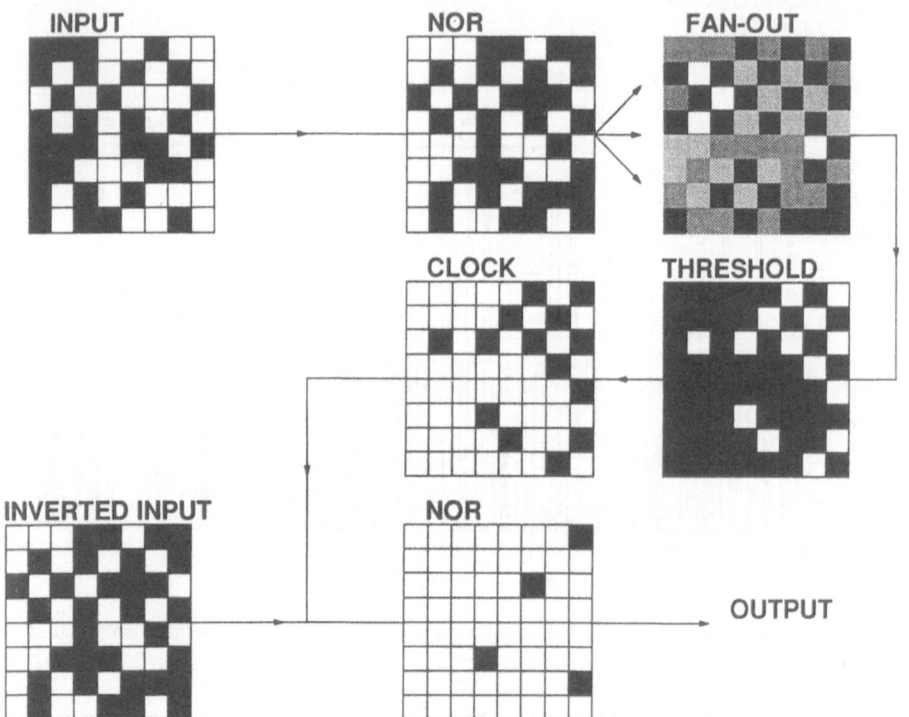

Fig. 3. Example of an algorithm implementable on O-CLIP II, a 16×16 near-est-neighbour interconnected processor. Target (cross) recognition is achieved in just two cycles

We have, since 1989, been able to show the transfer of 16×16 images between bistable etalons [25] and around loop circuits, now with transfer accuracies as high as 97%. O-CLIP power levels remain of order milliWatts per channel per etalon and timescales are in excess of milliseconds per etalon. Optothermal non-linear etalons typically operate individually with power-time products of a few nanojoules and this can be optimised in principle to the picojoule level. Tight tolerancing, and the need for physical pixellation to avoid thermal crosstalk, means that the implied 10^6 gates in parallel, operating at 1 μs, is most unlikely to be achieved. The merit of these etalons is that they have allowed us to obtain hands-on experience of the systems issues of optical circuitry and to develop architectures that are readily adapted to alternative technologies.

5 Fourth Generation

One such alternative is provided by optoelectronic Self ElectroAbsorption Effect Devices (SEEDs) [26]. With experience gained from interaction with AT&T Bell Laboratories at Naperville, where a series of symmetric-SEED based photonic switching fabrics have been constructed [27], we have now initiated an S-SEED O-CLIP module. A 16 × 8-parallelism circuit with NOR/NAND logic and 1 to 2 fanning has been achieved. Two S-SEED arrays are employed, operating with 850 nm GaAs diode lasers. Optical mounting is now on a dedicated baseplate.

6 Discussion

This paper has concentrated on architectural and hardware developments for a programme of all-optical demonstrators. The present status, nearest-neighbour interconnected logic gate arrays, corresponds topologically to the connectivity of electronic distributed array processors. It is of course recognised that the detailed functionality of the optical machines falls way below those of the electronic equivalents. The point has been to provide existence proofs for the optical components and to indicate the scalability once competitive all-optical devices (or optically-interconnected electronics) becomes available. To this end the fact that electronic machines of up to 256 × 256 parallelism, and with nearest-neighbour interconnection do already exist, must also be recognised. Electronics can accommodate both parallelism of a similar level to that proposed for optics, and local interconnects. The limitations of electronic communication do however become very apparent at the level of non-local interconnection.

We are therefore beginning to benchmark non-local algorithms that could be implemented on designs of extended CLIP-like machines that incorporate one or more shuffle interconnects. A first indication of the power of single cycle non-local reconfiguration of data lies in the sort procedure that plays a major role in computation. Time savings of up to 1,000 compared to dedicated software have been demonstrated for a 32 × 32 8-bit sort with and without a perfect-shuffle connection. Such interconnects have been demonstrated optically, using lens combinations or holographic techniques. Further detailed benchmarking is required, along with an understanding of a suitable set of interconnect options. Future circuits will need such options to be either available in separate loops or, far more acceptably, available as reconfigurable optical elements.

Much has to be improved yet in the yield and uniformity of optical components. Information degradation such as we observe will, however, always occur at some fault level so that fault tolerant algorithms and architectures that recognise the style of the optics must be developed. Future theoretical issues are adaptive programming and multiple instruction (MIMD). These are features that can be built into an optical machine; the question is whether the generalisation in performance is warranted by the increased hardware complexity.

With regard to the system components, the ongoing development of both optically and electrically addressed spatial light modulators is seen as key to fast parallel input, along with parallel data acquisition from disc-like optical storage. Complementary to the latter it is essential to have cache memory available that can be written into from the optical output of the circuit itself, on a timescale of a few system cycles. Some developments in the active (logic) devices have been described above. The use of SEEDs, and recent advances in pnpn structure optical thyristor components, leads to the question of smart pixels. How much logic should be incorporated into each optical cell? The answer would appear to be the maximum electronics, for both the cell itself and for local connections, that does not compromise the optical parallelism and hence the power of optics to carry out the non-local interconnection.

Finally, the issue of packaging is beginning to be addressed. Our own demonstrators have advanced from independent optical mounts and two-metre long gas lasers, through rail mounts and solid state lasers, to back-plates and diode lasers. The overall size of the latter circuit is of order 50 cm × 50 cm, still considerably too large.

Progress in these various component areas has not been described in this article, optical design issues are addressed in many of the referred publications and in the following paper. Our holograms and binary phase gratings are discussed elsewhere in this volume.

Acknowledgements

Over the past decade or so perhaps forty to fifty people at Heriot-Watt have made contributions to the wide programme described. Those who have been involved with some aspect of the O-CLIP activity are: G.S. Buller, R.G.A. Craig, M. Desmulliez, A. Kashko, N. McArdle, G. MacKinnon, D. McKnight, P. Meredith, M. Miller, S. Prince, I.R. Redmond, E. Restall, B. Robertson, G. Smith, S.D. Smith, J.F. Snowdon, M.R. Taghizadeh, F.A.P. Tooley, J. Turunen, S. Wakelin, A.C. Walker and R. Wilson. Funding for the project has been supplied by the U.K. Science and Engineering Research Council and by The Boeing Company Defence and Space Group at Seattle, Washington. Their support and foresight is very gratefully acknowledged. Support from the European Commission, for the EJOB project through the framework of the Stimulation Action, and for the exchange of information under the present Workshop on Optoelectronic Information Technology, is well appreciated.

References

1. Holah, G. D., Dempsey, J., Miller, D. A. B., Wherrett, B. S., and Miller, A.: Inst. Phys. Conf. Ser. **43** (1978) 505.
2. Miller, D. A. B., Smith, S. D., and Johnston, A.: Appl. Phys. Lett. **35** (1979) 658.
3. Walker, A. C., Tooley, F. A. P., Prise, M. E., Mathew, J. G. H., Kar, A. K., Taghizadeh, M. R., and Smith, S. D.: Phil. Trans. Roy. Soc. Lond. **A313** (1984) 249.

4. Wherrett, B. S.: J. Quantum Electron. **QE-20** (1984) 646.

5. From optical bistability towards optical computing. Proc. EJOB Project. Eds. Mandel, P., Smith, S. D., and Wherrett, B. S., Elsevier, Amsterdam (1987), pp. 362.

6. Smith, S. D., Mathew, J. G. H., Taghizadeh, M. R., Walker, A. C., Wherrett, B. S., and Hendry, A.: Opt. Commun. **51** (1984) 357.

7. Karpushko, F. V., and Sinitsyn, G. V.: J. Appl. Spect. (USSR) **29** (1978) 1323.

8. Wherrett, B. S.: Appl. Opt. **24** (1985) 2876.

9. Smith, S. D., Mathew, J. G. H., Taghizadeh, M. R., Tooley, F. A. P., and Walker, A. C.: Springer Proc. Phys. **8** (1986) 8.

10. Smith, S. D., Walker, A. C., Tooley, F. A. P., and Wherrett, B. S.: Nature **325** (1986) 27.

11. Taghizadeh, M. R., Redmond, I. R., Walker, A. C., Tooley, F. A. P., Smith, S. D., and Taylor, W.: Proc. SPIE **883** (1986) 245.

12. Wherrett, B. S., Smith, S. D., Tooley, F. A. P., and Walker, A. C.: Proc. Int. Conf. 'Frontiers in Computing', Elsevier-North Holland, Future Generation Computing Systems **3** (1987) 253.

13. Tooley, F. A. P., Craft, N. C., Smith, S. D., and Wherrett, B. S.: Opt. Commun. **63** (1987) 365.

14. Wherrett, B. S.: Opt. Commun. **56** (1985) 87.

15. Snowdon, J. F.: PhD Thesis, Heriot-Watt University, unpublished (1991).

16. Wherrett, B. S.: Workshop on Photonic Logic and Information Processing. Bowden, C. M., and Duthie, J. G., Eds., Proc. SPIE **769** (1987) 7.

17. Huang, A.: Proc. IEEE **72** (1984) 780.

18. Jenkins, B. K., Sawchuk, A. A., Strand, T. C., Forchheimer R., and Soffer, B. H.: Appl. Opt. **23** (1984) 3455.

19. Tanida, J., and Ichioka, Y.: Appl. Opt. **25** (1986) 371.

20. Wherrett, B. S., and Snowdon, J. F.: Int. J. Optical Computing 1 (1990) 41.

21. Craig, R. G. A., Buller, G. S., Tooley, F. A. P., Smith, S. D., Walker, A. C., and Wherrett, B. S.: Appl. Opt. **29** (1990) 2148.

22. Walker, A. C.: Opt. Commun. **59** (1986) 145.

23. Wherrett, B. S., Craig, R. C. A., Snowdon, J. F., Buller, G. S., Tooley, F. A. P., Bowman, S., Pawley, G. S., Redmond, I. R., McKnight, D., Taghizadeh, M. R., Walker, A. C., and Smith, S. D.: Proc. SPIE **1215** (1990) 264.

24. Craig, R. G. A., Wherrett, B. S., Walker, A. C., Tooley, F. A. P., and Smith, S. D.: Appl. Opt. **30** (1991) 2297.

25. McKnight, D. J., Redmond, I. R., Walker, A. C., Taghizadeh, M. R., Buller, G. S., Mathew, J. G. H., and Smith, S. D.: Opt. Comp. Proc. 1 (1991) 137.

26. Miller, D. A. B.: Opt. Quant. Electron. **22** (1990) 561.

27. McCormick, F. B., Tooley, F. A. P., Cloonan, T. J., Brubaker, J. L., Lentine, A. L., Morrison, R. L., Hinterlong, S. J., Herron, M. J., Walker, S. L., and Sasian, J. M.: OSA Proc. on Photonic Switching, Eds. Hinton, H. S. and Goodman, J. W., (Opt. Soc. Am., Washington) **8** (1991) 48.

Design of a S-SEED Cellular Logic Image Processor

S. Wakelin, F. A. P. Tooley, and G. R. Smith

Department of Physics, Heriot-Watt University,
Riccarton, Edinburgh EH14 4AS, UK

A cellular logic image processor was designed, constructed and successfully operated by interconnecting two symmetric self electro-optic effect device (S-SEED) arrays. This paper outlines some of the design issues associated with the implementation of a free-space digital optical system. It is argued that pixellated devices are more desirable than other devices and that if differential data representation is used, high contrast is not required.

1 Introduction

The circuit described in this paper is a looped S-SEED circuit [1]. It uses two S-SEED arrays. In each logic plane an array of 128 (16 × 8) devices are used. Each device used has two 10 μm by 5 μm windows. The input came from a spatial light modulator (SLM) to one device (S-SEED2). A binary phase fan-out element between S-SEED2 and S-SEED1 directed its output to two nearest neighbours. This loop allows logic to be performed on the output from S-SEED2 by S-SEED1 and on the SLM input and the output from S-SEED1 by S-SEED2. A schematic of the circuit is shown in Fig. 1. Beam combination of the output of one gate array with read beams of the other is via 50/50 beamsplitters, which also function as output ports. Input and output to the logic gate arrays is performed using polarisation optics.

The only previous looped S-SEED optical circuit was constructed by workers at AT&T Bell Laboratories in January 1990. It used four arrays of S-SEEDs. In each logic plane an array of 32 (8 × 4) devices were used. Each device used had two sets of two 10 μm diameter windows. There was no input spatial light modulator and the split shift interconnect (fan-out of 2) was formed using 4 polarising beamsplitters, two 160 mm long relays (4 lenses) and a thick glass block held at an angle [2]. In addition a ring counter was constructed by other workers which operates with a single channel at 50 Mb/s [3].

The circuit described in this paper is a partial implementation of a Cellular Logic Image Processor (CLIP) [4]. This architecture is capable of performing simple image processing tasks, such as noise removal, image tracking and maze solving. The design and construction of this demonstrator system has been carried out in order to investigate generic issues associated with the implementation

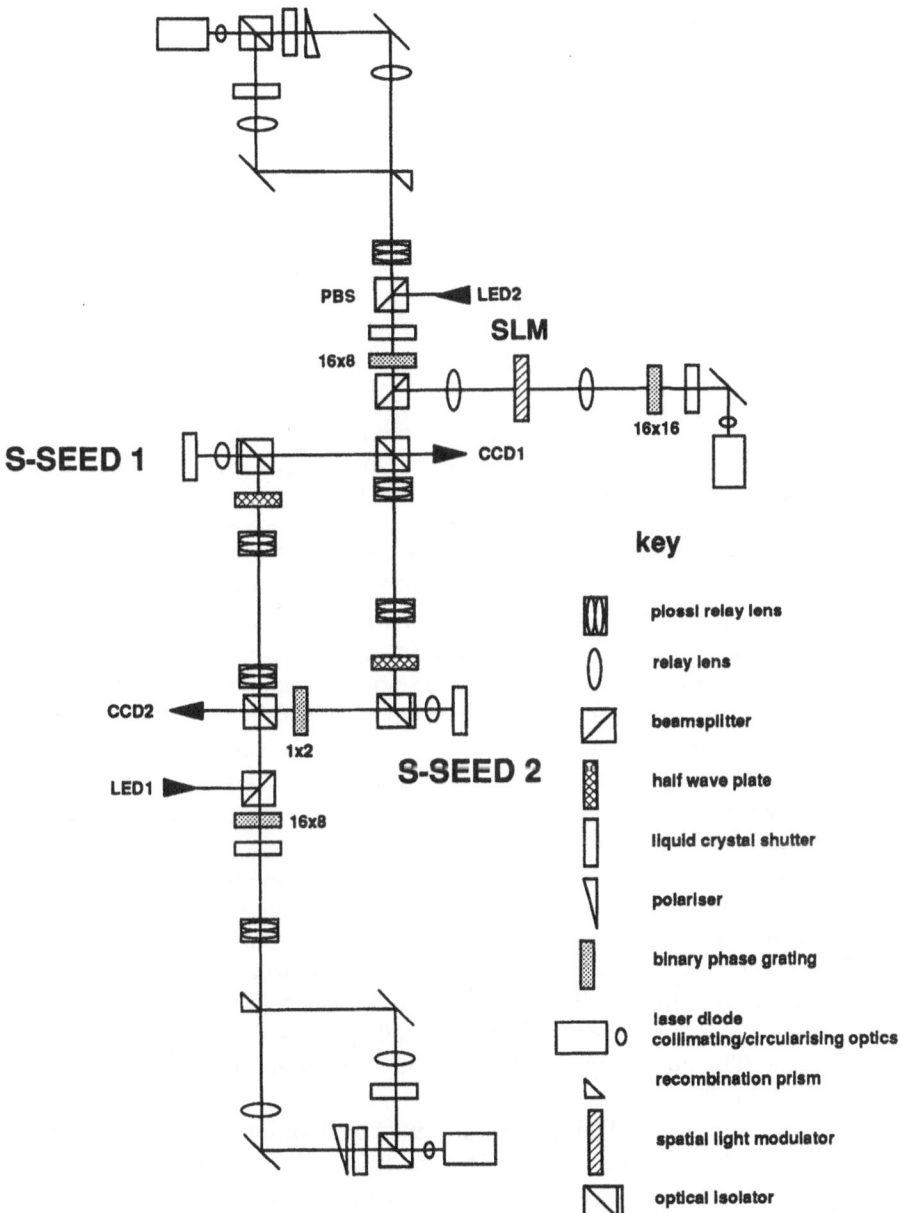

Fig. 1. Schematic of S-SEED loop circuit.

of free-space digital optical systems. The following sections of this paper describe some of the issues that have required investigation for the implementation of this particular circuit.

2 Optical Design

Good quality beam profiles throughout the system are necessary so that aberration accumulation does not limit the operational performance of the system. In particular, this system must be made tolerant to misalignment by ensuring that the depth of focus is large enough and the spot size small enough, that small displacements (1 μm) of the array would not decrease the coupling efficiency to the detriment of system performance. This is achieved using the optical system described below.

The images of both S-SEED arrays are relayed onto the other device array with unity magnification using pairs of 80 mm focal length achromats in a Plossl eyepiece arrangement, resulting in 42 mm focal length components. These relay lenses serve several purposes. As the manufacturers do not specify the focal length of the lenses to sufficient accuracy, use of pairs allows tuning of the focal lengths so that each pair is matched to the other pair in the relay. To keep the system telecentric, the aperture stop of the 7.79 mm focal length focussing lens for each device array must be imaged onto the aperture stop of the other. Direct 4-f imaging of S-SEED1 to S-SEED2 would not be possible without the use of relays because the stop of the focussing objective is within the compound lens. The use of the Plossl relay also allowed this to be done. The 1×2 binary phase grating could not be placed at an aperture stop due to the position of the stop within the objective. This reduction in telecentricity was of little consequence due to the low fan-out angle.

The SLM input is imaged onto S-SEED2 with a magnification of one tenth using an asymmetric Plossl pair of focal length 77.9 mm. This optical input is brought into the system using a 70/30 (T/R) beamsplitter. The SLM plane and the logic planes are illuminated by 850 nm LEDs to allow alignment when imaged onto CCD cameras. The input to the SLM is generated using a 16×16 binary phase grating with the focal spots on 200 μm spacing to match the SLM pixel spacing. The liquid crystal SLM is modulated under computer control to provide a dual rail input.

If S-SEEDs are used as logic gates, they must be illuminated by a preset beam before being written to with signals. The preset beam is incident on only one of the windows and determines whether the gate acts as a NOR or NAND gate. The preset and read beam arrays originate from the same laser source for each device array. After collimation and circularisation using pairs of Brewster prisms, the laser output is split using a 50/50 beamsplitter, focussed and recombined using a 'knife edged' gold coated mirror at the focal plane. An 8×16 BPG is positioned at the back focal point of the lens used to recollimate the beams. A 16×16 array is generated with independent control of the intensity of each 8×16 array.

The modulation of the beams required for the system to cycle is performed using liquid crystal shutters. These devices are effectively half-wave plates that may be switched on and off at a rate of 1Hz by the application of a 2.5 V a.c. Voltage (1 kHz). Three are used in each arm of the system and one in the SLM arm. Computer control of the voltages applied to the shutters allows the beam polarisation to be changed so that the polarisation optics discriminate against the beams not required during different parts of the cycle. By this technique, the two sets of read beams act also as a preset which determines which logic function is implemented.

The design of the system takes advantage of the principle that a symmetric system introduces no coma or distortion to off axis points. The regeneration of the signals by devices means that the optical path is from generation, reflection from one device array, to the other. The unfolded system is shown in Fig. 2, tracing the path from the phase grating to S-SEED2. The optical path for the largest field angle gives wavefront aberration at the exit pupil of less than 1/10 of a wave when the system is raytraced on CODEV. The centre of each spot must be focussed onto a position 2.5 μm from the edge of a 5 μm×10 μm window. The through focus encircled energy plots at the final position are shown in Fig. 3, showing greater than 90% coupling efficiency into a 5 μm diameter circle for on and off-axis points with a ±5 μm defocus. Half the S-SEED window is used for each focal spot to eliminate the problems associated with coherent fan-in.

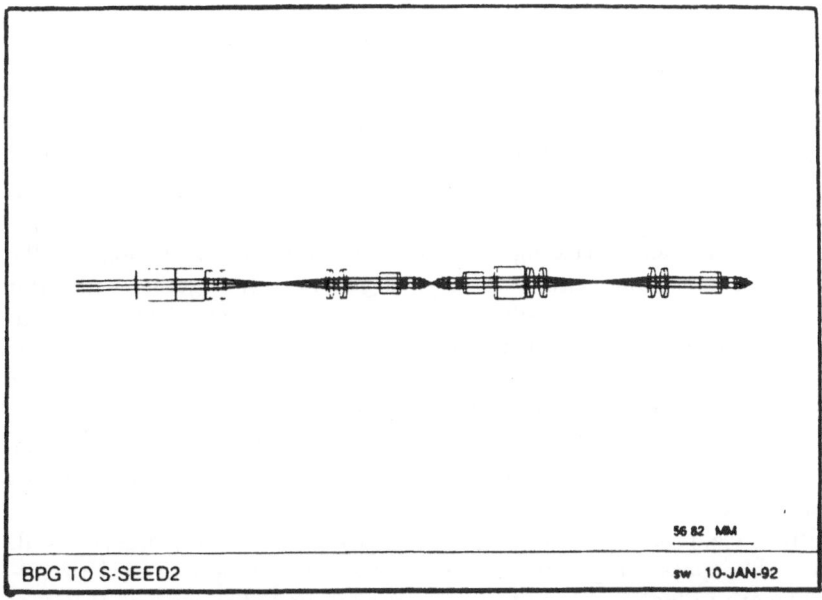

BPG TO S-SEED2   sw   10-JAN-92

Fig. 2. Unfolded layout of system

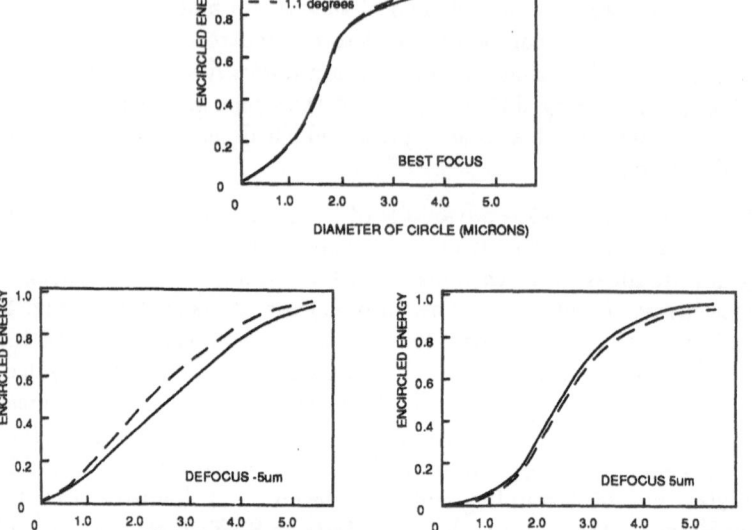

Fig. 3. Through focus encircled energy plots

3 Optomechanical Issues

All of the circuit components were mounted, after first placing them in custom built mounts, onto a machined base plate. A 1.25" thick low carbon, rolled steel sheet was machined with six slots 15 mm wide, 4.5 mm deep, parallel (and orthogonal) to a precision better than one arc-minute, with flatness better than 30 μm. The plate (and all the other steel components) were nickel plated to stop corrosion and to provide low friction.

The relay lenses and focussing objectives were mounted in two-part cells, consisting of an inner and outer ring. The rings were held together with grub screws thus allowing precise centering of the lens in the mount so that the beam deviation caused by decentration was less than 10 arc seconds. The outer diameter of the lens cell was 25 mm $\pm$ 5 μm to ensure that this centration is maintained when the lenses are used. The high uniformity in cell diameter was achieved by fabricating all of the components with the same set-up on a lathe. Other components mounted on the base plate were placed inside single steel cells of the same outer diameter.

The Plossl eye-piece was formed by simply using pairs of achromats in cells with a variable spacer ring between them. The spacer ring was used to correct focal length to high precision to ensure correct magnification and therefore registration. The range of 2 mm to 3.5 mm was found to be necessary.

A 28-pin flat-pack S-SEED mount was used, only two pins were used: 0V and 18V. This was screwed to a single-point flexure giving three angular movements which was bolted to a 3-axis Photon Control Microblock with micrometer drives to quantify alignment sensitivity. The use of the positioner is useful in initial system characterisation but unnecessary and limited the mechanical stability (1 μm drift over $\pm 20\,°C$ temperature range is measured). This 6-axis positioner was found to be easy to use (alignment took only a few minutes) and the setability was better than the required 1 μm.

The knife edged mirror used is a 4 mm thick optical flat coated with gold. This is mounted at 45 degrees by glueing it to a steel ring mount. The only degrees of freedom of this mount are roll and focus and these were used to correctly position the mirror. The powers of the 16×8 arrays of beams falling on the 2 windows (denoted R and S) are made equal to a precision of 1 part in 100 by adjusting an attenuator in the path of the more powerful beam (that which is not reflected from the combiner). The attenuator is a Polarcor filter.

The 3:1 ellipticity of the collimated laser output was corrected using anamorphic prisms. These were positioned and fixed on a steel plate. The fine control of aligning the optical axis (laser direction) with the mechanical axis was achieved using 7 minute and 15 minute deviation wedges. These were rolled in the slot with a manual setability of around 1 degree. This setability allows the direction of the beam to be set to a precision of 2 arc seconds. This precision is equivalent to 0.1 μm in the device plane which is not resolvable.

4 Alignment Procedure

The 1×2 BPG (interconnect) is removed until the final stage of the alignment. S-SEED 2 is observed with CCD2 (which has a 500 mm focal length lens). LED2 illuminates the device and the focus of the xyz mount is adjusted. The mount is then translated until the spots fall on windows. The tilts of the flexure are adjusted to ensure a flat image plane. The 16×8 BPG is rolled to ensure registration. Any magnification error is compensated for by adjusting the spacer of one of the Plossls. The laser is blocked and the pixels of the SLM input are aligned onto the centre of the windows. We now look at the reflection of LED1 from S-SEED1 and then 2 with CCD2 (a polariser must be used to discriminate against the s-polarised light coming directly from the LED). The focus of S-SEED1 is adjusted and it is translated and rotated until the windows of S-SEED1 and 2 overlap. The reflection of LED2 from S-SEED2 and S-SEED1 is observed with CCD1. The two arrays were found to be misaligned by only 15 minutes which was corrected with a wedge. The read beams for S-SEED1 were aligned using two 7 minute wedges. Finally, the 1×2 is inserted and rotated until the two arrays of 16×16 spots are aligned with one another. The stability of the arrangement may be assessed by noting that the system was fully operational 2 weeks after this procedure.

5 Image Rotation

In a loop circuit where two device arrays address each other, the problem of image rotation can occur if the loop is not confined to one plane. One way that this can occur is if the first beamsplitter has an error in orientation of the reflecting surface. This could be due to variation within manufacturers specification, or simply misalignment. This error will cause the reflected beam to be displaced in angle from the incident beam which results in a height change at the next component. Deviation through further wedged components can subsequently cause the image to rotate. To model this, the effect of the reflecting surfaces and refracting wedges may be represented as matrices, then an orientation vector representing the input plane is traced from one device to the other. The magnitude of the image rotation is then found by calculating the final rotation of the orientation vector.

A simplified version of the system was modelled for various wedge angles oriented to give maximum deviation. It is found that the magnitude of rotation angle resulting from the wedge is less than 1% of the wedge angle. This treatment may be extended to many reflecting surfaces and wedged components simply by transforming the initial orientation vector with matrices describing the extra components.

In the system we constructed, great care was taken to ensure that this image roll did not occur. If it had done so, it would have been impossible to remove with the adjustments designed into the system. We were successful since the amount of image rotation present when the system is aligned is less than that which could be observed (around 5 minutes).

6 Pixellation

Note that the alignment procedure used the pixels in the image plane to define when the system is aligned. The existence of pixels makes alignment easier to achieve rather than more difficult. The magnification problem is the easiest example of why pixellation helps. The magnification provided by the relay must be unity to a precision of a few parts in 1000. Since the specification of the focal length of the lenses used is only 1%, a "zoomable" compound lens must be used which is capable of being 'set' readily. Using a variable spacer between two achromats is an especially convenient way to achieve this since the performance of this form of lens is essentially independent of this spacing. The existence of a pixellated device ensures that the magnification (i.e. the ratio of the focal lengths of the two lenses) is matched to an absolute standard rather than a relative one. In the case of an unpixellated device, there is some scope for introducing small errors because there is no absolute standard. Great convenience is offered by the use of the pixels in setting the location of beams, image rotation, focus etc. It makes the difference between alignment being hard, arbitrary and time consuming to it being easy and something that one knows is correct to a measured precision. With pixels, the 3 sets of 16×16 beams that are incident on each device plane can all be judged to be in the same place on every pixel to a precision of better than 1 μm. With pixels, one is forced to take great care with distortion, whereas without pixels one can ignore distortion to some degree and consequently develop techniques that are not extensible.

Just as important as making alignment easy are two other advantages that pixels bring: carrier "confinement" and switching energy reduction. The mesas of a S-SEED are simply photodiodes. Any light that falls on them and that is absorbed creates photocurrent. The effect of light is therefore essentially independent of position in the 10 μm × 5 μm window. However, the response of the detector is slightly dependent on position due to surface recombination on the walls of the mesa. In contrast, when aligning a non-pixellated device the spot position is as important as the power in the spot. The two signal beams must have exactly the same effect as one another. This 'matching of effect' is extremely difficult to achieve in practise. If field division is used to achieve the fan-out in the interconnect or fan-in as it is in this experiment, the two signals are mutually coherent. If they are incident on the same area of the diode, they will interfere which will seriously decrease the effective uniformity. With non-pixellated devices, the signals must therefore be incident at a position where their effect is not a maximum (overlapping the read beam).

In addition, the use of pixels enables the first steps to be taken toward smart pixels which will use electrical gain to lower the effective switching energy of the device and increase its functionality. The S-SEED is a device which has a high effective nonlinearity because of the electric field across it. It would be much more difficult to fabricate such low energy devices without the use of optoelectronics. More importantly, the logic function is made about a factor of 10 times more tolerant to non-uniformity through the use of differential data representation. A true differential mode of operation would be impossible without pixels. A pseudo-differential mode is possible by coupling two NOR gates to make a flip flop. However, such a device is still essentially a threshold gate which operates on the difference of two signals over a limited range. The S-SEED operation mode uses the ratio of two inputs over a large range.

It is appropriate to discuss the disadvantages that use of a pixellated device brings. The most serious is that an extra fabrication step is required. If a dust particle is in a window, that device becomes unusable. In addition, three sets of beams need to be aligned with the windows whereas with non-pixellated devices, they only need to be aligned with one another. In practise, this means that the non pixellated devices do not have to be mounted such that they can be rolled whereas the pixellated devices do. The use of a precision cylinder mounted in a slot allows roll to be achieved at no extra cost. It takes around 30 seconds to rotate the mount to align it with the other image. In addition, an extra xy alignment is needed with the pixellated device, this is achieved in our system using two wedges each which take about 1 minute to align. In conclusion, the drawback of using a pixellated device are outweighed by the advantages that a pixellated device offers.

7 Spatial Noise

A differential detector is ideally insensitive to the optical power level, the output being dependent only on the ratio of two powers used to represent a bit. S-SEEDs have two spatially separated but electrically connected detectors. The input is a 'dual-rail' pair composed of two beams. An important measure of the performance of any logic gate is its tolerance to input signal level variation.

The operation of a S-SEED is determined by the form of the power transfer characteristic shown in Fig. 4. The power reflected from the S window ($\bar{Q}$) is plotted against the ratio of the total signal powers to the R and S windows (input contrast). The power incident on the S window is fixed. If the input contrast is increased from zero, the reflectivity of the S window switches from a low to high at an input contrast T. The transition point (ideally) occurs at an input contrast of $1/T$ when the input contrast is reduced from a high value. The ratio of the two reflectivities, C, is the output contrast. For an S-SEED, C is typically 2-6, the reflectivity of the low absorbance window is 30% and $T = 1.4 \pm 0.2$.

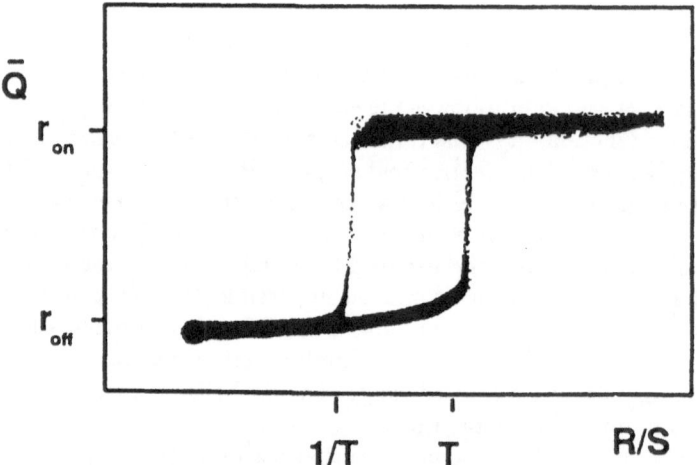

Fig. 4. S-SEED power transfer characteristic

Considering a S-SEED used to implement a two input logic gate. This input requires two sets of beam-pairs. If all the beams presented in such a system had only powers P or P/C, there would be no opportunity for such a gate to make an error. However, typical measured variation in power levels in demonstrator systems is such that the brightest beam may be up to two times as powerful as the dimmest beam. Even with such a large variation, differential logic gates can work correctly.

A power, P, is defined as the mean power of the brightest and dimmest pixels in the array. ΔP_G is half the difference between the brightest and dimmest pixels, and $\Delta P_G/P$ is referred to as the 'global non-uniformity'. The dual rail pair that has the largest difference in power ($2\Delta PL$) defines the "local non-uniformity", $\Delta P_L/P$ and $\Delta P_L = \beta \Delta P_G$ where $\beta < 1$. Tolerance to non-uniformity of the signals incident on a S-SEED may be estimated by considering the occurrence of the 'worst case' combination of signals. This worst case is found by examining the various possible combinations of two signal inputs with the powers of the four beams constrained by ΔP_L and ΔP_G.

The variation of allowable non-uniformity with T for various values of C and β are shown in Fig. 5. It is seen that tolerance is much higher when the local non-uniformity is low. For the case where the local non-uniformity is zero, $\beta = 0$ (Fig. 5a), $\Delta P_G/P = 1$ for a $(1, 1)$ input. For the case where $\beta = 1$, note that $\Delta P_G/P$ is independent of C. The other point to note is that for other values of β there is a very weak dependence of the tolerance on the output contrast, and that in most cases of interest, low contrast is desirable.

A preset beam is used to determine the logic operation performed by the S-SEED. If the preset beam is 'left on' during the 'write' cycle, the tolerance to non-uniformity will vary for the $(0, 1)$ and $(1, 1)$ cases according to its magnitude. The tolerance of the $(0, 1)$ signal input can be increased at the expense of the $(1, 1)$ input if $C > T^2$. From comparison of these tolerances to non-uniformity with and without a bias, it can be seen that the allowable non-uniformity can be greater than doubled for most values of C and T. For example, $\Delta P_G/P = 30\%$ with $C = 4$ and $T = 1.3$ compared with 13% with $B = 0$ for $\beta = 1$, and $\Delta P_G/P = 55\%$ compared with 22% with $B = 0$ when $\beta = 0$. Note that this value of tolerance means that the logic gate would still work even if the brightest beam is more than twice as powerful as the dimmest beam.

Switching speed is proportional to differential power, so use of a bias beam to improve non-uniformity tolerance will affect the speed of the device. The slowest gate is that with the lowest differential power which is related to the worst signal level non-uniformity. The ultimate tolerance to spatial noise is important, but equally of concern is that the non-uniformity be kept low so that the system speed is as fast as possible.

A similar analysis may be carried out with four input logic. This would be required if a 2-D nearest neighbour interconnection scheme were implemented. Assuming totally uniform input power levels, for correct operation to occur the values of C and T must fall within the unshaded region in Fig. 6. For values of the input contrast T that are usually associated with S-SEEDs, the devices must be operated with output contrasts C approximately equal to two which will give lower speed than a higher contrast. When considering worst cases of signal level non-uniformity, the limiting cases give $\Delta P_G/P$ between 14% and 3.5% depending on the relative proportion of local non-uniformity when $T = 1.5$ and $C = 2$. These values and the severe restrictions on C and T indicate that unrealistically strict tolerances must be applied to achieve correct operation of a four input gate performing AND or OR functions and suggest that fan-in of four could only be achieved by cascading gates with fan-in of two.

The analysis presented here is based on a simplified model of the true response of S-SEEDs. In particular, the characteristic is assumed to be perfectly 'square' and independent of the optical power. In addition, by considering the worst case, we exaggerate the problems associated with spatial noise. In practise, it is unlikely that the two inputs to a gate will consist of the four beams with the worst possible combination of variation. Thus, it is likely that a fully working system could be successfully constructed with spatial noise higher than the limits quoted here. The tolerance to spatial noise can be made high by the correct

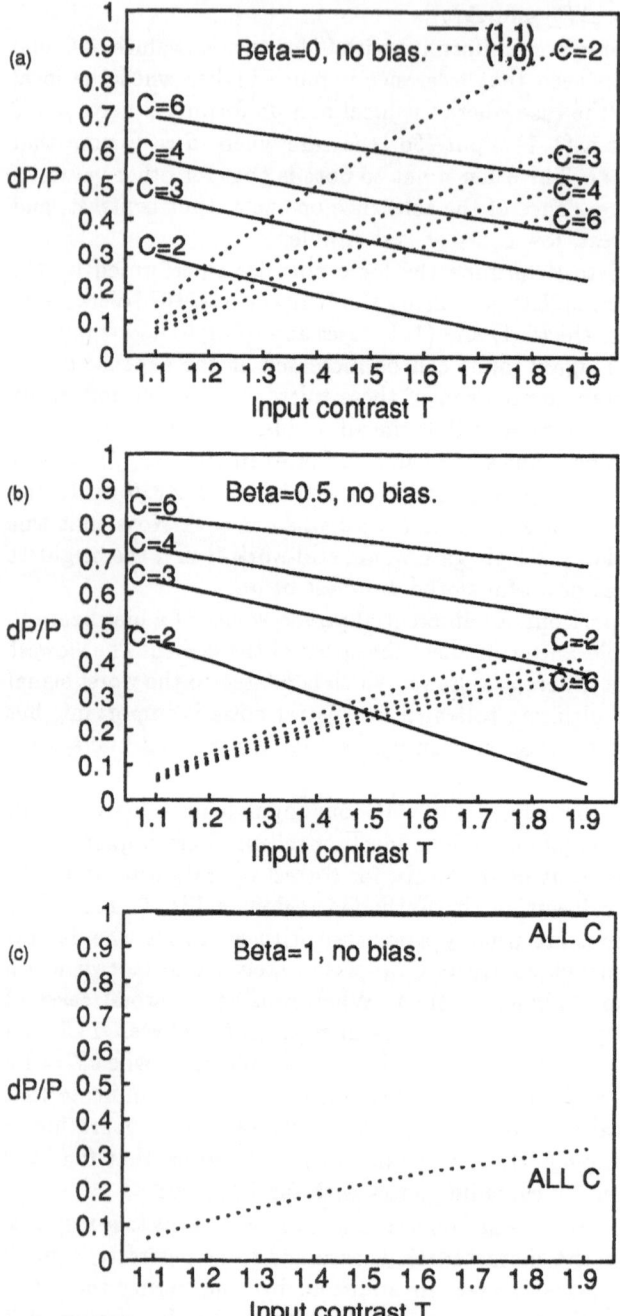

Fig. 5. Tolerance to non-uniformity of two input differential logic gates

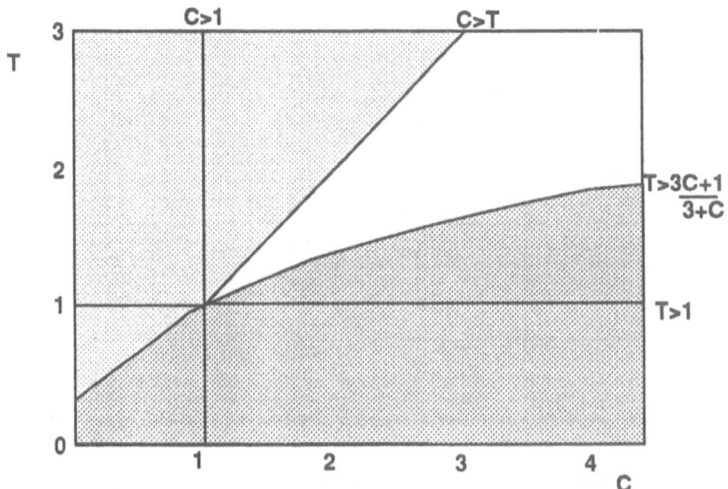

Fig. 6. Operating region for four input differential logic gate

choice of input and output contrast, using a fan-in of two and a bias beam. A full discussion of this tolerance analysis is to be found in [5].

8 Results

A free-space digital optical system has been described which uses two S-SEED arrays to form a cellular logic image processor. The system was fully operational. Programmable logic was performed by each of the gate planes. The full truth table (4 entries) was checked at each of the 128 pixels of both gates in both NAND and NOR operation, a total of 1024 logic operations. Of these, about 1% did not work correctly due to dust on the windows of the S-SEEDs or some other as yet unidentified reason. An algorithm for noise removal was programmed into the system and was shown to correctly operate. The input (SLM) and output (reflection from S-SEED2) images are shown in Fig. 7. The dual rail representation is provided by having the R and S windows horizontal. The algorithm identifies 'on' pixels which do not have 'on' pixels above or below them and removes them. The line of 'on' pixels is removed and the 4 × 4 block is converted into a 4 × 6 block. The switching speed of this system cannot be accessed with the existing configuration due to the slow speed of the liquid crystal devices. However, the measured switching speed of similar systems was accurately equal to the ratio of switching energy (2 pJ) and differential power (2 μW). This system should therefore operate at around 1 MHz clock rate. The implementation of this system has allowed investigation into various issues in optical and opto-mechanical design, the results of which will be used in further investigations of more complex systems utilising optical interconnections.

INPUT IMAGE   OUTPUT IMAGE

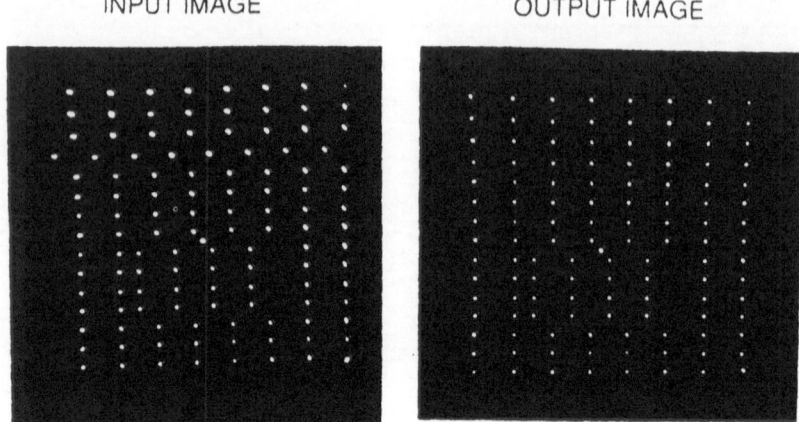

Fig. 7. Input and output images obtained when a noise removal algorithm is implemented

Acknowledgements

The experimental work was funded by SERC and Boeing. The 16×16 and 1×2 BPG was etched by Mike Miller and Neil Ross using a mask and other facilities provided by Mo Taghizadeh. Doug. McKnight designed and fabricated the SLM and shutters. Robert Craig and Ian Redmond wrote the software that controls the SLM and shutters. Much technical input was forthcoming from AT&T Bell Laboratories and in particular conversations with F.B. McCormick and A.L. Lentine proved to be particularly useful. This work would have been impossible without the unreserved support of Professors Smith, Walker, and Wherrett.

References

1. McCormick, F. B., Tooley, F. A. P., Sasian, J. M., Cloonan, T. J., Brubaker, J. L., Lentine, A. L., Morrison, R. L., Hinterlong S. J., and Walker, S. L.: Demonstration of cascaded operation of 2K arrays of S-SEEDs. Electr. Lett. **27** (1991) 1869.
2. Prise, M. E., Craft, N. C., Lemarche, R. E., Downs, M. M., D'Asaro, L. A., and Chirovsky, L. M. F.: Cascaded operation of arrays of S-SEEDs. Appl. Opt. **30** (1991) 2841.
3. Lentine, A. L., Chirovsky, L. N. F., and D'Asaro, L. A.: Opt. Lett. **16** (1991) 36.
4. Duff M. J. B. and Fountain T. J.: Cellular Logic Image Processing. Academic Press (1986).
5. Wakelin S., and Tooley F. A. P.: The tolerance to spatial noise of differential logic gates. accepted for publication in Opt. Comp. Proc. (Sept 1992).

Part V

Thermo-Optic Devices

Thermo-Optical Logic Gate Array Using SOS Waveguide

H. Gualous, A. Koster, W. Chi, N. Paraire, and S. Laval

Institut d'Electronique Fondamentale - CNRS URA 22, Université Paris-Sud, Bât. 220, F-91405 Orsay Cedex, France

1 Introduction

Silicon on sapphire (SOS) has been studied at IEF for many years as a nonlinear waveguide at $\lambda = 1.06$ μm. Resonant excitation of guided waves in submicron undoped SOS films can be obtained with a Nd:YAG laser beam for precise values of the incidence angle by means of a submicron period grating coupler etched in silicon. Sharp and contrasted angular resonances can be observed near the fundamental TE_0 mode excitation on either the reflected or transmitted beam. The wavelength being slightly smaller than the silicon absorption edge, fast optical and electrical switchings have been observed and studied at IEF in the nonlinear pulsed regime (20 ns and 200 ps pulse durations). These switchings can be useful for either optical or electrical pulse shaping [1]; they result, at first from electronic nonlinearities induced by a carrier density in excess of the order of 10^{18} cm^{-3} due to high optical excitation and later (with 20 ns pulses) from the competition between these effects and thermal nonlinearities of opposite sign due to the fast relaxation (lifetime < 1 ns) of the electron/hole pairs in excess.

When the devices operate under cw illumination, the predominant nonlinearity has a thermal origin; it can be greatly enhanced by Joule effect, thanks to a voltage applied between electrodes deposited on the silicon film. Studies carried out at Münster and Brussels with Joule enhanced sensitivity devices using thick silicon substrate as Fabry-Pérot resonator and Schottky photodiode structures have shown interesting results in optical bistability with individual devices [2,3], although the non-uniformity of the silicon plate thickness has not allowed realization of 2D arrays. The good spatial uniformity of the TE_0 resonance on the SOS plate characterized during the studies carried out on an individual device recently led to the realization of a 6×3 gate array. The transfer characteristic in transmission allows NOR logic gate operation in an off-axis geometry: the holding beam is exciting the TE_0 mode with an incidence angle close to $14°$, the control beams are exciting the TE_1 mode near $1°$.

2 Performance of the First Individual Device

2.1 Description

The device under study is represented in Fig. 1, the SOS sample is from Kiocera and the undoped silicon film (thickness ≈ 0.6 μm) was heteroepitaxied by CVD on a sapphire substrate. The rear surface of the substrate (thickness ≈ 300 μm) has been optically polished not parallel to the silicon surface (angle $\approx 0.3°$) in order to avoid guided mode excitation by reflection on this face. The grating coupler is realized by holographic lithography with an argon laser at $\lambda = 0.4579$ μm and Ar$^+$ ion milling of silicon. At $\lambda = 1.064$ μm, its 330 nm period allows excitation of only one diffracted order propagating in the silicon film. The grating modulation depth is about 100 nm leading to a high coupling efficiency and a short growth or attenuation length for the guided mode (≈ 50 μm).

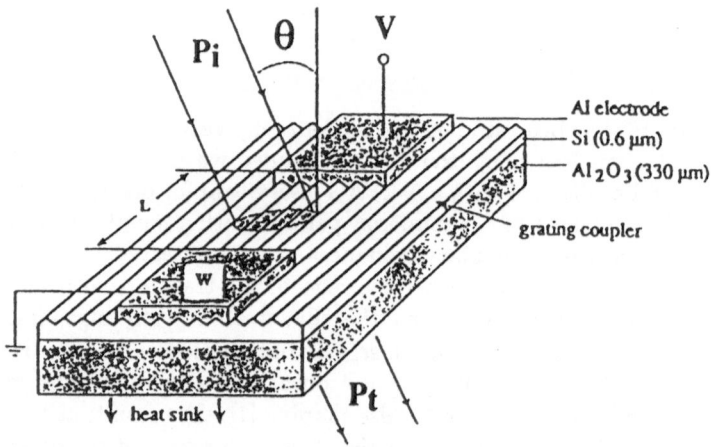

Fig. 1. Schematic view of the optothermal SOS waveguide device

Four test devices with square active region $L = W = 200$ μm or 400 μm have been made: aluminium electrodes have been deposited by a lift-off process; the best results have been obtained with the 200×200 μm^2 device, the grating grooves of which are parallel to the electrode strips. The sapphire is in backing contact with a copper heat sink at a controlled temperature T_0 and the experimental studies are made on the transmitted beam thanks to a hole made in the heat sink.

2.2 Linear characteristics

The incident gaussian TEM$_{00}$ beam from a cw Nd:YAG laser is linearly polarized, attenuated ($P_i \approx$ mW) and focused on the guiding film with a beam waist diameter $\Phi_{1/e^2} \approx 750$ μm. Figure 2a gives the transmissivity measured on the whole beam, at the centre of the 10×9 mm^2 SOS plate, before electrode deposition; this curve is plotted versus the incidence angle θ. In the angular range from $0°$ to $40°$, three TE guided modes are excited. Repeating such measurements in nine different points of a 6×6 mm^2 region in the centre of the plate, the following extreme values of the resonance angle and of the minimum transmissivity are obtained for the TE$_0$ mode: $\theta_0 = 14.19° \pm 0.03°$, $T_{0\,min} = 7.5\% \pm 0.5\%$. These sharp TE$_0$ resonances (FWHM $\Delta\theta = 0.20°$) are fairly well fixed on a large region of the plate and this allows gate array realization. After electrode deposition, the dark current versus applied voltage shows linear evolution though the contacts between aluminium and silicon may be non-ohmic. Then in the linear regime, the photocurrent I_p has been studied versus θ with a low applied voltage of 5 V: such a characteristic is given in Fig. 2b for the 200×200 μm^2 device enlightened by an incident optical power of 3 mW and a smaller beam waist $\Phi_{1/e^2} \approx 400$ μm. The difference between the photocurrent and the dark current is proportional to the absorption rate A in the interelectrode region. The parasitic TE$_0$ resonance (P) near $12°$ (see Fig. 2b) comes from the wedge shaped sapphire substrate. The optimum positions of the incident beam relative to the electrodes which give maximum photocurrent at resonance are quite different for the TE$_0$ and the -TE$_1$ modes principally as these two modes are counter-propagating.

Figure 3 shows the spatial evolutions of a gaussian incident beam and the corresponding guided wave intensity of two counter-propagating modes at resonance (for example + or −TE$_0$ excited with $\theta = +$ or $-\theta_0$), such guided wave intensity profile can be observed behind a shield limiting the incident beam and in the wave guide deexcitation region thanks to a 50 mm diaphragm attached to the power-meter which is close to the SOS plate (incident beam is displaced relative to the shield and power-meter). In this experiment (+ or - TE$_0$ excitation) the maximum photocurrent is obtained for two different positions 100 μm apart of the active region relative to the incident beam. Measurements of the locally transmitted beam at resonance (without a shield and displacing the power-meter) show that a very small transmitted power can be observed in a limited region close to the guided wave maximum (point a). In fact, $T_{0\,min}$ observed on the total transmitted beam strongly depends on the beam dimension; with this 400 μm beam waist diameter, $T_{0\,min} = 25\%$.

Operation in the nonlinear regime: The experimental conditions are shown in Fig. 4.

B_h and B_c are the holding and the control (input) beams respectively. B_h is used alone to observe the bistable evolution of the locally transmitted power (50 μm diaphragm) versus the incident beam power in Fig. 5 (observation conditions corresponding to point a defined in Fig. 3). The applied voltage is $V = 400$ V and different positive angular detunings $\delta\theta = \theta - \theta_0$ from $0.05°$ to $0.25°$ are chosen. Larger detuning leads to higher threshold power for switching down, but transfer characteristics are steeper; the use of $V = 600$ V with $\delta\theta = 0.25°$ reduces the optical threshold by a factor ≈ 2 predicted by the modelling.

NOR logic gate operation has been performed in the off axis geometry. The more convenient incidence angle for the control beam corresponds to the -TE$_1$ guided wave excitation ($\theta \approx +1$) whose FWHM of resonance curve is $\Delta\theta_1 =$

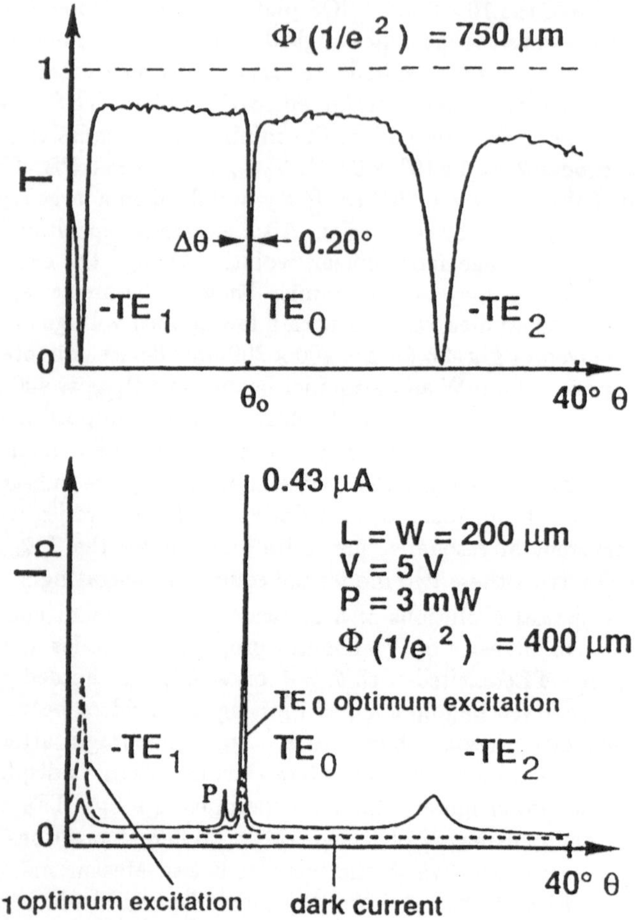

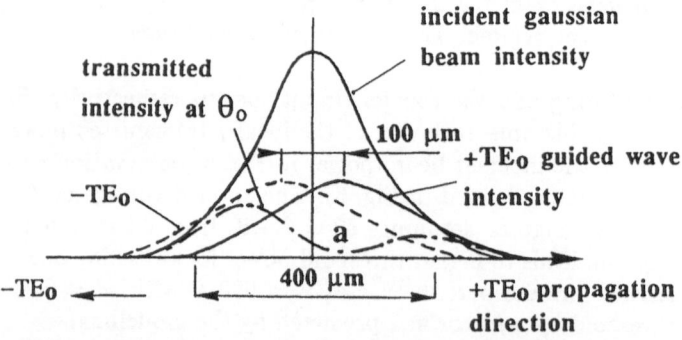

$\Phi(1/e^2) = 750\ \mu m$

1

T

$\Delta\theta \rightarrow |\leftarrow 0.20°$

(2a)

$-TE_1$ TE_0 $-TE_2$

0

θ_0 $40°\ \theta$

0.43 μA

I_p

$L = W = 200\ \mu m$
$V = 5\ V$
$P = 3\ mW$
$\Phi(1/e^2) = 400\ \mu m$

TE_0 optimum excitation

(2b)

$-TE_1$ TE_0 $-TE_2$

P

0

$-TE_1$ optimum excitation   dark current $40°\ \theta$

Fig. 2. (a) Transmission coefficient versus incidence angle θ, (b) photocurrent versus θ

incident gaussian
beam intensity

transmitted
intensity at θ_0

100 μm

$+TE_0$ guided wave
intensity

$-TE_0$

a

$-TE_0 \leftarrow$ 400 μm $+TE_0$ propagation
direction

Fig. 3. Spatial evolutions of the $-TE_0$ and $+TE_0$ guided wave intensities excited at resonance by an incident gaussian beam, and the evolution of the transmitted intensity

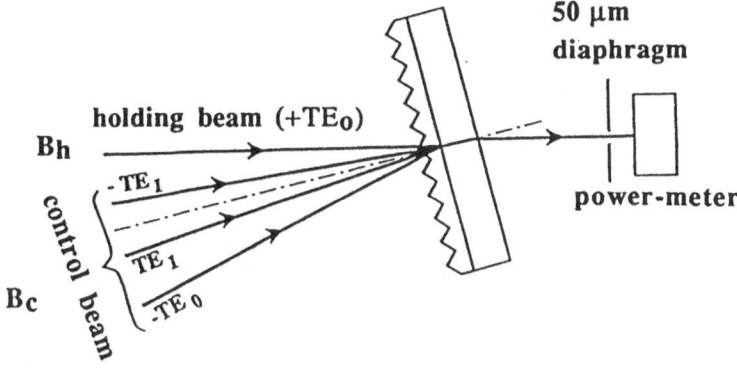

Fig. 4. Experimental conditions for the NOR gate operation in off-axis geometry

0.9° and so prescribes not too accurate incidence conditions. With an angular detuning $\delta\theta = 0.20°$ and a power of 70 mW for the holding beam and $V = 400$ V (corresponding to point b in Fig. 5), dynamical NOR gate operation has been demonstrated but with rather slow switching times, the control beam power being modulated between 0 and 110 mW by an acousto-optic modulator.

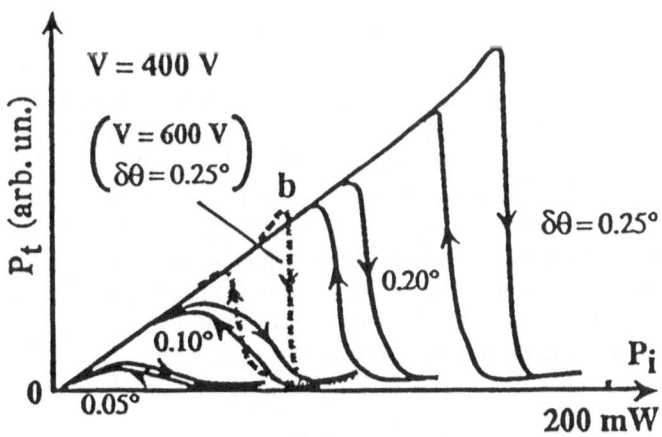

Fig. 5. Evolution in the nonlinear regime of the locally transmitted power versus the incident power for different angular detunings and a 400 μm beam waist

Due to the gaussian profile of the incident beam and to the device geometry, there is a large non-uniformity in the temperature of the enlightened region as the heating power density is proportional to the guided wave intensity in the interelectrode region (see Fig. 3). Thus, only a small part of the incident beam can switch down and the transfer characteristic determined on the whole transmitted beam is not much contrasted and leads to a fan-out smaller than 1.

2.3 Nonlinear operation modelling

The refractive index n of silicon is temperature dependent and dn/dT is approximately equal 2×10^{-4} K^{-1}. Thus the incidence angles corresponding to the excitation of guided modes are also temperature dependent. Experimental measurements give $d\theta_0/dT = +0.013$ K^{-1} in the $18° - -35°$C range for the fundamental mode and lead to a temperature rise of the order of 15°C to compensate an angular detuning $\delta\theta = 0.20°$. Silicon film temperature can be changed by heating. If uniform excitation of the TE$_0$ mode is assumed in the interelectrode region, neglecting dark current, the heating power P_h induced by optical absorption and associated Joule effect is:

$$P_h = WL \left(1 + \frac{V^2}{V_0^2}\right) \cos\theta A(\theta, n) I_i ,$$

where A is the resonant absorption rate, V the voltage applied between the electrodes, I_i the incident optical intensity. Neglecting electrical field boundary effects, the characteristic voltage V_0 is:

$$V_0 = L \sqrt{\frac{\hbar\omega}{q(\mu_n + \mu_p)\tau}}$$

where τ is the photo-carrier lifetime and μ_n, μ_p the electron and hole mobilities. If V is greater than V_0, the Joule heating can rapidly become more important than optical absorption. The characteristic voltage has been measured and was found to be 118 V. This value is significantly higher than the calculated one $V_0 = 70$ V (calculated with $\mu_n + \mu_p = 400$ cm^{2}V^{-1}s^{-1} and $\tau = 400$ ps taking into account carrier lifetime degradation in SOS after ion milling of the grating coupler) [4]. The non uniform excitation of the guided wave by a gaussian beam in the nonlinear regime could explain this difference, which is close to a factor of two. Indeed, this modelling assumes plane wave excitation and predicts, as usual, optical bistability for an angular detuning greater than a critical one of the order of FWHM of the resonant absorption rate A.

3 First Results with the Logic Gate Array

3.1 Description of the gate array

The spatial uniformity of the TE$_0$ resonance being demonstrated, a 6×3 gate array has been realized using microelectronic processes on the SOS plate used previously. Figure 6 gives the device sizes and electrode shapes.

The distance between electrodes is 20 μm and the pixel length 350 μm. In order to get an efficient energy tranfer to the guided mode, each pixel is limited by total etching of the silicon film by ionic milling in the interelectrode region. The

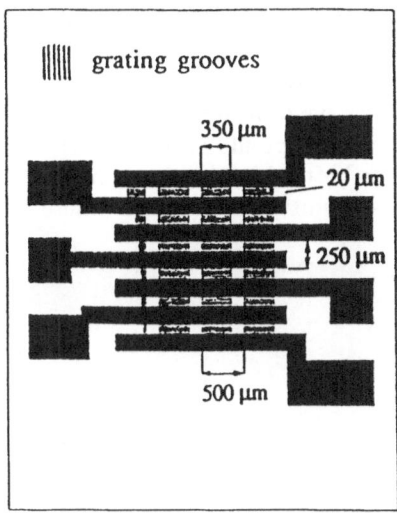

Fig. 6. Schematic view of the 3 × 6 gate array

seven contacts allow independant bias of the six lines. Heat diffusion studies with the rear sapphire substrate temperature fixed suggest the following dimensions: 250 μm between two lines and 500 μm between two devices on a line. Six test devices of various shorter lengths have been added, the total device area now being 1.5×1.5 mm^2 in the centre of the SOS plate.

At first, the heat sink was kept as previously; so only experiments on isolated devices were feasible, but offering the advantage of a more direct comparison of one element of the array with the individual device.

3.2 Pixel illumination conditions

When a device is illuminated by a gaussian beam with a beam waist diameter of the order of L or W, this inhomogeneous illumination induces non-uniformity of the thermal nonlinearity in the interelectrode region leading to fan-out of the NOR gate less than 1. With the present device, experiments are made with an illumination as uniform as possible to reduce the spatial effects. The small distance (20 μm) between electrodes allows good illumination uniformity and avoids the electrical field non-uniformity present in the first device. The pixel being delimited by etching, there is no electrical field boundary effect and the beginning of the guided mode growth is well defined. To secure sufficient optical power for the nonlinear studies, the incident beam is focused on one element of the array with a 600 μm beam waist. In the linear regime, the array is adjusted with respect to the incident beam to achieve the maximum photocurrent at TE$_0$ resonance. Measurements of photocurrent and transmitted beam are made via an imaging system; the transmitted beam is measured locally with a spatial resolution of 20 μm in the image plane (6 μm in the object plane). The FWHM of the photocurrent resonance is $\Delta\theta_p = 0.33°$ while the corresponding width of the transmitted power is $\Delta\theta = 0.20°$. Such measurements of the locally transmitted power have been repeated regularly along the pixel length image: Fig. 7 shows the evolution curves of the locally transmitted power far from resonance ($\theta = 13°$, which is proportional to the incident beam intensity profile on the pixel) and at resonance. These two curves demonstrate the better illumination uniformity and at resonance the larger region where the transmitted beam is minimum as compared with a gaussian beam excitation.

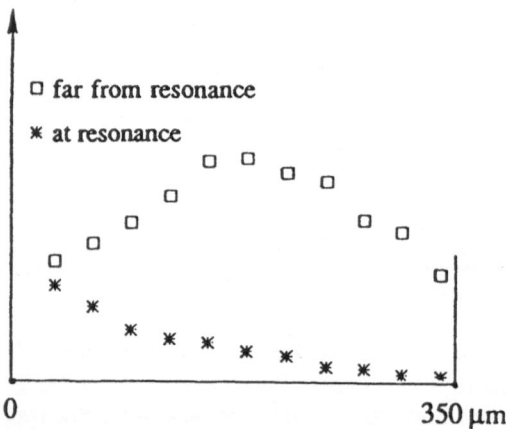

Fig. 7. Evolution in the linear regime of the locally transmitted power along the pixel at and far from resonance

3.3 First Results in the Nonlinear Regime

Figure 8 displays a plot of the locally transmitted power measured in the centre of the pixel image versus the incident power, for $\delta\theta = 0.20°$ and a voltage bias $V = 30$ V ($V/V_0 \approx 2$). The maximum transmitted power is obtained with an incident power of the order of 10 mW. As with the first studied device, the minimum transmitted power is close to zero, but here there is no bistable or hysteresis cycle. The width of the switching region is notably larger than with the individual device, and allows the observation of a well contrasted transfer characteristic of the total transmitted power. Such a characteristic is shown in Fig. 9, for $\delta\theta = 0.35°$ and $V = 30$ V. For this large angular detuning (larger than $\Delta\theta_p$), there is no bistable cycle although it is predicted by a simple plane wave model.

3.4 Prospects

A modification of the device illumination using a cylindrical lens should lead to uniform and higher intensity on several pixels with the object of achieving a

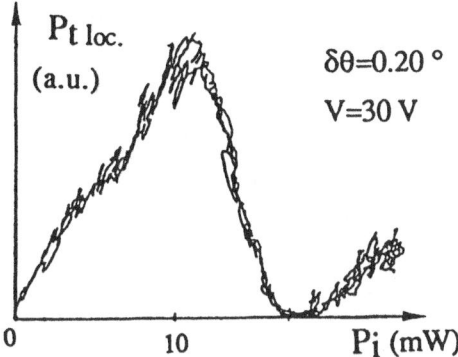

Fig. 8. Evolution in the nonlinear regime of the locally transmitted power versus the incident power in the centre of the pixel

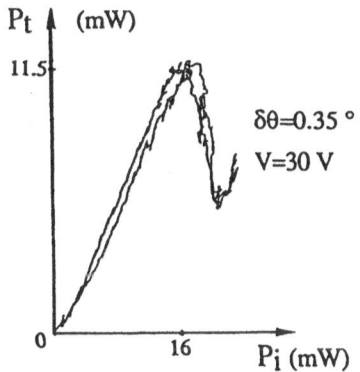

Fig. 9. Transfer characteristic measured on the whole transmitted beam

minimum fan-out of 2. Interconnection between several pixels will be studied, the goal being to demonstrate elementary image processing.

Acknowledgments

This work has been supported by CNET and ETCA research contracts.

References

1. Berard, D., Paraire, N., Chi, W., Koster, A.: Using silicon nonlinearities in waveguides for passive nanosecond optical pulse shaping at $\lambda = 1.064$ mm. Annales de Physique (Colloques) **16** (1991) 63–72.
2. Jäger, D., Forsmann, F.: Optical, optoelectronic and electrical bistability and multistability in a silicon Schottky SEED. Solid State Electr. **30** (1987) 6771.

3. Thienpont, H., Vanholder, S., Ranson, W., De Tandt, C., Vounckx, R., Veretenni-coff, I.: Thermo-optic SEED arrays for optical information processing. ECO3, The Haag, Proc. SPIE **1280** (1990).
4. Chi, W.: Thesis, Orsay, February 1991 (unpublished).

Optical Switches and Oscillators Based on Thermally Induced Optical Nonlinearities in II–VI Semiconductors

J. Grohs, S. Apanasevich*, F. Zhou,** H. Ißler, A. Schmidt,***
and C. Klingshirn

University of Kaiserslautern, Department of Physics, W–6750 Kaiserslautern, Germany

In this contribution we will focus on thermally induced dispersive and absorptive optical nonlinearities in II-VI semiconductors. We demonstrate an optical oscillator by introducing the sample into a hybrid ring resonator with long round trip time. In the case of a bistable input–output characteristic (IOC) of the nonlinear element (we use an interference filter in reflection consisting of a glass matrix doped with CdSSe) we get strong mode locking to the resonator round trip time with a complex mode structure including Farey tree [1,2] transitions and mode coexistence. For a barely bistable IOC (realized with a CdS single crystal) we find a completely different kind of dynamics. Relaxation oscillations become important and stabilize the system to prevent a transition to chaos. Further investigations are concerned with the electrooptic bistability in ZnSe single crystals. By applying an electric field perpendicular to the light beam we observe changes of the optical properties that are due to the temperature change and to the Franz–Keldysh effect.

1 Introduction

In a previous paper we reported on thermally induced absorptive optical bistability (OB) in CdS single crystals and the self oscillations achieved by introducing the sample into a hybrid ring resonator with long round trip time [3]. Due to the flat input–output characteristics (IOC) of the lower branch of this optically bistable element we could not observe the descending steps in the resonator intensity as a function of time predicted by theory [4,5]. The aim of this paper is threefold: We will describe a system with two nearly parallel branches of the bistable loop to investigate the mode structure in presence of descending steps. Then we will show how the dynamics of our feedback system changes if the width of the bistable region of the IOC is decreased up to the nearly monostable case. Finally, we will present results on the optical nonlinearities of II-VI single crystals with additional electric field. The resulting optoelectronic elements

---

* permanent address: Division for Optical Problems of Information Technology and Academy of Sciences, Minsk, Republic of Belarus
** permanent address: Changchun Institute of Physics, Changchun, PR China
*** permanent address: University of Würzburg, Institute of Physics, W–8700 Würzburg

shall be used in future to achieve optical oscillators with and without additional feedback.

2 Experimental Setup

The setup of the experiment is shown in Fig. 1. The light of an Ar^+ laser is incident onto the sample. This may be a high reflectivity coated edge filter from CdS_xSe_{1-x} doped silicate glass (made by Changchun Institute of Optics, for the microscopic structure see e.g. [6]) or of CdS or ZnSe single crystals. With the filter we work in reflection (as indicated in Fig. 1) and with $\lambda = 496.5$ nm. The single crystals are operated in transmission and with $\lambda_{CdS} = 514.5$ nm, $\lambda_{ZnSe} = 476.5$ nm.

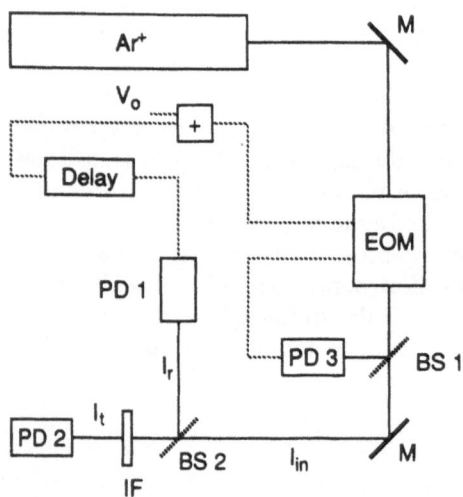

Fig. 1. Schematic diagram of the experimental setup. Ar^+ Argon laser, M mirror, BS beamsplitter, IF interference filter, PD photodiode, EOM electro optical modulator

The voltage corresponding to the reflected (transmitted) light can be digitized with 8 bit resolution and stored in the delay–generator. After the round–trip (or delay–) time τ_R it is decreased by the feedback factor R (R<1), fed back to the electro optical modulator (EOM) and added to the constant intensity I_0 given by the constant voltage V_0.

3 The Optical Nonlinearity Without Additional feedback

With R=0 we have no feedback at all and by varying I_0 slowly with a triangular pulse shape we can investigate the quasistatic IOC in reflection and transmission. The results of this procedure are shown in Fig. 2 for the interference filter (Fig. 2a) and the CdS single crystal (Fig. 2b). The filter shows bistable behaviour

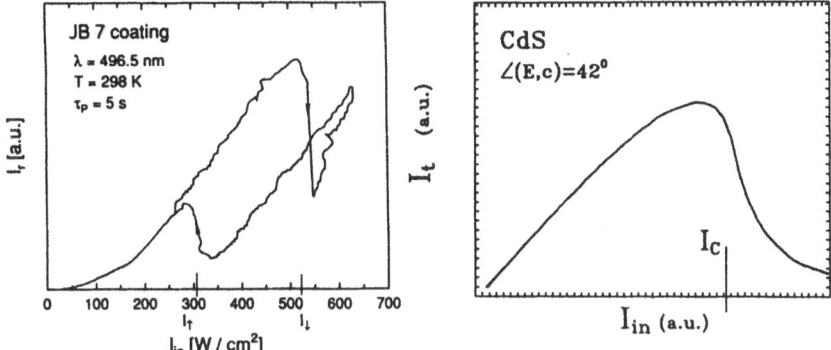

Fig. 2. Quasistatic input–output characteristics of the semiconductor doped glass filter JB7 in reflection (left) and of the CdS single crystal platelet in transmission (right)

due to changes of the refractive index [7] with two parallel stable branches. In CdS we can choose an induced absorptive characteristic which can be mono- or bistable using the dichroism of this material by varying the angle between **E** and the crystallographic **c** axis. In the polarisation configuration chosen here $\angle(\mathbf{E}, \mathbf{c}) = 42^0$, the sample exhibits a strongly nonlinear but just not bistable behaviour due to thermally induced absorption [8].

4 The Bistable Element in the Ring Resonator

From the simplest point of view in the case of $R>0$ and large round trip times in the resonator, one can expect an iteration on the quasistastic IOC of the nonlinear element [3,4]. Because of the fact that the hybrid cavity produces a retarded positive feedback and the IOC a partly negative feedback we get oscillatory outputs in some region of the parameter space. In this section we use the longest possible delay time in our experiment, namely $\tau_R = 2$ s and a feedback of $R = 0.75$. Figure 3 shows the oscillation modes with the bistable filter as a function of time that arise for different values of the control parameter I_0. All the modes show influences of dynamical effects but the sharp change of the intensity after every delay time can be clearly seen. Nevertheless, the locking into multiples of τ_R is not perfect because of the thermal inertia of the sample.

We will denote the oscillation modes according to the following rule: A mode with k ascending and l descending steps in the intensity will be denoted as (k, l) mode.

The mode of Fig. 3a is the one arising first after leaving the stationary regime when increasing the value of I_0 up to $I_0 = 0.29\ I_\downarrow$ where $I_\downarrow$ is defined in Fig. 2a. It is a $(6,2)$ mode. For lower values of I_0 the system reaches a stationary state on the high reflecting branch of the IOC and by increasing I_0 the oscillation arises when the fixed point on the high reflecting branch reaches the switching value. Modes with higher numbers of steps don't arise because of noise induced

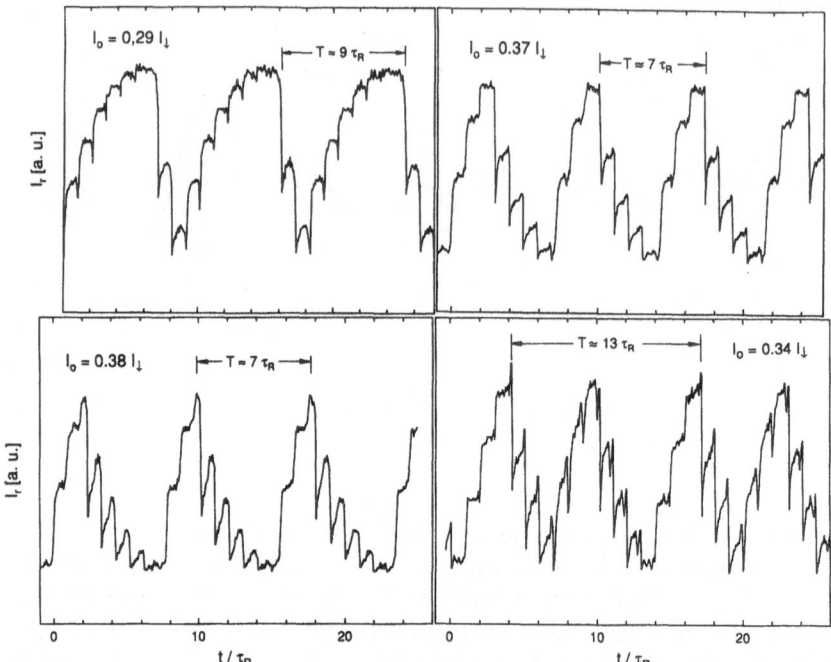

Fig. 3. Experimentally observed oscillation modes as a function of time for different values of the intensity I_0 incident onto the ring resonator: (6,2) mode, (3,4) mode, (2,5) mode, (6,7) mode (for the definition of the modes see text)

switching processes [9,10]. Then the descending steps in the output intensity occur because after every round trip the constant intensity I_0 is added to the reflected intensity I_r and the sum of these two values is lower than the value of the incident intensity one round trip before. When the lower switching value is reached the system jumps to the high reflecting branch again and here the opposite is true: The sum of I_0 and I_r is higher than the incident intensity one round trip time before and this leads to the ascending steps.

Following these arguments it is quite clear that higher values of I_0 are causing less ascending steps and more descending steps and vice versa. This can be seen from the other modes of Fig. 5. Changing I_0 to the value of 0.37 $I_\downarrow$ we get less ascending steps (namely only 3) and more descending steps (namely 4), as can be seen from the (3,4) mode shown. With even higher values of I_0 we come to the other border of the oscillation region and just before reaching stability again, now on the lower branch of the IOC, one can observe the (2,5) mode arising for $I_0 = 0.38\ I_\downarrow$.

Until now we have shown only modes with one single maximum. An example of a more complicated mode is shown in Fig. 3 for $I_0 = 0.34 I_\downarrow$ where one can see a (6,7) mode with 2 different maxima during the whole oscillation period $T=13\tau_R$. The intensity interval in which this higher mode exists is much smaller than that of the fundamental modes. The (6,7) mode is only stable for a few minutes before it is destroyed by fluctuations of the laser or long time scale changes of the optical properties of the sample.

In [3,4] a very similar iterated bistable characteristic has been investigated except that the lower branch of the IOC was a line through the origin but with smaller slope than that of the upper branch. It has been pointed out that the transition between the possible modes is given by a Farey–tree structure. This means that in the intensity interval between a (k,l) and a (m,n) mode there will exist a (k+m,l+n) mode, with which this construction can be repeated producing higher generations of the Farey tree. An additional feature of the map investigated by [4] was the phenomenon of mode coexistence. Here we shall investigate what this structure looks like in our special case. In order to do so we approximate the IOC of the nonlinear element using two parallel straight lines as indicated in Fig. 4a. Here, the switching down intensity $I_\downarrow$ has been scaled to unity and $I_\uparrow$ is given as a fraction of $I_\downarrow$. R is defined as the feedback ratio in the loop like in the experiment. The adiabatic eliminiation of the temperature, i.e. the use of a map instead of a differential equation is singular [11,12] and we cannot expect that this approximation will explain all features of the experiment, but Figs. 4b-e show that the principle mode structure can be simulated quite well, of course without the transient features of the experimentally observed modes.

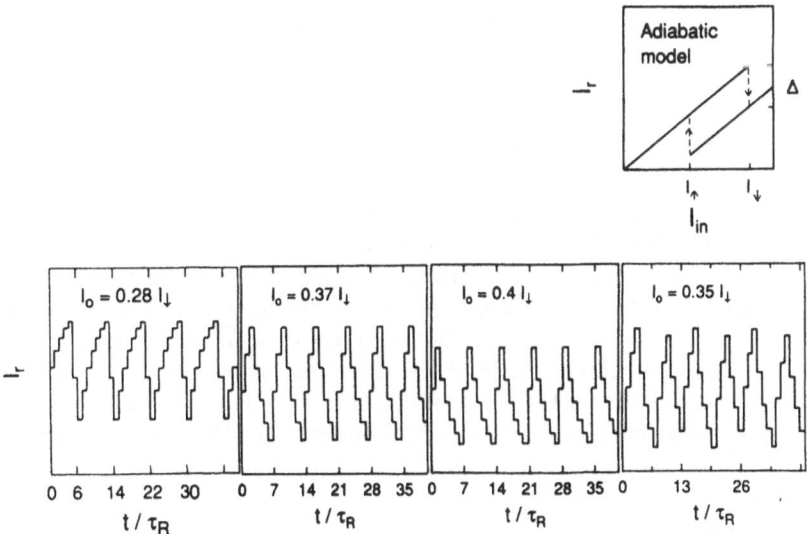

Fig. 4. Adiabatic Input–Output characteristics and oscillation modes evaluated by its iteration: (6,2), (3,4), (2,5), (6,7) modes

To get an overview of the mode structure we show in Fig. 5 the average period $\bar{T} = T/n$ of the oscillation arising as a function of I_0. Here, T is the whole period of the oscillation in units of τ_R and n is the number of different maxima in this period. So for the first generation of the Farey tree n is equal to one, for the higher generations n increases.

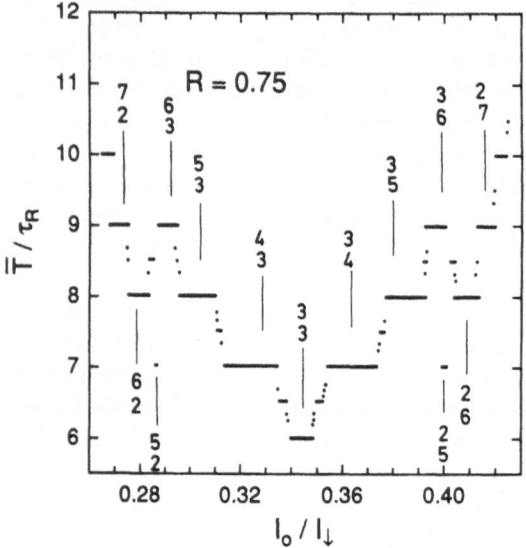

Fig. 5. The average period $\bar{T}$ of the oscillation as a function of the input intensity of the resonator

If we look at Fig. 5 we recognize the following: First, the scenario is quite symmetric with the number of ascending and descending steps inverted on both sides of the oscillation region. Then, for very high and low values of I_0, i. e. at the borders of the oscillation region, $\bar{T}$ increases very strongly due to the fact that the system needs very many ascending (low I_0) or descending steps (high I_0) to reach the switching points on the upper or lower branch, respectively. Another characteristic feature of Fig. 5 is the oscillating structure of $\bar{T}$ in the region of the (2,5) and (3,6) mode. Looking closely at the transition between different modes it turns out that e. g. between the (3,3) and (3,4) mode a Farey tree sequence at least up to the fourth generation exists. Additionally to the Farey–tree structure we found the phenomenon of mode coexistence in the simulation in accordance with [4]. It can be recognized in Fig. 5, for example, between the (3,6) and (2,5) mode. These modes are not connected by higher modes of the Farey–tree but the system jumps abruptly from one mode to the other. When changing the input intensity into the other direction this jump occurs at a slightly different value of I_0 giving rise to mode coexistence in this I_0 interval. This mode coexistence instead of the Farey tree modes is most prominent at the transition between two families of the Farey tree. The $(2, n)$–mode family exists with monotonously decreasing $\bar{T}$ beginning at the border of the oscillation region to the (2,6) mode.

The transition between $(2, n)$ and $(2, n-1)$ modes is mainly given by the Farey tree as well as the transitions between modes of the $(3, m)$ family $(3, m)$ and $(3, m-1)$ with $3 < m \leq 6$. But in the same I_0 interval in which the mode of the $(3, m)$ family with the longest $\bar{T}$, namely (3,6), should exist, also the mode of the $(2, n)$ family with the shortest period time, namely (2,5) can arise. Their intervals of existence overlap to a certain extent. So we get this complicated structure in the transition between one family and another where mode coexistence as well as Farey tree transitions occur. This phenomenon can also be found in a more complex way in higher modes of the Farey tree.

Fig. 6. The average period $\bar{T}$ of the oscillation as a function of the normalized input intensity I_n in the interval $[0.05\ldots0.5]$ and of the feedback factor R in the interval $[0.6\ldots0.98]$

To investigate the influence of the reflectivity R on the mode structure we repeated the simulations for different values of this parameter. I_n is defined in such a way that the values of I_n in the interval $[0 \ldots 1]$ cover the whole oscillation region ranging from $I_{0\,min}$ to $I_{0\,max}$. The results can be seen in Fig. 6, where $\bar{T}$ encoded in grey tones is shown as a function of R in the interval from 0.6 to 0.98 and of I_n in the interval from 0.5 to 0.95.

For each point we determined the mode 5 times, starting from different random initial conditions. So, the dotted areas indicate regions with mode coexistence. Additionally, one clearly recognizes that with increasing R the structure gets more fractal because the width of the mode locking intervals decreases. Additionally, the oscillating structure of $\bar{T}$ with very small mode locking intervals in the region of transitions between families of modes is more pronounced then.

First investigations have shown that in this system there exist also forms of mode coexistence that cannot be detected by smoothly changing the value of I_n but only by beginning each iteration with random starting conditions. It also turned out that there exist two topologically different kinds of mode coexistence: in one case the coexistence interval is of nearly constant width in I_0 and is purely shifted when the feedback factor R is changed, in the other case the coexistence interval is generated by a crossover of some neighbouring modes when changing R and thus the I_0 interval is of more complicated shape. Details about this phenomenon will be published elsewhere.

This complex mode structure was not seen in the experiment. The reason is that the most complex modes are damped out because of the finite relaxation time of the nonlinearity. Additionally, the noise in the system increases with increasing feedback R preventing a precise variation over small intensity intervals.

5 The Nearly Not Bistable Element in the Ring Resonator

In a former paper we described the mode structure of the oscillations arising when a bistable induced absorber made from a CdS single crystal is incorporated into the hybrid ring resonator [3]. We observed Farey tree like transitions between different modes without any descending steps. Changing the absorption of the sample by rotating the polarization by 90° leads to a monostable IOC and a period-doubling route to optical chaos has been observed [3]. In this contribution we want to concentrate on the intermediate case of nearly not bistable behaviour and use a polarization yielding the IOC shown in Fig. 2b. The value of the delay time is chosen to be 1 s and is thus even larger than in [3].

First we want to look at the overall behaviour of the system when the input intensity onto the resonator is changed in 15 min from $I_0 = 0.9I_\downarrow$ to $I_0=0$. In Fig. 7 one can see a photo of an oscilloscope with I_0 shown on the x-axis and I_t on the y-axis. Bright points on the photos correspond to values of I_t that are stable for a long time i. e. they correspond in the adiabatic limit to the plateaus in the intensity. For high values of I_0 the output of the system is stationary and the fixed point moves to higher transmitted intensities with decreasing I_0. At a

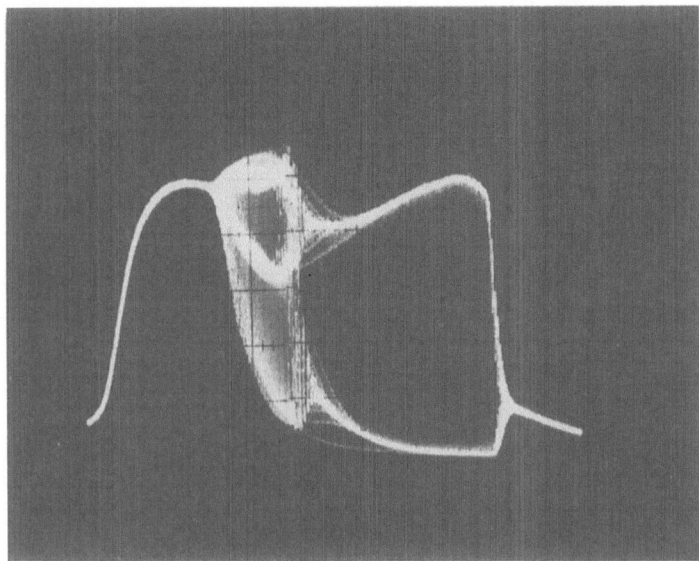

Fig. 7. Overall behaviour; x-axis: intensity incident onto the ring resonator, y-axis: transmitted intensity

critical value the first bifurcation occurs and an oscillation between two values of I_t arises. By decreasing I_0 further these values split again, not into discrete levels but into a band which gets broader with decreasing I_0. This behaviour becomes clear when we look in detail at the oscillation modes as a function of time arising for specific values of I_0. The basic mode is the $(1,1)$ mode shown in the first line of Fig. 8. It is locked into the round trip time nearly perfectly. An adiabatic iteration of the IOC of Fig. 2b similar to the preceeding section yields that one should find a period doubling route to chaos with decreasing I_0. Definitely, the observed scenario in Fig. 8 is different from that. It exhibits dynamics on a lower time scale that gets more prominent with decreasing intensity. By comparing with the numerical simulation (see the right hand side of Fig. 8) that is based on an ordinary differential equation for the temperature in the laserspot with a temperature relaxation time τ we find that the new oscillation has a period of approximately 10τ.

The origin of this relaxation oscillation is the momentary mismatch of light intensity falling on the sample and the temperature at the switching point to high absorption of the crystal which leads to a short intensity peak. The feedback for this peak one round trip time later is given by the overall feedback factor R times the product of the slope of the IOC at the points of oscillation (the plateaus in the adiabatic limit). Now its obvious why the amplitude of the fast oscillations increases with decreasing input intensity leading to the splitting bands in the overall picture of Fig. 7: the points of oscillation move towards the steepest region of the IOC and thus towards the most effective feedback for the relaxation oscillations. The frequency of these oscillations is much lower than the frequency

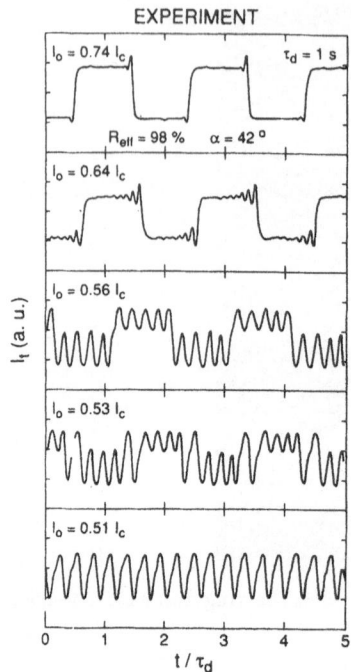
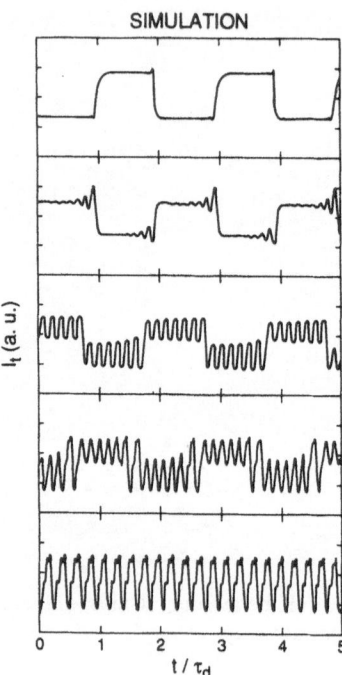

Fig. 8. Oscillation modes as a function of time: Comparison between experiment and simulation

given by the inverse temperature relaxation time because the dynamics of the system gets slower when the involved states come closer to the critical point of the IOC with the largest slope [3,10,13].

When the amplitude of these oscillations is so large that the oscillations on the upper and the lower level of the basic mode start to overlap a switching between these levels can occur: the up to now strictly separated oscillations on upper and lower branch start to mix until a series of full amplitude oscillations with high frequency can be observed.

This self induced transition is hysteretic: now the intensity can be increased again and because of the phenomenon of mode coexistence [3] the system remains in this oscillation mode. The fast oscillations show now a (2,1) structure but the relevant time base is not the delay time τ_d but the period of the relaxation oscillations. This means that during this transition the temporal intensity distribution with 3 peaks remains qualitatively the same. Furthermore, the minima of these oscillations indicate very weakly a (2/1) structure aslo.

The reason for these unexpected results compared to the adiabatic simulation that leads to a series of period doublings can be found in the fact that the self induced relaxation oscillations in turn influence the IOC. This can be seen from Fig. 9 which shows modulated and unmodulated quasistatic IOCs of the CdS crystal without additional feedback. The slowly varying input intensity now additionally has been modulated [13,14] in a sinusoidal fashion with frequency f and amplitude A that is given by the relaxation oscillations. The average of the transmitted intensity during one period of the sinusoidal modulation has been plotted as the solid line in Fig. 9. A comparison with the unmodulated IOC (broken line) shows that the nonlinearity of the system is dynamically reduced by the fast oscillations which leads to the loss of deterministic chaos in this case.

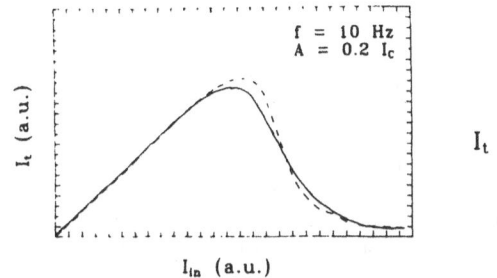 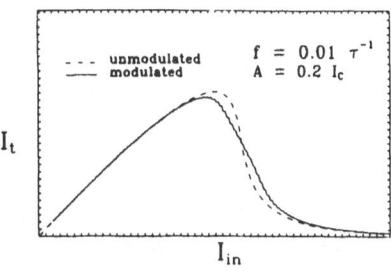

Fig. 9. Quasistatic IOC of the CdS crystal with and without additional modulation (left: experiment; right: simulation)

6 Optoelectric Nonlinearity

The photo–thermal nonlinearity can be influenced by an applied electric field e.g. via heat produced by photoconductivity. Results for CdS, ZnSe and ZnTe have been published e.g. by [15,16] In this section we present results on optically bistable ZnSe single crystals with additional electric contacts made from aluminium. The influence of the electric field applied throughout these contacts is twofold: first, Joule heating leads to an additional temperature rise in the crystal and second, the electrostatic potential induces a band tilt, the so called Franz–Keldysh or three dimensional Stark effect, that both influence the optical properties of the sample. These effects can be seen in the quasistatic IOCs of Fig. 10: without electric field the ZnSe crystal exhibits thermally induced dispersive optical bistability with the second loop corresponding to a light power of 53 mW. With an applied voltage of 100 V (corresponding to an electric field of about 8 kV/cm) we observe the second loop at an approximately 3 mW lower input power and even a third dispersive loop appears. Calculating the electrical power that is dissipated into the crystal by the current of 0.15 μA leads to a value of 150 μW. Thus the change of the switching power cannot only be due to the temperature rise induced by Joule heating but the field must have directly influenced the optical properties. To confirm this interpretation, modulation experiments will be useful to separate the effects of thermal changes with time constants in the ms range and the Stark effect with a very much shorter time constant.

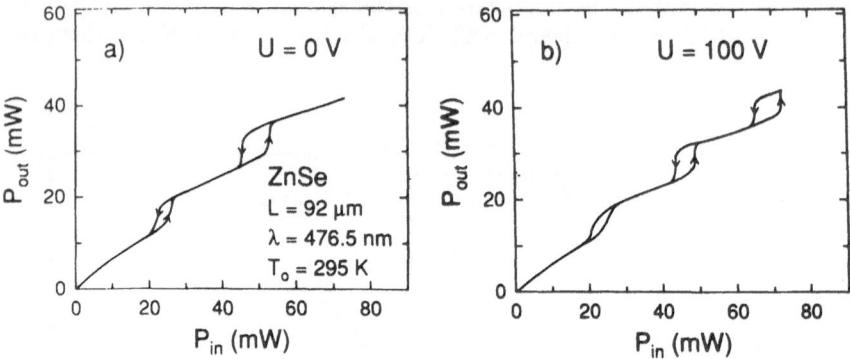

Fig. 10. Quasistatic IOC of the ZnSe crystal with and without electric field

7 Conclusions

We investigated the thermally induced optical nonlinearity of a semiconductor doped glass in a FP with additional feedback provided by a hybrid ring resonator with long round trip time. Without this feedback we observed dispersive optical bistability. The transition between two oscillation modes in the limit $\tau_R \gg \tau$ is partly given by a Farey–tree structure which is pronounced within one mode family (m, n) with m or n constant. In the region between these families mode coexistence may also occur. In the case of a nearly not bistable CdS element in the resonator we achieve two main transitions in the structure of the output intensity: the transition from stationary behaviour to an oscillation in the form of the basic mode with only two alternating plateaus and the self induced formation of a new temporal structure which is a higher frequency relaxation oscillation that is superposed to the basic mode for high input intensities and that has full amplitude for low input intensities. The transition between these two states is hysteretic. Additionally, the optoelectronic properties of ZnSe crystals have been investigated. Here, the electric field directly as well as temperature changes due to Joule heating influence the optical bistability.

Acknowledgments

This work has been financially supported by the "Stiftung Volkswagenwerk" and by the "Forschungsschwerpunkt Materialwissenschaften der Universität Kaiserslautern". We thank Dr. H. J. Korsch for valuable discussions. Thanks for help in the linear experiments are due to H. Gießen.

References

1. Vitanovic, P.: In Optical instabilities. Boyd, R. W., Raynor, M. G., and Warducci, L. M., Eds. Cambridge University Press, Cambridge, 1986, p. 151.
2. Hardy, G. H., and Wright, E. M.: Theory of numbers. Oxford University Press, Oxford, 1938.
3. Wegener, M., Klingshirn, C.: Phys. Rev. A 35 (1987) 1740; *ibid.* Phys. Rev. A 35 (1987) 4247; Wegener, M., Klingshirn, C., Müller–Vogt, G.: Z. Phys. B 68 (1987) 519.
4. Lindberg, M., Koch, S. W., Haug, H.: J. Opt. Soc. Am. B 3 (1986) 751; Haug, H., Koch, S. W., Lindberg, M.: Phys. Scripta T13 (1986) 178.
5. Ikeda, K.: Opt. Commun. 30 (1979) 257; Ikeda, K., Daido, H., Akimoto, O.: Phys. Rev. Lett. 45 (1980) 709.
6. Ekimov, A. I., Onushchenko, A. A.: JETP Lett. 34 (1981) 345; Efros, Al. L., Efros, A. L.: Fiz. Tekh. Poluprovodn. 16 (1982) 1209; Uhrig, A., Oberhauser, D., Dörnfeld, C., Klingshirn, C., Neuroth, N.: Proc. SPIE 1127 (1989) 101; Uhrig, A., Banyai, L., Hu, Y. Z., Koch, S. W., Klingshirn, C., Neuroth, N.: Z. Phys. B 81 (1990) 385;
7. Karpushko, F. V., Sinitsyn, G. V.: J. Appl. Spectrosc. 29 (1978) 1323; Miller, D. A. B., Smith, S. D., Johnston, A.: Appl. Phys. Lett. 35 (1979) 658; Gibbs, H. M., McCall, S. L., Venkatesan, T. N. C., Gossard, A. C., Passner, A., Wiegmann, W.: Appl. Phys. Lett. 35 (1979) 451; Gibbs, H. M.: Optical Bistability: Controlling light with light. Academic Press (1985); Mandel, P., Smith, S. D., and Wherrett, B., eds.: From optical bistability towards optical computing. North Holland (1987); Wherrett, B. S., Tooley, F. A. P., eds.: Optical computing. Proc. of the 34th Scottish Universities Summer School in Physics 34 (1988).
8. Lambsdorff, M., Dörnfeld, C., Klingshirn, C.: Z. Physik B 64 (1986) 409.
9. Lugiato, L. A., Broggi, G., Colombo, A.: In Frontiers in quantum optics. Ed. R. Pike, Hilger, London (1987); Mitschke, F., Deserno, R., Mlynek, J., Lange, W.: IEEE JQE 21 (1985) 1435; Koch, S. W.: Dynamics of first order phase transitions in equilibrium and nonequilibrium systems. Lecture Notes in Physics 207, Springer (1984);
10. Grohs, J., Schmidt, A., Kunz, M., Weber, C., Daunois, A., Rupp, a., Dötter, W., Werner, F., Klingshirn, C.: Proc. SPIE 1127 (1989) 39.
11. Berre, M., Ressayre, F., Tallet, A., Gibbs, H. M.: Phys. Rev. Lett. 56 (1986) 274.
12. Abraham, N. B., Firth, W. J.: J. Opt. Soc. Am. B 7 (1990) 951.
13. Grohs, J., Ißler, H., Klingshirn, C.: Opt. Commun. 86 (1991) 183.
14. Boden, C., Mitschke, F., Mandel, P.: Phys. Rev. Lett. 65 (1990) 1873.
15. Witt, A., Wegener, M., Lyssenko, V. G., Klingshirn, C., Wingen, G., Iyechika, Y., Jäger, D., Müller–Vogt, G., Sitter, H., Heinrich, H., Mackenzie, H. A.: IEEE Journ. Quant. Electr. 24 (1988) 2500; Kazukauskas, V., Grohs, J., Klingshirn, C., Wingen, G., Jäger, D.: Z. Physik B 79 (1990) 149; Schmidt, A., Kunz, M., Lacis, I., Daunois, A., Klingshirn, C.: Z. Physik B 86 (1992) 337.
16. Ullrich, B., Bouchenaki, C., Zielinger, J. P., Nguyen Cong, H., Chartier, P.: J. Appl. Phys. 69 (1991) 7357.

Thermally Induced Optical Bistability in GaAs/(AlGa)As Multiple Quantum Wells for Application as a Temperature Sensor

U. Zimmermann,[1] *K.-H. Schlaad,*[1] *G. Weimann*[2] *and C. Klingshirn*[1]

[1] University of Kaiserslautern, Department of Physics, W-6750 Kaiserslautern, FRG
[2] Walter-Schottky-Institut, W-8000 München, FRG

With the aim of realizing a temperature sensor based on thermally induced optical bistability that can be operated by a conventional diode laser, we investigated the absorptive behaviour of GaAs/(AlGa)As multiple quantum well structures. Even at room temperature, these structures exhibit a steep excitonic absorption edge in the near infrared spectral region. With rising temperature, this absorption edge shifts strongly enough to the red to lead to a sharp increase in the absorption of suitable photon energies below the absorption edge of the quantum wells. This opens the possibility of observing thermally induced optical bistability in the sample with moderate pumping intensities. The bistability is observed by focusing an infrared laser beam on the multiple quantum well structure. The transmitted laser intensity shows the switching of the sample between the two possible absorptive states. The incident laser intensities at which this switching occurs are extremely sensitive to the temperature of the material surrounding the sample. Using this effect, an optical temperature sensor can be realized in a very practical design using an optical fiber to guide the incident and reflected laser beam that can possibly be provided by a diode laser.

1 Introduction

Even though temperature is one of the more common quantities in nature, it can be a very crucial parameter in scientific experiments and industrial production. Consequently, a large variety of techniques for measuring this quantity has been developed for all kinds of applications. However, it is still interesting to explore alternative techniques that may also have useful applications.

One promising method is to use the temperature dependence of thermally induced optical bistability (TOB) to realize a temperature sensor. This sensor works on an all optical principle, it could for example be used in places where even the small electro-magnetic fields caused by thermocouples etc. are not tolerable. This method of measuring temperature has already been investigated by J. Grohs et al. [1]. They performed successful experiments on CdS excited by an Ar$^+$ laser, showing that, in principle, sensing temperature by means of TOB is possible.

In this contribution, we will review similar but more detailed measurements we performed on *GaAs/(AlGa)As* multiple quantum well structures (MQWS). We investigated these structures for the following reasons:

- the absorption edge of the MQWS lies in the near infrared. This allows the use of the small and comparably inexpensive diode lasers to induce TOB, making this temperature sensor more attractive for application. Also, it is relatively easy to find optical fibers with high transmission in this spectral region,
- because of the quantum confinement, excitonic effects in MQWS become stronger than in bulk material, leading to a steep absorption edge even at room temperature [2] which is quite essential for observing TOB at moderate pumping powers.
- This sensor may also have a very useful application in optical computing. Since many optical modulators such as Miller-SEED's, envisioned for use in an optical computer, are constructed of MQWS similar to the ones used for our sensor, it should be possible to incorporate the sensor directly on an 'optical-computer-chip'. By operating the SEED and the sensor with the same laser, a shift in the position of the laser wavelength relative to the MQWS absorption, caused by temperature changes of laser or SEED, would immediately show itself in the temperature signal of the sensor. Adjusting the laser wavelength, the temperature drifts could then be compensated, ensuring proper performance of the SEED-elements.

2 Theory

To qualitatively understand the basics of thermally induced optical bistability, it is sufficient to consider the following simplified model: an area Q of a sample at temperature T, for example a MQWS, is connected to a heat reservoir at the constant temperature T_0 through a medium of thickness d with heat conductivity κ_m. A *cw* laser beam with total power P_0 is sent through the sample, a fraction A of P_0 is absorbed, creating e–h pairs in the sample. When these recombine, a fraction η of their energy, assumed to be independent of T and P_0, is transferred to the phonon field, thus heating the sample. We assume for simplicity that the heat conductivity of the sample is much larger than κ_m; the temperature T can then be regarded as constant over the sample. The heat flow out of the sample is given by

$$J = \frac{Q\kappa_m}{d}(T - T_0) \ .$$

Under stationary conditions, the energy flows into and out of the sample are equal, leading to

$$A = \frac{Q\kappa_m}{\eta P_0 d}(T - T_0)$$

If A depends on T and $A(T)$ is known, the above equation can be solved graphically: the intersections of $A(T)$ with the line given by $Q\kappa_m/(\eta P_0 d)\cdot(T-T_0)$

give the solutions [3,4,5,6]. TOB is observed if there is a P_0 for which the equation above has two stable solutions. This is demonstrated in Fig. 1 (left) for a typical $A(T)$–curve: there exists a line having three intersection points with $A(T)$, the outer ones marking the two possible stable states the sample can occupy.

The switching process works as follows [7]: for $P_0 = 0$, the sample is in the high–transmitting state. As P_0 increases, T increases, increasing A. This further increases T since more power is absorbed, again increasing A. For P_0 small enough, this process still converges to the high–transmitting solution. However, at an input power marked P_{down}, the feedback becomes regenerative and the sample switches to the low–transmitting state. When P_0 is then decreased, the sample stays in the 'low' state until it switches back to 'high' at input power $P_{up} < P_{down}$. For P_0 with $P_{up} < P_0 < P_{down}$ we find two possible stable states of absorption, thus the name bistability. Plotting the transmitted power P_t against P_0 shows the characteristic hysteresis loop (Fig. 1 (right)).

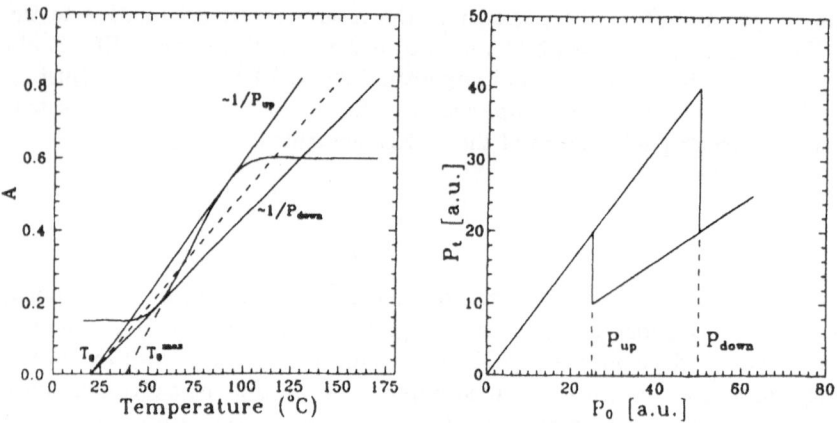

Fig. 1. Graphical solution for typical $A(T)$ (left) and resulting hysteresis loop (right)

If the temperature T_0 is increased, the values of P_{down} and P_{up} both decrease and converge to an identical value determined by the slope of the line tangent to the steepest part of $A(T)$. At the temperature T_0^{max}, marked by the intercept of this line with the T-axis, the bistability disappears [1].

This simplified model gives us a criterion for the existence of TOB when $A(T)$ is known. We also see why a steep absorption edge of the used material is important: the steeper the absorption edge, the greater the slope of the $A(T)$–curve. This increases the minimum slope of possible lines with the required three intersections, thus reducing the laser power necessary for inducing the switching process at a given T_0.

3 Experiment

We performed a pump- and probe beam experiment, using an Ar^+ laser pumped Ti:Al_2O_3 laser tunable from 1.6 to 1.4 eV for the pump beam and a tungsten lamp for the white probe spectrum (I_0). Laser intensities were measured with photo diodes, the probe light transmitted (I_t) and reflected (I_r) by the sample was detected by an optical multichannel analyser (OMA) behind a spectrometer. The sample was placed in an oven.

The sample used in the following measurements is a GaAs/($Al_{0.3}Ga_{0.7}$)As MQWS consisting of 60 periods (well width: 103 Å, barrier width: 150 Å) grown by molecular–beam–epitaxy (direction of growth: [110]). The MQWS is glued to a sapphire substrate with a transparent adhesive, the $GaAs$ substrate on which the MQWS was grown is removed by a selective etching technique to prepare the samples for transmission experiments [8]. The sapphire substrate serves as the heat reservoir at T_0 as described in Sect. 2, the adhesive is the medium of thickness d with heat conductivity κ_m. The temperature T_0 is adjusted through the oven.

The laser beam was focused on the sample with a spot diameter of $50\mu m$, the probe beam, being much smaller in diameter, could be scanned through the laser spot. With this setup, we are able to induce TOB with different photon energies of the pump beam at various substrate temperatures T_0. At the same time, we can record the transmission spectrum of any part of the sample, making it possible to deduce the temperature of this part of the sample from the relative red–shift of the spectrum. The red–shift of the absorption edge of the MQWS, being responsible for the temperature dependence of A, is caused by the decrease of the band gap energy with rising temperature [9]. This shift of the spectra is shown in Fig. 2.

The MQWS was heated with the oven. Since we are only interested in the fraction of incident light really absorbed in the sample, we also detected the relatively strong reflected intensity (I_r) and define A by $A = 1 - (I_t + I_r)/I_0$.

We recorded a set of spectra at temperatures from 25°C to 200°C. Out of these data, we get $A(T)$ curves for different photon energies below the absorption edge (Fig. 3). As one can see, TOB should occur for T_0 around 25°C at the photon energies stated in Fig. 3.

In the experiment, we detected the bistability by monitoring the incident (P_0) and the transmitted laser power (P_t). The dependence of width and position of the hysteresis loop on T_0, as predicted in section 2, is also observed experimentally (Fig. 4). In order to extract the value of T_0 from these hysteresis loops, we use their relative width W defined by $W = (P_{\text{down}} - P_{\text{up}})/P_{\text{down}}$[1]. W has the advantage of being dimensionless and independent of the units of P_0 (voltage signal of a photodiode). Measured values for W at different T_0 and various photon energies of the pump laser are shown in Fig. 5 (the solid lines represent a square fit to the measured values). Using this diagram, we can now sense the temperature T_0 of an object connected to the sapphire substrate by measuring P_{down} and P_{up} in arbitrary units, calculating W and deducing T_0 from Fig. 5.

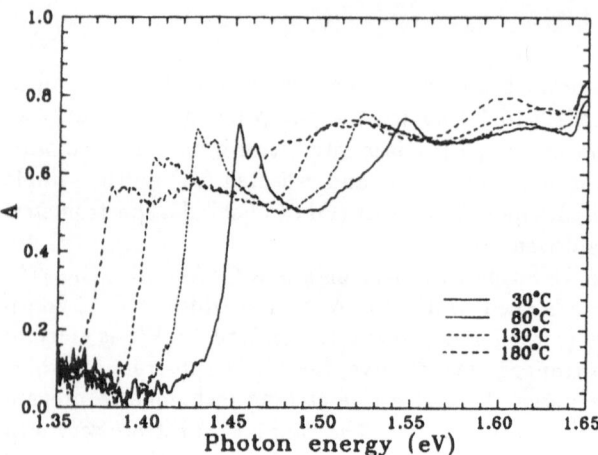

Fig. 2. Red–shift of the absorption spectrum with rising temperature

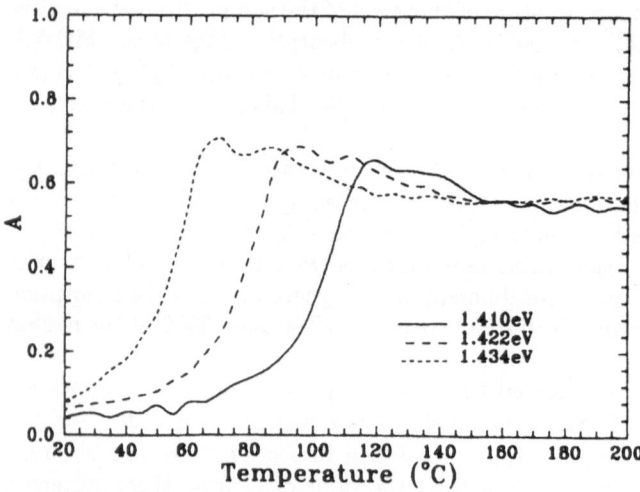

Fig. 3. $A(T)$–curves for different photon energies

The temperature range that can be covered by using just one photon energy lies around 35°C. It is limited by $T_0^{\max}$ on the upper side, by destruction of the sample because of overheating of the adhesive on the lower side since P_{down} increases strongly with decreasing T_0. This range can be extended by varying the photon energy. The reproducibility of the measured values is good, being limited only by the accuracy with which the switching powers and the photon energy can be determined. Of course, P_0 has to be varied on a timescale long enough to exclude effects of switching dynamics on the switching intensities. At present, an accuracy of about 1°C is obtained.

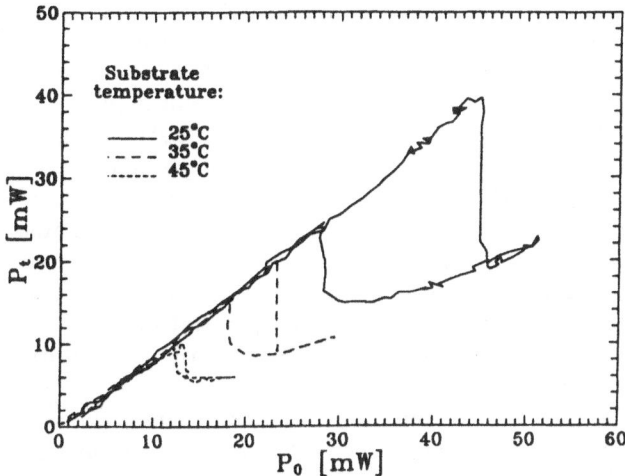

Fig. 4. Hysteresis loops for different temperatures T_0 at photon energy 1.424 eV

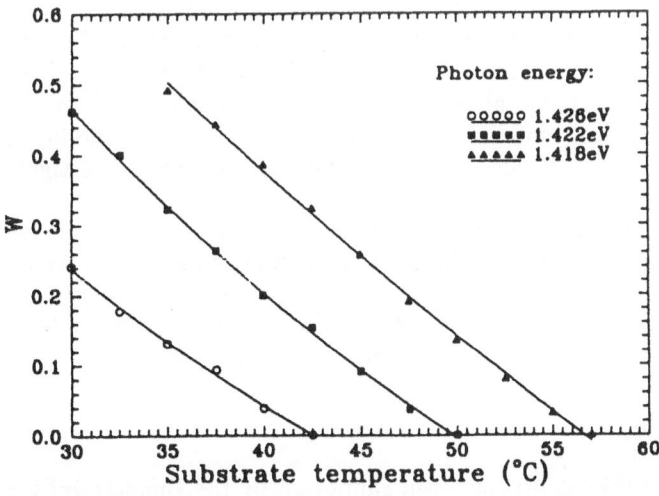

Fig. 5. Relative width of the hysteresis loop as a function of the substrate temperature at different photon energies

Comparing the measured W with values obtained from the recorded $A(T)$–curves using the graphical solution method described in section 2, we find the agreement to be quite poor. The main reason for this is the following: in our simple model, we made the assumption that the temperature T of the sample is a constant over its entire area. In the experiment, however, this does not hold: by measuring T at different points of the sample as described above, we find that T varies strongly over the area of the sample [10]. This is quite intelligible since our sample is much larger in size than the Gaussian laser spot.

With the aim of realizing a miniature temperature sensing device, we used a thick optical fiber (diameter: 100 μm) to guide the laser power onto the sample by placing one end of the fiber directly on the surface of the MQWS. This has the great advantage of being easy to adjust and guaranteeing a laser spot of fixed size. The switching of the MQWS could be seen in the laser light reflected from the sample through the fiber. However, the reflected signal was too weak to precisely measure the switching powers. To increase the reflected laser power, we are trying to prepare a sample with an aluminum-mirror evaporated on the surface of the MQWS opposite to the fiber. By fixing a small slab of this sample to the end of the fiber, we would have a very practical miniature temperature sensing device.

The system can also be used as an temperature threshold sensor [1]: when illuminating the sample with a constant laser power P_0, switching from 'high' to 'low' is induced when T_0 is raised above the temperature at which P_0 equals P_{down}. The desired threshold temperature can be varied by adjusting P_0.

4 Outlook

As mentioned above, the model used here is too simple for quantitative statements. We are currently working on computer simulations of an extended model including an intensity profile of the laser beam, lateral heat conduction and the resulting temperature profile in the plane of the sample, and dynamic effects. We recently integrated an optical modulator in our setup. By using it to modulate P_0 on a timescale down to μs and also gating the OMA, we intend to investigate the dynamic behaviour of the TOB.

Our future work regarding the application of the sensor will focus on realizing the design with the optical fiber as explained in Sect. 3. Should this be successful, we will try to further improve our sensor by using thinner optical fibers which should reduce the laser power needed. We also intend to connect a diode laser to our sensor in order to get a compact temperature sensing system.

Acknowledgements

The project underlying this report has been supported by the Bundesminister für Forschung und Technologie (Förderungskennzeichen TK 0575). We thank the SIEMENS AG for their financial support. Stimulating discussions with H. Bartelt are acknowledged. The authors are responsible for the contents of this contribution.

References

1. Grohs, J., Müller, M., Schmidt, A., Uhrig, A., Klingshirn, C., Bartelt, H.: Optics Comm. **78** (1990) 77.
2. Schmitt–Rink, S., Chemla, D. S., Miller, D. A. B.: Advances in physics **38** (1989) 89–188.
3. Miller, D. A. B.: J. Opt. Soc. Am. B **1** (1984) 857.
4. Lambsdorff, M., Dörnfeld, C., Klingshirn, C.: Z. Physik B **64** (1986) 409.
5. Wegener, M., Klingshirn, C.: Phys. Rev. A **35** (1987) 1740.
6. Henneberger, F.: Phys. Stat. Sol. (b) **137** (1986) 371.
7. Miller, D. A. B., Gossard, A. C., Wiegmann, W.: Opt. Lett. **9** (1984) 162.
8. LePore, J. J.: J. Appl. Phys. **51** (1980) 6441.
9. Panish, M. B., Casey, Jr., H. C.: J. Appl. Phys. **40** (1969) 163.
10. Wegener, M., Klingshirn, C., Daunois, A., Bigot, J.-Y., Cherkaoui Eddeqaqi, N., Grun, J. B.: Appl. Phys. Lett. **52** (1988) 685.

Chances for Nonlinear Optical Switching Elements

H. Bartelt

Siemens AG, Corporate Research ZFE BT PE 51, Paul-Gossen-Str. 100,
W-8520 Erlangen, Germany

The development of optical switching elements has been directed, until now, mostly to applications in the field of signal processing with very high data capacity. But optical switching elements could also offer great advantages if used in optical or fibre optical sensor systems. The discussion of two relatively simple examples is intended to demonstrate the chances of nonlinear optical switching elements for applications as sensor elements or as sensor specific signal processing elements.

1 Introduction

Optical switching elements are being mainly developed for applications in highly parallel or very fast signal processing systems [1]. Another attractive field of applications could be found in optical sensor systems (especially fibre optic sensor systems) even with relatively limited performance concerning parallelism and switching speed. Fibre optic sensor systems are mostly used for applications where electric sensor systems are problematic or even not applicable (e.g. high voltage levels, interference with electro-magnetic fields). Possible further applications are limited until now because there exist almost no optical modules for optical processing of the sensor signals. Transformation into an electronic signal and further electronic signal processing would mean a loss of many of the specific optical advantages. In addition, new sensing principles are desirable in order to enlarge the field of possible applications. In both directions, nonlinear optical switching elements could be useful. Although purely optical switching elements would be preferable, elements requiring an electric voltage may also be acceptable, as long as only low electric power levels are necessary which could be supplied optically by an optical fibre [2].

In the following a possible application as a temperature switch or temperature sensor based on thermal bistability will be described and a concept of a reference signal switch will be discussed as a simple signal processing application.

2 Temperature Switch Based on Thermal Bistability

Many types of bistable optical elements are based on thermal effects. The illuminating intensity may either change the absorption or the refractive properties of the spacer material in a Fabry-Perot structure. The switching effect will be induced by an appropriate illumination intensity. These thermal effects may also be applied for an optical temperature sensor or temperature switch. In this case, the illumination intensity will be kept constant and the switching process is induced thermally by a varying ambient temperature. We consider here, as an example, a BEAT-element (Bistable Etalon with Absorbed Transmission) where the refractive properties of the spacer material within the etalon are changing with temperature [3,4]. An additional absorber layer allows for a relatively good optothermal conversion efficiency adapted to wavelengths in the near infrared region. The transmission of the Fabry-Perot is described by the Airy formula. In addition, one may describe the temperature change by a relaxation equation which gives, with a steady-state assumption a temperature dependent transmission:

$$T_t = (T - T_0)\kappa/(\tau A P_i) \qquad (1)$$

with T_t: transmittance

T: temperature

T_0: ambient temperature

κ: heat capacity

τ: thermal relaxation time

A: absorptance

P_i: illumination power.

Both conditions, the Airy formula and equation (1), have to be fulfilled simultaneously for the switching element. The solution can be obtained from the intersection of both curves. In Fig. 1 the calculated thermal characteristics for transmittance depending on relative ambient temperature are shown for three different illumination intensities [4]. For a fixed illumination intensity, the element can be used as a switch changing the transmission properties strongly at a given temperature. The relative width of the hysteresis curve (switch up intensity divided by switch down intensity) is independent of the absolute illumination intensities at the switching element. Therefore it is possible to detect the temperature (by varying the illumination intensities) without knowing these actual intensity levels. This is a very useful property for a fibre-optic sensor in order to become independent from the attenuation of the fibre-optic transmission line.

In an experimental setup, it was shown that with such a BEAT-element (provided by the Technical University of Berlin [3]) it is possible to obtain thermally induced switching with a laser diode as a light source [3,4,5]. In Fig. 2, the switching effect is shown with a constant illumination intensity of 39 mW at a wavelength of 788 nm by varying the ambient temperature.

In a very similar way, absorptive bistability can be used for a temperature sensor [6].

3 Switchable Reference Channel

As a simple example for a possible signal processing use of optical switching elements in a sensor system, a concept for a switchable reference channel will

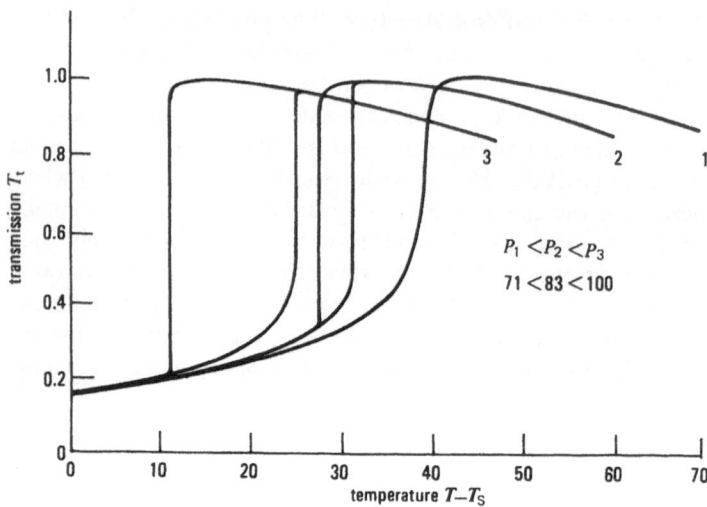

Fig. 1. Calculated thermal characteristic of a BEAT-element with reflectivity 0.8 at three different illumination intensities (temperature relative to a starting temperature T_s, intensities in arbitrary units)

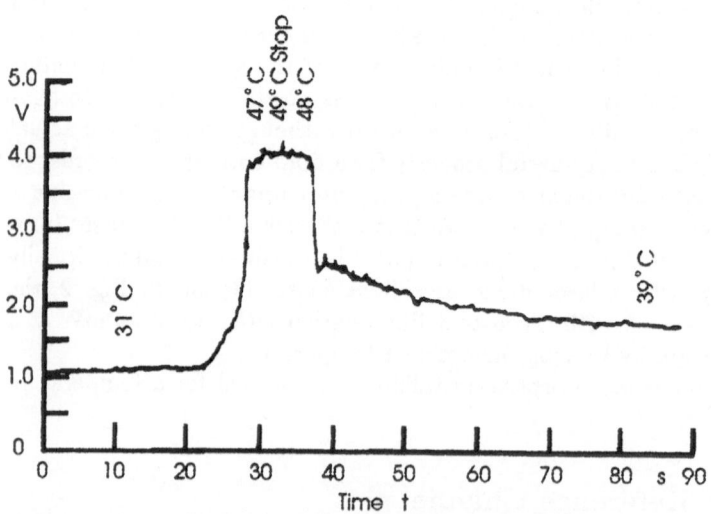

Fig. 2. Thermally induced switching of a BEAT-element with constant illumination intensity

now be described. A major problem in intensity modulated fibre-optical sensor systems is caused by additional attenuation effects e.g. due to bending losses, to connector elements and to variations in the properties of light emitting and light detection elements. Therefore in many cases a good measuring accuracy can only be obtained with a reference channel. Typical methods for obtaining a reference signal are wavelength multiplexing, a time delayed reference signal or a bridge system [7]. In the following, the concept of optically controlled active switching between the sensor signal and a reference signal will be discussed, which has the advantage that identical wavelengths and fibres are used for sensor signal transmission and reference signal transmission. In Fig. 3, possible concepts either for a reflective or for an absorptive switch are shown. In case of a reflective switch (switching the transmission and reflection coefficient) the sensor signal I_1 and the reference signal I_2 are described as:

$$I_1 = I_0 \left(R_r + T_r^2 R_s \right) \tag{2}$$
$$I_2 = I_0 \left(R_t + T_t^2 R_s \right)$$

with I_0: illumination intensity
$\quad R_r$: reflectivity in reflection state
$\quad R_t$: reflectivity in transmission state
$\quad T_t$: transmission in transmission state
$\quad T_r$: transmission in reflection state
$\quad R_s$: reflectivity of the sensor element

From both signals the sensor effect can be obtained independent from the actual intensity at the sensor head:

$$R_s = \frac{R_t - V R_r}{V T_r^2 - T_t^2} \text{ with } V = I_2/I_1 \tag{3}$$

A high stability of the switching levels is of great importance for good accuracy of the sensor system. The possible error ΔE of the system is described by:

$$\Delta E = (dR_s/dR_t)\Delta R_t + (dR_s/dR_r)\Delta R_r + (dR_s/dT_r)\Delta T_r + (dR_s/dT_t)\Delta T_t \tag{4}$$

or

$$\frac{\Delta E}{R_s} = \frac{R_t}{R_t - V R_r}\frac{\Delta R_t}{R_t} - \frac{V R_r}{R_t - V R_r}\frac{\Delta R_r}{R_r} - \frac{2V T_r^2}{V T_r^2 - T_t^2}\frac{\Delta T_r}{T_r} + \frac{2T_t^2}{V T_r^2 - T_t^2}\frac{\Delta T_t}{T_t} \tag{5}$$

Depending on the switching levels, the stability of the reflection and transmission coefficients should be kept better than 1% in order to achieve an accuracy of the sensor system in the range of some percent. Therefore, bistable elements with well defined switching levels would be required.

In a similar way, an absorptive switch (changing the level of attenuation) and a mirror could be used for a reference switch.

Other signal processing examples may be added such as an optically address-able bus system for sensor systems or an optical analog to digital converter.

(a)

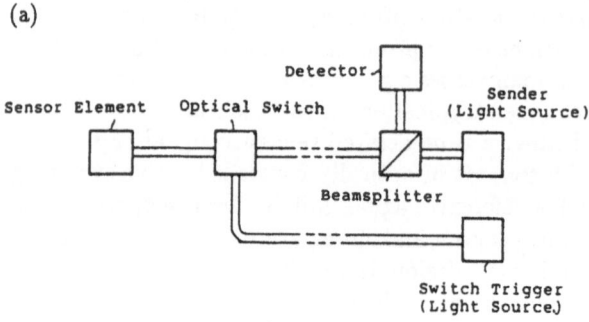

(b)

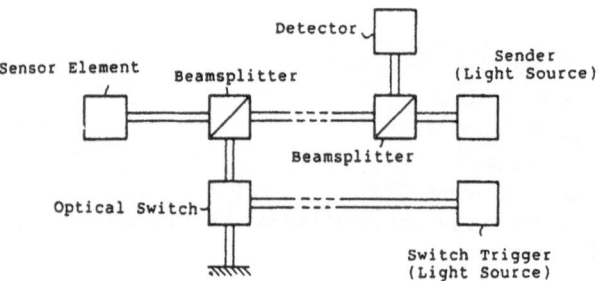

Fig. 3. Concepts of an optical reference switch in a fibre optic sensor system: (a) with a reflection/transmission switch, (b) with an absorptive switch

4 Conclusions

Fibre-optical sensor systems offer great potential for the application of non-linear optical switching elements. For practical reasons, a working wavelength in the near infrared wavelength region, where solid state light sources can be used, would be very important. In addition, compatibility of the required light intensities for switching with available solid state light sources as well as high stability with environmental conditions are required. Fast switching speed and high parallelism are of lower importance for these applications. As shown with the BEAT-element, an application with a solid state laser source is, in principle, possible. The concept of the optical reference switch gives an example that even with a rather simple use of such elements, the functionality of an optical sensor system can be enhanced. On the other hand, there is certainly a great effort necessary to improve the properties of optical switching elements, even for relatively simple technical applications.

5 Acknowledgements

The cooperation with Prof. Eichler and Dr. Kumrow for obtaining the BEAT-element and fruitful discussions with L. Frank and Dr. Bosselmann are gratefully acknowledged.

References

1. Gibbs, H.M.: Optical bistability: Controlling light with light. Academic Press, Orlando (USA) 1985.
2. Groß, W.: Optical power supply for fiber-optic hybrid sensors. Sensors and Actuators A **25-27** (1991) 475–480.
3. Bartelt, H., Bosselmann, T.: Refractive optical bistable switch as temperature sensor. Optik **89** (1992) 169–480.
4. Eichler, H.J., Glaw, V., Kumrow, A., Penschke, V., Wahi, A.: Optical bistable thin film devices using wide gap II-VI compounds, J. Cryst. Growth **101** (1990) 695–698.
5. Buller, G.S., Paton, C.R., Smith, S.D., Walker, A.C.: Optical bistable nonlinear interference filter for use with near-infrared laser diodes, Opt. Commun. **70** (1989) 522–528.
6. Grohs, J., Müller, M., Schmidt, A., Uhrig, A., Klingshirn, C., Bartelt, H.: On the possible use of photo thermal optical bistability as a temperature sensor, Opt. Commun. **78** (1990) 77. See also: Zimmermann, U., Schlaad, K.-H., Weimann, G, Klingshirn, C.: Thermally induced optical bistability in GaAs/(AlGa)As multiple quantum wells for application as a temperature sensor. In these proceedings.
7. Dakin, J., Culshaw, B., Edts.: Optical Fiber Sensors, Artech House. London 1988/89.

Part VI

III-V Bistability and Devices

Subnanosecond Switching and Recovery in a Fabry-Perot Etalon Based on Bulk Heavily Doped n-GaAs

D. J. Goodwill, F. V. Karpushko,* S. D. Smith, and A. C. Walker

Department of Physics, Heriot-Watt University Riccarton, Edinburgh, EH14 4AS, UK

Shaping of 1.9 ns laser pulses and optical switching behaviour with about 100 ps switching on/off times are measured in a bulk n-GaAs Fabry-Perot etalon. The etalon structure is not optimised, having a finesse of about 4 at low incident power. The impurity-related fast nonlinearity causes optical switching at wavelengths slightly longer than that of the fundamental absorption edge with incident intensities of $\sim 10^6$ W/cm^2, corresponding to a switch energy of ~ 1 pJ/μm^2.

1 Introduction

A strong refractive index nonlinearity of impurity origin which can arise in thin film interference structures was suggested and discussed in [1,2]. It was thought to an electro-optic process arising from photo-induced changes in the electric fields within the space charge domains at the boundaries between layers of the interference structure. Experimental evidence consistent with this proposition was obtained for the case of heavily doped n-GaAs by a four-wave mixing technique with nanosecond [3] and picosecond [4] laser pulses. The experiment in [4] had 10 ps time resolution and showed the lifetime of the nonlinearity to be about 100 ps.

To excite a similar electro-optic interaction inside a heavily doped bulk semiconductor layer without applying any external electric field, a field can be created by means of carrier diffusion processes when spatially non-uniform absorption occurs in a semiconductor which has significantly different mobilities for the electron and hole subsystems [3]. In the case of a Fabry-Perot interferometer with an absorbing spacer layer, the non-uniform absorption can take place as a result of the standing wave light distribution within the spacer. With this situation of the creation of a standing-wave electric field by the optical beam itself, heavily n-doped bulk GaAs Fabry-Perot etalons were investigated [5] and negative nonlinear refractive index changes of up to -5×10^{-3} were obtained. At incident intensities of $> 5 \times 10^5$ W/cm^2 in pulses of 10ns duration, the peaks of the interferometer transmission spectra (averaged over each pulse) were sharply

---

* Permanent address: Division for Optical Problems in Information Technology, Byelorussian Academy of Sciences 220072, Minsk, P.O. Box 1, CIS

asymmetrical and shifted to short wavelengths compared to those of the low intensity spectra. This change in the transmission spectra of the etalons is explained in Ref. [5] as being due to bistable switching on and off.

2 Experiment/Discussion

In the present paper we confirm the above results for one of those etalons by the direct time-resolved measurements of shape changes in 1.9 ns incident pulses partially transmitted and partially reflected by the device. The ~ 35 μm thick Fabry-Perot etalon was made by thinning a bulk sample of n-GaAs with Te doping density of $\sim 2 \times 10^{18}$ cm^{-3} grown by liquid-transport. This was then coated on both sides by dielectric mirrors of 70% reflectivity and placed on a glass substrate. The contrast ratio of the etalon in the linear transmission regime was however only about 5 due to a slight wedge in the spacer layer. Optical pulses were generated by a Lumonics Hyper-Dye SLM laser, pumped by 45 mJ pulses from a Lumonics Excimer-500 XeCl laser. The gain medium of the dye laser was LC8630 (1.0g/l in methanol), which gave a peak pulse energy of 230 μJ after two amplification stages and a tuning range of 880-920 nm. The laser produced single longitudinal mode pulses with a linewidth of 500 MHz and a smooth temporal profile of 1.9 ns duration (10%-10% level). The output beam of the dye laser was spatially filtered to produce a Gaussian profile and then focussed onto the Fabry-Perot etalon at normal incidence with a spot diameter of ~ 30 μm. Part of the incident, reflected, and/or transmitted beams were picked off, delayed between each other by 6 ns and steered onto a single Antel ARS-1 silicon photodiode, connected to a Tektronix 7912 digitizing storage oscilloscope. Only one detector was used so as to reduce the pick-up of electrical noise from the excimer laser discharge. The rise and fall time of the detection system were determined experimentally using pulses with a length of only 200 fs from a separate dye laser, and are both about 200 ps.

Changes in the temporal profiles of the transmitted and reflected pulses were observed between 880 nm and 895 nm, which is slightly below the GaAs band edge and is the wavelength range over which the impurity related nonlinearity shows the maximum refractive index changes [5]. The most dramatic pulse shaping takes place at the short wavelength side of the etalon transmission peaks, due to the negative sign of the nonlinearity, when the incident pulse intensity exceeds $\sim 10^6$ W/cm^2. Figure 1 shows an example of switching in the reflected pulse at a wavelength of 894.3 nm. This wavelength is detuned to the short wavelength side of the low-power cavity transmission peak at 895.5 nm. Figure 1(a) shows the incident and reflected intensities as a function of time, while Fig. 1(b) shows the reflectivity as a function of time. As the incident intensity increases on the leading edge of each pulse, the negative refractive nonlinearity pulls the cavity into resonance with the laser wavelength and so more of the incident light is transmitted. This is evident in Fig. 1 as a switch-down in the reflectivity. There also appears to be a switch-up in the reflectivity as the incident intensity decreases on the trailing edge and the cavity recovers, although at this point

(a)

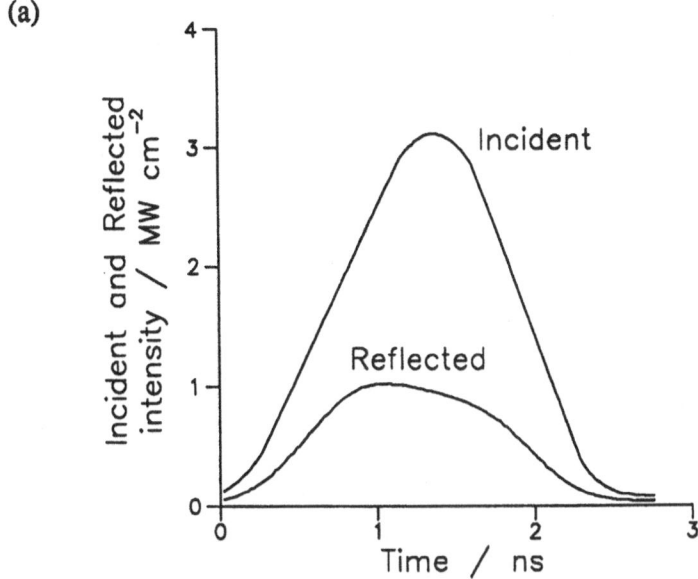

(b)

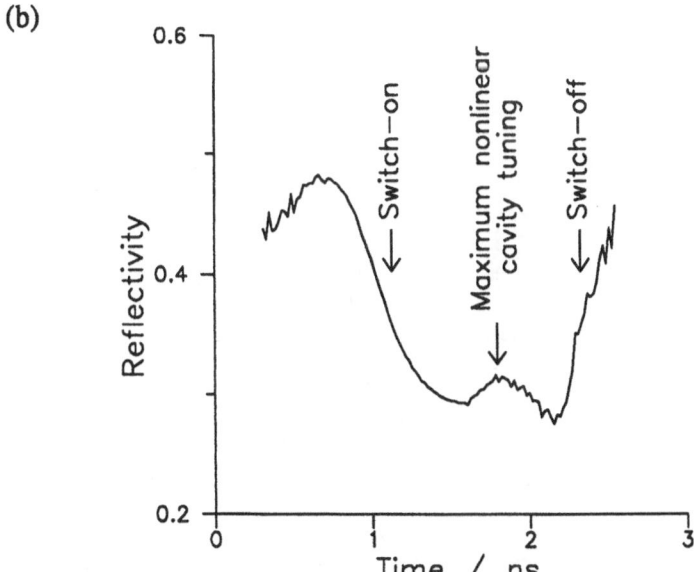

Fig. 1. (a) Pulses incident on and reflected from a nonlinear n-GaAs Fabry-Perot etalon. The laser wavelength is 894.3 nm, which is detuned to the short wavelength side of the low-power cavity transmission peak at 895.5 nm. (b) Reflectivity versus time for the same pulses as (a), showing switch-on and switch-off. Note that the 200 ps risetime and falltime of the detection system have not been deconvoluted from the measured

(a)

(b)

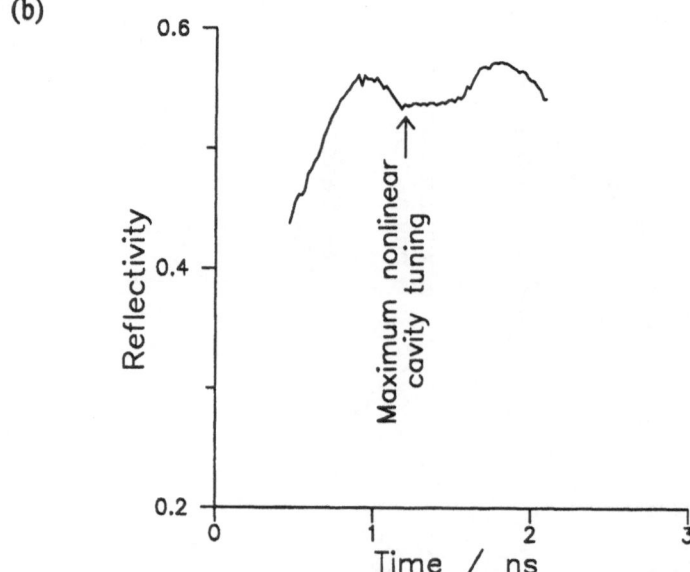

Fig. 2. (a) Pulses incident on and reflected from a nonlinear n-GaAs Fabry-Perot etalon. The laser wavelength is 893.0 nm, which is detuned to the long wavelength side of the low-power cavity transmission peak at 892.6 nm. (b) Reflectivity versus time for the same pulses as (a). Note that the 200 ps risetime and falltime of the detection system have not been deconvoluted from the measured signals

(a)

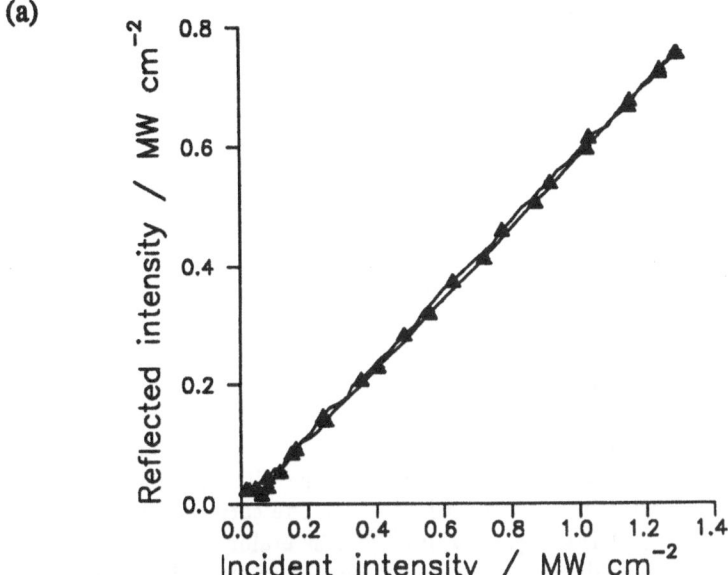

(b)

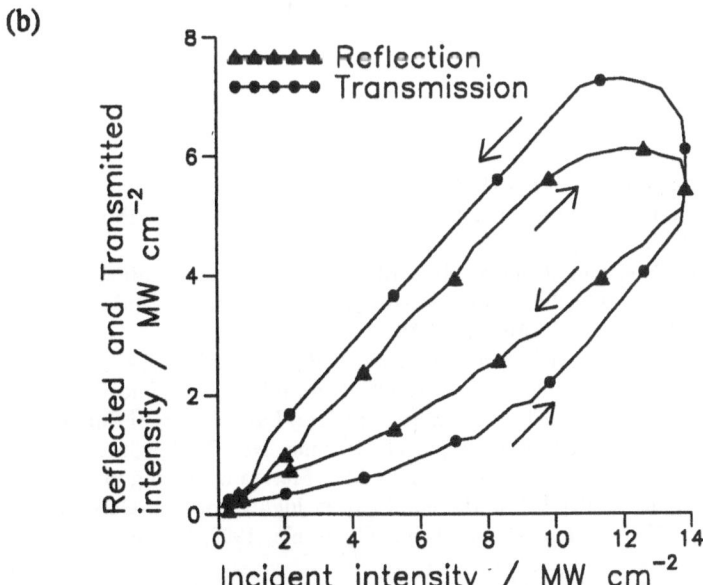

Fig. 3. Optical switching in a bulk n-GaAs Fabry-Perot etalon with 1.9 ns pulses.
(a) Reflected intensity versus incident intensity for a low incident power, linear case.
(b) Transmitted and reflected intensities versus incident intensity for a high incident
power, strongly nonlinear case Points are highlighted at 200 ps intervals. Arrows indicate increasing time. Note that the 200 ps risetime and falltime of the detection system

the signal to noise ratio is poor. The nonlinear effects observed at 893.0 nm are shown in Fig. 2 for comparison. This wavelength is on the long wavelength side of the cavity transmission peak at 892.6 nm. The reflectivity is seen to increase on the leading edge of the pulse as the cavity is pulled further from resonance, and then to recover on the trailing edge of the pulse. The contrast is not as great as in Fig. 1.

The transmitted and reflected intensity versus incident intensity for a low energy pulse (a) and a high energy pulse (b) at a single wavelength close to that of Fig. 1 are plotted in Fig. 3. In the low power case (Fig. 3(a)), no nonlinearity is observed, as expected. This also demonstrates the linearity of our detection system. In the high power case (Fig. 3(b)) a clear switch-on is observed and there is apparently a switch-off at a later time, although the recovery point is again close to the noise level. However, since this recovery was observed at the same point for many pulses, we are confident that the switch-off was genuine. From the measured data we estimate the recovery time to be about the same as the response time of our detection system (i.e. 100–200 ps). The switching-on time appears to be even shorter in this over-driven case. The corresponding value of the threshold switching energy is ≤ 1 pJ/μm^2 for the etalon used.

In conclusion, we have demonstrated sub-nanosecond switching on/off in a non-quantum well semiconductor Fabry-Perot etalon, based on the impurity related nonlinearity in heavily doped n-GaAs. Values of ~ 100 ps for both switching on and off times and switching energies of ≤ 1 pJ/μm^2 were obtained even in this unoptimised structure.

One of the authors (F. K.) is grateful to the Royal Society for support of a visiting research fellowship at Heriot-Watt University.

References

1. Karpushko, F.V.: Nonlinearity of thin-film semiconductor interferometers due to interlayer boundary photoemf and electrooptic processes. Journal de Physique **49** (1988) C2-87 – C2-90.

2. Karpushko, F.V.: Fast and strong nonlinearity of thin-film semiconductor interferometers due to intrinsic photo- and electrooptical effects of the semiconductor interlayer. Phys. Stat. Sol. B **150** (1988) 791–796.

3. Karpushko, F.V., Bystrimovich, S.A., Morozov, V.P., Porukevich, S.A., Sinitsyn, G.V. and Utkin, I.A.: Fast photorefractive effect in high-doped semiconductors, applications to optical bistable interferometric devices. Laser Optics of Condensed Matter **2**, 351-362, Edited Garmire, E., et al., Plenum, New York, 1991.

4. Karpushko, F.V., Garmire, E., Kost, A. and Ching-Mei Yang: Strong picosecond refractive index nonlinearity below the band gap in highly doped n-GaAs. Post-deadline paper, IQEC conference, Anaheim, California, 1990, Technical Digest, p. 359.

5. Karpushko, F.V., Bystrimovich, S.A., Goncharenko, A.M., Porukevich, S.A., Sinitsyn, G.V. and Utkin, I.A.: Direct measurement of the strong refractive nonlinearity of impurity origin in heavily-doped GaAs. J. Mod. Opt. **39** (1991) 1593–1598.

Optical Bistability in Quantum Well Semiconductor Devices

R. Kuszelewicz, B. G. Sfez, D. Pellat, and J. L. Oudar

C.N.E.T - Laboratoire de Bagneux, France Telecom, 196, avenue Henri Ravera, F-92220 Bagneux, France

1 Introduction

Epitaxial microcavity devices, such as surface emitting lasers [1,2], vertical cavity modulators [3,4], and all-optical bistable microresonators [5,6], are currently subject to a lot of interest for their potential application in photonic switching. Due to their small active volume and their optical access perpendicular to the surface, they can be integrated as two-dimensional arrays, with a device density of 10^6 devices/cm^2 or larger. Such arrays of all-optical bistable devices could be useful as opto-optical modulators, wavelength converters, optical logic gates, or fast access digital optical memories.

Since our first observations of optical bistability [7] in all-epitaxial étalons, much effort has been devoted to the improvement of the nonlinear characteristics and the decrease of bistability thresholds. This paper reports on the main results that have been obtained in order to improve the operating characteristics of such microresonators and to improve the density of elements that can be operated simultaneously. This has been achieved following three main directions. The first one consists in optimizing the optical cavity parameters resulting in considerably reduced power density thresholds. Then, we discuss the origin of this saturation relating this mechanism to the saturation of the band-tail absorption [9]. Finally, an additional decrease of the optical power density could be demonstrated [10] through the reduction of the lateral dimensions of individual pixels arranged in high density arrays. The technological process, involving spatially-resolved, thermally-induced alloy-mixing of the multi-quantum well active layer, is described and the non linear characteristics of these pixels is analysed.

2 Cavity Optimization

In order to keep the switching threshold of bistable talons as small as possible, it is well-known that high finesse cavities are desirable. This enhances the intra-cavity field at resonance, and minimizes the single-pass phase-shift required from the nonlinear medium. This can be achieved by using high reflectivity mirrors, and keeping absorption losses in the nonlinear medium at a level comparable to that of the mirror losses. In order to keep the optical response less sensitive to temperature changes and to angular misalignment or beam divergence, it is of a great importance to reduce the nonlinear medium thickness when increasing the mirror reflectivity. This has the additional advantage of keeping the overall thickness of the epitaxial structure at reasonable values.

A recent analysis[11] of cavity optimization, taking the saturation of nonlinear refractive index into account [12], shows that the overall epitaxial thickness goes through a minimum when the active layer thickness is reduced proportionally to the mirror transmission losses $\tau = 1 - R$. This minimum overall thickness lies in the vicinity of 5 μm when using $AlAs/Al_{0.1}Ga_{0.9}As$ mirrors centered at a wavelength $\lambda \sim 0.85$ μm. In addition it is found that a simultaneous optimization of threshold and contrast in the reflective mode can be achieved when the front- and back-mirror losses τ_F and τ_B are such that $\tau_F = 3\tau_B$. Finally, as these mirror transmission losses vary exponentially with the number of layer pairs, the appropriate ratio τ_F/τ_B is obtained for a specific value of the difference ΔN between the layer pair numbers of the two mirrors. For the above-mentioned mirror composition, one finds $\Delta N = 7$, taking into account the surface contribution to the front mirror reflectivity.

Based upon these considerations, the following sample ($n°$FB17) has been grown by molecular beam epitaxy on a [001] GaAs substrate. The structure consists of 23.5 periods of $AlAs/Ga_{0.9}Al_{0.1}As$ quarter-wave thick layers acting as a back mirror, a 18.5-period MQW active medium consisting of 10 nm-thick $Ga_{0.5}Al_{0.5}As$ barriers and 10 nm-thick GaAs wells, and 16 periods of the same $AlAs/Ga_{0.9}Al_{0.1}As$ alternate layers acting as a front mirror. At the wafer center its reflectivity spectrum exhibits a narrow resonance at 0.859 μm of $\delta\lambda = 0.34$ nm (FWHM) and a cavity finesse $\mathcal{F} = \Delta\lambda/\delta\lambda = 250$, significantly higher than the critical finesse needed for achieving bistability [13], estimated as 100. From this experimental finesse, we deduce that the back mirror reflectivity is larger than 99.5%.

The nonlinear response of this sample was studied using an Ar^+-pumped continuous-wave Ti-sapphire laser operated around 860 nm and modulated by an acousto-optic modulator. The experimental set-up is described in details in ref.8. The pulses typically have a full duration of less than 2 μs and a repetition rate of 10 kHz in order to avoid any heating of the sample. When tuning the laser on the high energy side of the Fabry-Perot resonance peak, optical bistability was observed due to the negative refractive index shift of electronic origin.

By adjusting lenses in the optical set-up, the spot size could be varied in the range of 4 to 30 μm. Bistability could only be observed for spot diameters larger than 5 μm. It is thought that for smaller spot sizes, the diffraction losses reduce the effective étalon finesse below the critical finesse, so bistability cannot be observed due to the saturation of nonlinear refractive index. For spot sizes in the range 5 to 9 μm, the minimum bistability threshold power did not vary significantly. Figure 1 shows the hysteresis loop observed at a wavelength of 858.26 nm, with a spot size of 6 μm and a triangular shape input pulse of 1 μs duration. The center of the bistability region is at 1.2 mW, with the two switching thresholds at ±6% of this value. A typical switching contrast of 10:1 was observed. True

bistability was confirmed by the observation of the memory effect for a time duration at least two orders of magnitude larger than the characteristic switching
times. For spot sizes larger than 9 μm, the minimum bistability threshold power
increased only moderately. With a spot diameter of 29 μm, it was measured as
4.1 mW, which corresponds to an incident power density of only 500 W/cm^2.
Much wider hysteresis loops could be observed than with spot sizes below 10 μm.
Lateral carrier diffusion cannot itself explain this fact [14]. One has to account
for diffraction effects since, in high finesse cavities, the minimum Fresnel number
at normal incidence, for a diffraction-free operation is proportional to the cavity
finesse [15].

However, when this effect is minimized, very low power density thresholds are
observed. With a laterally restricted microresonator, waveguiding would allow
to avoid this diffraction loss. Our measured threshold power density suggests
that a threshold power of 5 μW could be obtained for an effective device area of
1 μm^2.

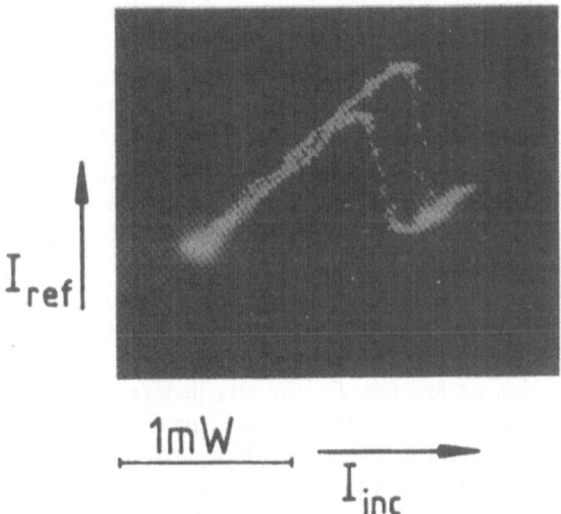

Fig. 1. Hysteresis loop observed at 858.26 nm with a 1 μs triangular input pulse

3 Origin of the Non Linear Index Saturation

The optical bistable properties observed in these structures are obtained using
the large excitonic-resonant nonlinear refractive index δn_{NL} of GaAs MQWs
at room temperature [16], which involves the photoexcitation of electron-hole
pairs, initially created in the low energy tail of the excitonic and band-to-band
optical transitions (Urbach's tail), and subsequently thermalized in higher energy
states through interaction with the lattice vibrations. In the course of our studies

of such bistable microcavities, it was found that, due to the short nonlinear medium length (typically 100 QWs with a 20 nm periodicity), optical bistability could only be observed in the highest finesse cavities [17], which suggested that the available nonlinear phase-shift was limited by some saturation mechanism. Recently we developped a technique for directly measuring the nonlinear index of semiconductors embedded in a Fabry-Perot cavity, and we indeed observed a clear saturation behavior of the nonlinear index change δn_{NL} as a function of intracavity power P [18]. The experimental data were found to fit very well with a saturation law:

$$\delta n_{\mathrm{NL}}(P) = \delta n_S \frac{P/P_S}{1 + P/P_S} \ . \tag{1}$$

This saturation behavior has been studied in a 10 nm/10 nm GaAs/AlGaAs MQW structure, as a function of the detuning from the excitonic resonance. The absorption dependence of the saturation parameters suggests a very simple physical interpretation of the origin of the observed saturation, in terms of the saturation of the Urbach's tail absorption.

The investigated sample ($N°$ 1896) consists of 14 periods of AlAs/$Al_{0.1}Ga_{0.9}$ As quarter-wave-thicklayers as the back mirror (reflectivity 97%), a 130-period GaAs/ $Al_{0.3}Ga_{0.7}$As MQW active layer with 10 nm/10 nm nominal thicknesses, and finally 7 periods of the same AlAs/$Al_{0.1}Ga_{0.9}$As alternate layers for the front mirror (reflectivity 92%), grown on a semi-insulating GaAs substrate. The reflectivity of the mirror is chosen in order to obtain simultaneously a high finesse (up to 50 at the lowest absorption) and a good contrast (typically 30 to 50, depending on the spot location on the sample). With this sample we could observe a large hysteresis loop and a low bistability threshold (typically 2.8 mW for the incident power).

In order to investigate the absorption dependence of the saturation parameters, the energy difference between the Fabry-Perot resonance and the active layer band-gap was modified by tuning the sample temperature between 13°C and 40°C. On Fig. 2 the saturation parameters P_S and δn_S deduced from this fit have been plotted as a function of the linear absorption. This shows that the saturating power is inversely proportional to the linear absorption whereas δn_S remains practically constant (within our experimental accuracy).

Taking into account the values of spot size and diffusion length relevant to our experiment, we calculate [19] an effective area for the carrier distribution $S_{\mathrm{eff}} = 190 \ \mu\mathrm{m}^2$, so an estimate for the plane-wave saturation intensity I_S is $\alpha_0 I_S = 1.8 \times 10^7 \ \mathrm{W/cm}^3$.

Several mechanisms can contribute to a departure from a linear relationship between δn_{NL} and the intensity I. They can be discussed by considering that in our conditions the nonlinear index is dominated by population effects, so that in thermal equilibrium and at a given wavelength, δn_{NL} is a function of the electron-hole pair density N only. In steady-state, N is itself a function of I, so that one can write:

$$\delta n_{\mathrm{NL}}(I) = \delta n_{\mathrm{NL}} \{N(I)\} \tag{2}$$

One possible limiting mechanism for $\delta n_{\mathrm{NL}}(N)$ would be the saturation of the excitonic contribution, which has been discussed by a number of authors [20].

However, these limiting values of $\delta n_{\mathrm{NL}}^{\mathrm{excitonic}}$ are typically 10^{-3}, much below our measured value of δn_S. It is now recognized that optical bistability needs the additional contribution of band filling effects. This contribution to $\delta n_{\mathrm{NL}}(N)$ may have a sublinear dependence for the following reason [21]: as the lower energy levels become completely populated, additional carriers can only occupy higher energy levels, which are less resonant with the incoming light and contribute to a lesser extent to the nonlinear index change. However this effect should not lead to a limiting value of δn_{NL} but only to a sublinear dependence.

Therefore the limiting behavior of $\delta n_{\mathrm{NL}}(I)$ can only come from the $N(I)$ relationship. In steady-state this dependence can be expressed by the following implicit equation:

$$N = \frac{\alpha(N)\tau(N)}{\hbar\omega}I \, , \tag{3}$$

where the carrier density dependence of the absorption α and the carrier lifetime τ are written explicitly, and $\hbar\omega$ is the photon energy. The only possibility for the solution of (3) to display an asymptotic behavior like the right-hand side of (1) is that either $\tau(N)$ or $\alpha(N)$ vanishes for a finite value of N. Although τ is known to decrease with N due to bimolecular or Auger recombination, it does not vanish at a finite value of N.

We are thus led to the conclusion that the observed saturation behavior comes from the vanishing of the absorption at a finite carrier density. In the simplest case of a two-level system, the absorption could be written as:

$$\alpha(N) = \alpha_0 \left(1 - \frac{N}{N_t}\right) \, , \tag{4}$$

where N_t is the carrier density at transparency, i.e. when gain starts to occur at the wavelength of interest.

If one ignores the $\tau(N)$ dependence, (3) and (4) give the following relation between the carrier density and the intensity:

$$\frac{N}{N_t} = \frac{I/I_S}{1 + I/I_S} \, , \tag{5}$$

with $I_S = \hbar\omega N_t/\alpha_0\tau$. Although (4) is only an approximation in the case of semiconductors, it predicts correctly the asymptotic behavior of (1). We note that it also predicts that I_S is inversely proportional to the linear absorption coefficient, in agreement with our experimental results.

With the present data, (5) leads to $N_t = 0.8 \times 10^{18}$ cm^{-3}. In 2D units this corresponds to $N_t^{\mathrm{2D}} = 1.5 \times 10^{12}$ cm^{-2} per quantum well. The typical nonlinear index per electron-hole pair is $\delta n_S/N_t^{\mathrm{2D}} = 1.6 \times 10^{-14}$ cm^2. These two numbers are in good agreement with calculations using many-body theory [22,23]. Therefore, we believe that the vanishing of α at a finite carrier density N_t is the right mechanism to explain the nonlinear index saturation observed in our experiments. This parameter is also of crucial interest in the field of semiconductor lasers, as gain starts to occur for $N > N_t$. We note that the present experiment also provides a new method for the systematic measurement of N_t in semiconductors.

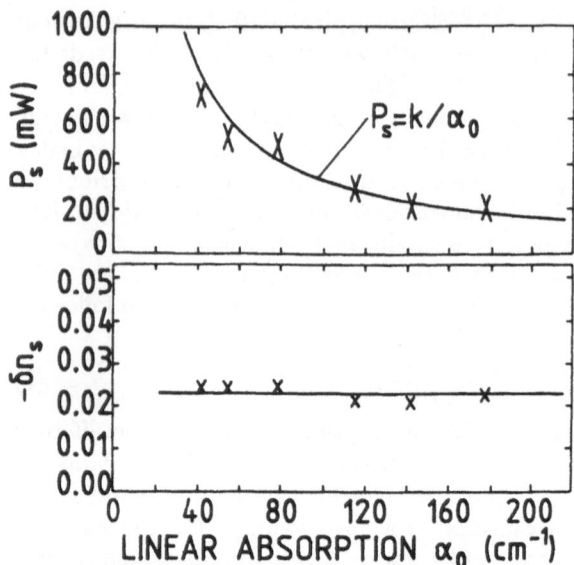

Fig. 2. Top curve: plot of the saturation power P_S as a function of the linear absorption coefficient α_0. The crosses are the fitted parameters used in Fig. 1, and the plain line is a fit showing that P_S is inversely proportional to the linear absorption coefficient. Bottom curve: plot of the maximum nonlinear δn_S as a function of α_0. δn_S remains constant within experimental accuracy

4 Pixellation Through Alloy-Mixing Processes

The operating characteristics of all these devices greatly improve with lateral confinement of both light and carriers. Thus several etching techniques have been developped in order to pixellate the original device into arrays of micro-components. However these techniques induce non-radiative recombination centers which ultimately limit the device efficiency to a time scale not larger than a few picoseconds. Passivation [24] as well as epitaxial regrowth have been tried to contain this problem. However passivated microresonators retrieve their original properties for only a few hours due to chemical interactions between the passivation agent and the water coming from the air [25]. Epitaxial regrowth is a difficult process due to the nonplanar geometry of the etched samples and requires several technological steps although it is of a great potential for achieving efficient optical and electrical confinement. An effective pixellation of nonlinear microresonators using alloy-mixing techniques, has been successfully achieved, which leads to the improvement of the nonlinear properties of the device, even in the continuous regime.

Pixellation of nonlinear resonators has the two-fold objective of confining carriers within a diffusion length radius [26], and guiding light so as to act against the effects of diffraction. Therefore a minimum barrier energy of typically 25 meV is required together with an index step greater than 10^{-2}. The alloy-mixing process modifies the structure of the multiple quantum well active region to create an equivalent alloy with different electronic and optical properties. It is accomplished by a local implantation of isoelectronic ions in the near-surface region of the sample. The disordering of the quantum well region is obtained by the propagation of defects through the structure [27], which leads simultaneously to an increase of the band-gap and a decrease of the refractive index.

After the growth of the back mirror (14 pairs of $AlAs/Al_{0.1}Ga_{0.9}As$ with a nominal width of 73.8 nm/59 nm) and of the active layer (130 pairs of $GaAs/Al_{0.3}Ga_{0.7}As$ with a nominal width of 10 nm/10 nm) by Metal-Organic Chemical Vapor Deposition, the sample is masked (resin mask) in order to protect arrays of square microresonators of various sizes ranging from 50 μm × 50 μm to 2 μm × 2 μm. It is then implanted with Sb^+ ions (energy 350 keV, dose 5×10^{14}). After thermal annealing (4 hours, 850°C), a SiO_2/Si front mirror is evaporated at the surface of the sample in order to complete the Fabry-Perot structure. Secondary Ion Mass Spectroscopy profiles of the Aluminium concentration showed a mixing efficiency of roughly 40%, which was uniform over the whole active region (2.6 μm). This alloy-mixing led to a luminescence peak blue-shift of 110 meV, which was uniform over the whole structure.

The linear and nonlinear behavior of the new structure have been characterized in a manner similar to the one used in experiments reported up above, but the modulator rather forms 1 μs long triangular pulses at a rate of 10 kHz. The index step δn created by the alloy-mixing process was evaluated by illuminating the pixellated region at a wavelength around the resonance peak of the device i.e. $\lambda = 840$ nm. Since the refractive index inside the microresonator differs from outside (where interdiffusion took place), the Fabry-Perot resonance occurs at a different wavelength. The difference $\delta\lambda$ between these wavelengths could be related to the index change δn through $\delta n \sim n\delta\lambda/\lambda$, where n is the group velocity index. The index change was found to be -1.2×10^{-2}.

Using the simultaneous visualization of the microresonators arrays (background illumination) and of the operating laser beam spot, it has been possible to analyze the nonlinearities of individual microresonators. Figure 3(a) shows the partial image of a 25 × 25 array of 5 μm × 5 μm microresonators. The microresonators appear to be brighter than the background. It is because the laser wavelength is chosen to be at the resonance of the alloy-mixed region. The laser is focused onto one of the pixels. Figure 3(b) shows the response of this particular pixel. A hysteresis loop is obtained in the quasi-continuous regime, showing the nonlinear properties of the device.

Actually, these properties are greatly improved, as has been demonstrated by the comparison between microresonators of different sizes ranging from 20 to 5 μm -size. The comparison of the slopes at the origin, between the microresonators response and the off-resonance response shows that the linear reflectivity has not been modified by the process. It shows also that the effect of the nonlinearity increases significantly when the microresonator size decreases. It is thus possible to observe the hysteresis behavior for the 5 μm square microresonators whereas it is not observable for the 20 μm square microresonators.

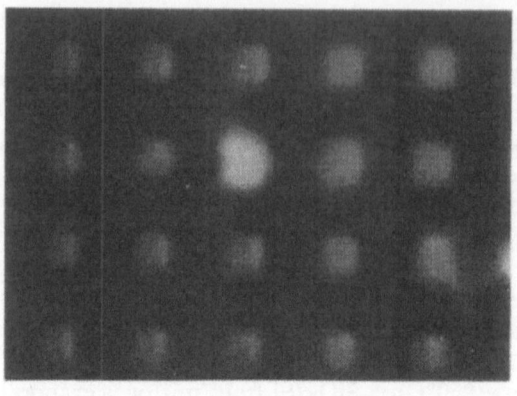

(a)

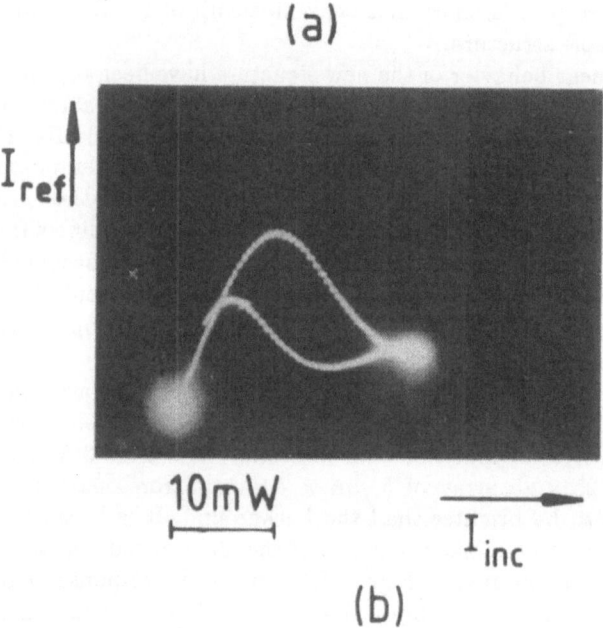

(b)

Fig. 3 (a) Partial micro-cartography of a 25×25 array of 5 μm $\times$ 5 μm microresonators using the set-up of Fig. 1. The laser spot is located on a microresonator. (b) Hysteresis loop obtained on this particular microresonator

The pixellation process has thus improved the characteristics of the device. However clear bistability was not observed in this sample. We attribute this to the slight absorption in the Si layer of the front mirror, which can be avoided by changing the composition of this layer. Moreover a larger nonlinear enhancement is expected for smaller microresonator sizes. Different runs demonstrated similar results, which evidenced the reproducibility of this technique. The good definition of the 5 μm microresonators shows that a higher resolution is possible using this alloy-mixing technique.

5 Conclusion

Various solutions have been proposed in order to reduce the individual power consumption of integrated microresonators, and optimize their non linear response. In particular, at least one order of magnitude is expected to be gained in the threshold intensity by simply optimizing the structure geometry. Up to now power density thresholds of 500 W/cm^2 have actually been observed. The experimental analysis of the mechanism of the non linear saturation has shown that it originates from the saturation of the band-tail absorption, leading to refined criteria in the structure design. Finally, we have demonstrated the possibility of fabricating nonlinear microresonator arrays whose CW nonlinear characteristics improve after pixellation. The alloy-mixing technique which has been used allows a good confinement of both carriers and light.

Acknowledgments

The results presented here have been obtained in close collaboration with R. Azoulay, J. C. Michel, Y. Nissim, E. V. K. Rao, from Centre National d'Etudes des Télécommunications and with R. Planel from CNRS/Laboratoire de Microélectronique et Microstructures (L2M).

References

1. Gourley, P. L., Drummond, T. J.: Appl. Phys. Lett. **50** (1987) 1225
2. Jewell, J. L., Scherer, A., McCall, S. L., Harbison, J. P., Florez, L. T.: Electron. Lett. **25** (1989) 1123.
3. Simes, R. J., Yan, R. H., Geels, R., Coldren, L. A., English, J. H., Gossard, A. C.: Appl. Opt. **27** (1988) 2103.
4. Whitehead, M., Parry, G.: Electron. Lett. **25** (1989) 566.
5. Sahlen, O., Olin, U., Masseboeuf, E., Landgren, G., Rask, M.: Appl. Phys. Lett. **50** (1987) 1559.
6. Acklin, B., Bagnoud, C., Dupertuis, M. A., Martin, D., Morier-Genoud, F.: O.S.A. Proceedings on Photonic Switching, Salt Lake City, Utah, 1991.
7. Kuszelewicz, R., Oudar, J. R., Michel, J. C., Azoulay, R.: Appl. Phys. Lett. **50** (1988) 1559.
8. Oudar, J. L., Kuszelewicz, R., Michel, J. C.: submitted to Electron. Lett.
9. Sfez, B. G., Oudar, J. L., Kuszelewicz, R., Pellat, D.: Appl. Phys. Lett. **60** (1991) 1163.
10. Sfez, B. G., Rao, E. V. K., Nissim, Y. I., Oudar, J. L.: Appl. Phys. Lett. **60** (1992) 5.
11. Oudar, J. L., Kuszelewicz, R., Sfez, B. G., Michel, J. C., Planel, R.: Opt. and Quantum Electron. **24** (1992) 5193.

12. Sfez, B. G., Oudar, J. L., Michel, J. C., Kuszelewicz, R., Azoulay, R.: Appl. Phys. Lett. **57** (1990) 324.
13. Oudar, J. L., Sfez, B. G., Kuszelewicz, R., Michel, J. C., Azoulay, R.: Phys. Stat. Sol. (b) **159** (1990) 181.
14. Sfez, B. G., Padjen, R., Oudar, J. L.: QELS'91, paper QTuI34, Technical Digest Series, 1991, 11, p. 96 (Optical Society of America, Washington, DC).
15. Olin, U.: J. Opt. Soc. Am. B **7** (1990) 35.
16. Special issue on Excitonic Optical Nonlinearities, J. Opt. Soc Am. B **2** (1985) 7.
17. Oudar, J. L., Sfez, B. G., Kuszelewicz, R., Michel, J. C., Azoulay, R.: Phys. Stat. Sol.(b) **159** (1990) 181.
18. Sfez, B. G., Kuszelewicz, R., Oudar, J. L.: Opt. Lett. **16** (1991) 11.
19. Weare, D., O'Carroll, C., Wickham, C.: Europhysics Lett. **8** (1989) 25.
20. Miller, A., Manning, R. J., Milsom, P. K., Hutchings, D. C., Crust, D. W., Woolbridge, K.: J. Opt. Soc. Am B **6** (1989) 567, and references therein.
21. Poole, C. D., Garmire, E.: Opt. Lett. **8** (1984) 356.
22. Haug, H., Koch, S. W.: Phys. Rev. A **39** (1989) 1887.
23. Bava, G. P., Debernardi, P.: Electron. Lett. (to be published). From the absorption spectra of Fig. 4 at Eg - 35meV, the transparency is achieved for $N_t = 1.0 \times 10^{12}$ cm^{-2}.
24. Yablonovitch, E., Sandroff, C. J., Bhat, R., Gmitter, T.: Appl. Phys. Lett. **51** (1987) 439.
25. Oudar, J. L., Sfez, B. G., Kuszelewicz, R., Michel, J. C., Azoulay, R.: Phys. Stat. Sol. (b) **159** (1990) 181.
26. Sfez, B. G., Padjen, R., Oudar, J. L.: Quantum Electronics and Laser Science Conference 91, paper QTuI34
27. Rao, E. V. K.: Proc. SPIE **866** (1987) 24, and references cited therein.

All Optical Bistability in a Type II Heterostructure

R. Teissier, R. Planel, and F. Mollot

Laboratoire de Microstructures et Microélectronique,
196, av. Henri-Ravera, F-92220 Bagneux, France

In "type II" heterostructures, potential profiles of conduction and valence bands are such that electrons are separated from holes, in real space. As a consequence, illumination induces charge separation and thus internal electric fields, which, in turn, modify the properties of the structure. On the other hand, it is now well established that such type II heterostructures may be obtained from various associations of GaAs, GaAlAs, and AlAs [1,2,3]. This happens as far as X-point conduction states are concerned. As a consequence, new tools for "band gap engineering" are available in this well controlled family of material; for instance, an indirect-to-direct transition has been observed in a GaAs/AlAs superlattice under external electric field [4], with consequences on the photoluminescence intensity. As a further example, in a more sophisticated and asymmetric structure, spatial separation of charges has been demonstrated to create an internal electric field with subsequent action on the photoluminescence emission energy of the structure [5]. More recently, Zrenner and coworkers [6] have observed an intrinsic optical bistability under external electric field, thought to be due to hot carriers.

We describe here a structure which exhibits intrinsic all-optical bistability in photoluminescence, with huge variation of intensity and spectrum. It is observable without any external field, in as-grown samples. However, electrical measurements and switchings are also possible. On the basis of luminescence spectrum analysis, and of electrical measurements, we propose an explanation different from the one of Ref. [6]. Each state corresponds to a different quasi static charge repartition within the heart of the structure; and switching as well as bistability are explained in terms of charge accumulation and associated electric fields.

The structure, which was grown by Molecular Beam Epitaxy, consists of a n-i-n sequence, in which the intrinsic region is made of four layers, as described and labelled in Fig. 1. It may be viewed as one type-II "unit cell", embedded into a $Ga_{0.65}Al_{0.35}As$ barrier for Γ- or X-electrons as well as for holes. In Fig. 1, are plotted the effective potential profiles pertinent for these three sorts of electronic states, as well as relevant energies, due to confinement. These quantities have been obtained using a transfer matrix calculation, based on the Envelope Function Approximation [7,8]. This type of calculation may account for electric fields which are present in the structure; thus, through the self-consistent coupling with Poisson equation, we get a reasonable quantitative description of the carrier population effects on the electronic levels of the structure. Although it cannot be presented and discussed in details here, this calculation was a great help for discussion of the experimental results. According to this calculation, the central B- and C- layers (respectively 2.5 nm thick GaAs and 10 nm thick AlAs) consist in a type- II structure, at least in moderate population conditions: more precisely, the hole ground states are of G-symmetry, confined in the GaAs layer; whereas the electron ground state is of $X_{x,y}$ symmetry [9], and is confined in the AlAs layer (the X_z-symmetry level is situated some 10 meV above). It is of interest to recall that, in GaAs/AlAs heterostructures, typical type-II recombination time has been found to lie in the 10^{-4} s range for indirect structures ($X_{x,y}$ conduction state), and in the 10^{-6} s range for pseudo-direct structures (Xz conduction states). This is to be compared with direct and type I recombination time, in the 10^{-9} s range. The calculation also indicates that we may obtain a 6 meV blue-shift of the X-HH energy, due to the photocreated electrostatic potential, for both electron and hole densities equal to 10^{11} cm^{-2}. Another consequence of the optical excitation is the filling of the two-dimensional state densities (3.8×10^{11} cm^{-2}/meV for $X_{x,y}$ electrons and 0.39×10^{11} cm^{-2}/meV for holes), which leads to consider the two associated pseudo Fermi levels (PFL).

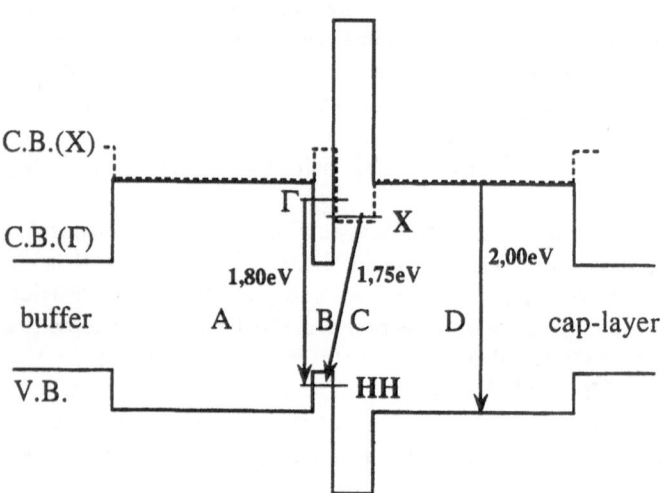

Fig. 1. The n-i-n structure is made of the following sequence: substrate and buffer $n+$, GaAs; A, Ga$_{0.65}$Al$_{0.35}$As, 100 nm; B, GaAs, 2.5 nm; C, AlAs, 10nm; D, Ga$_{0.65}$Al$_{0.35}$As, 100nm; cap-layer $n+$, GaAs, 100nm. The potential profiles for Γ-electrons and holes (solid line) and X-electrons (dashed line) are sketched, as well as calculated energy levels. The resulting transition energies are indicated

We have studied the photoluminescence of this sample at low temperature (5 K), under an all-line Ar-laser excitation. (Excitation spectra of PL have been performed to confirm the assignment of main lines, but will not be detailed here). The exciting beam was focused over 100 μm on the sample surface. The exciting power P_{ex} used ranged from 10^{-4} to 10^3 W/cm^2. Electrical measurements under optical excitation have been made, with simultaneous recording of the PL. For that purpose, 300 μm $\times$ 300 μm mesa-type diodes were fabricated by conventionnal lithography and etching; a peripheral ohmic contact was deposited, made by Au-Ge-Ni-Au deposition and alloying.

The PL consists of two main lines (see Fig. 2): one (at about 1.80 eV) is attributed to recombination between the B- and C-layers, the other (at about 2.00 eV) to recombination in the Ga$_{0.65}$Al$_{0.35}$As layers. Two different stable states are found in the 40 to 250 W/cm^2 P_{ex} range, as shown in Fig. 2. The most prominent distinction is found on the high energy line of the spectrum. In the low power state, labelled hereafter as the OFF state, the barrier recombination is weak. In the high power state, the ON state, it is some 50 times more intense, and is broadened towards lower energies. The low energy line does not shift, but is more intense in the ON state typically by a factor of two.

Excitation spectroscopy of luminescence, as well as theoretical estimations of the transition energies, allow us to identify the luminescence lines of Fig.2. The low energy part of the spectra is attributed to carrier recombination in the central structure (B- and C-layers); and the high energy part is attributed to recombination in the Ga$_{0.65}$Al$_{0.35}$As barrier. For very weak excitation ($P_{ex} < 1$ W/cm^2), one observes the spectrum caracteristic for a type II indirect recombination, involving electrons with $X_{x,y}$ symmetry: a weak zero-phonon recombination (1740 meV), associated to three more intense phonon replica at lower energies. For $P_{ex} \sim 1$ W/cm^2, two other lines appear, associated with the Γ-HH direct transition of the GaAs quantum well (1820 meV) and with recombination in the Ga$_{0.65}$Al$_{0.35}$As barrier (2012 meV). For $P_{ex} > 5$ W/cm^2, we observe the luminescence associated to the population of the X_z electronic level. As P_{ex} reaches, typically, 100 W/cm^2, the type II luminescence merges into the Γ-HH line and the total intensity of radiative recombination from the central cell of the structure increases rapidly. The theoretical description presented above allows us to associate these shifts with charge densities present in the layers. Under both actions of photocreated electrostatic potential and level filling, the X PFL in C-layer is lifted up to the Γ level in B-layer. The saturation appears, for P_{ex} around 100 W/cm^2, when the Γ level is populated, allowing fast Γ-HH recombination. We estimate the required densities at 7×10^{11} cm^{-2}. Let us point out that this blue shift is characteristic for a type II transition.

An hysteresis loop is also found on the photocurrent, which is typically smaller by 20% in the ON state. But the key features of mesa type samples are i) that the PL of open-circuited diodes resembles the ON state PL for all P_{ex}, even low; ii) and that the photo-tension (in open circuit) is positive on the cap layer, which indicates a significant hole accumulation also in the D-layer. As a consequence, the bistability is not found in open circuit diodes, but is restored when a load resistor R_{ext} smaller than 10 kOhm is inserted in the external circuit. Increasing R_{ext} leads to a diminution of both low and high commutation thresholds down to, respectively, 1.0 and 1.2 W/cm^2 for $R_{ext} = 10$ kOhm. In short-circuited

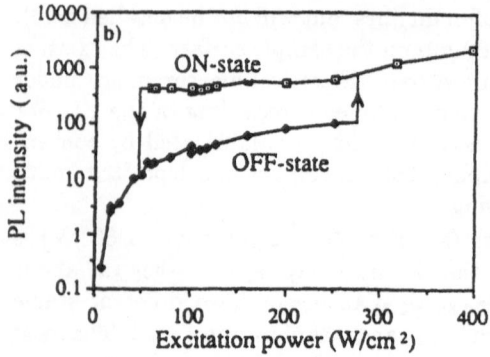

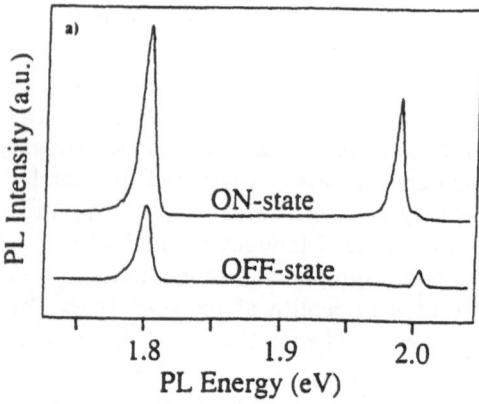

Fig. 2. a) PL OFF- and ON-spectra of as-grown sample for $P_{ex} = 100$ W/cm^2. In the 1840–1980 meV range, no PL is observed. b) Hysteresis loop versus P_{ex} of integrated PL in the 1950 to 2050 meV range

diodes, the same behaviour as in as-grown samples is found (which illustrates the well established feature that as-grown standard MBE structures are likely to be short circuited in the direction perpendicular to the layer, by oval defects and similar imperfections).

In view of these results, we are led to the following proposal: The OFF and ON states correspond respectively to weak or strong accumulation in the D-layer, most probably in the triangular attractive potential created by the accumulation of electrons in the C-layer; this triangular well may be compared to those appearing in modulation-doped heterojunctions, and explains the low energy broadening of the barrier luminescence in the ON state.

To explain the open-circuit behaviour of the structure we have to point out that electrons can be collected in the C-layer from both sides of the structure, whereas holes photo-created in the D-layer cannot be collected in the B-layer because of the C-AlAs barrier in the valence band. It gives rise to an excess hole population in the D- or cap-layer, and then to a positive voltage. The high PL intensity at energy close to 2 eV allows us to think that hole accumulate in the D-layer where they recombine with continuously created electrons. In the A-layer, both electrons and holes can move freely, then any electric field in this layer would be screened as far as charge accumulation is possible at its extremities. Figure 3a shows the resulting potential profile.

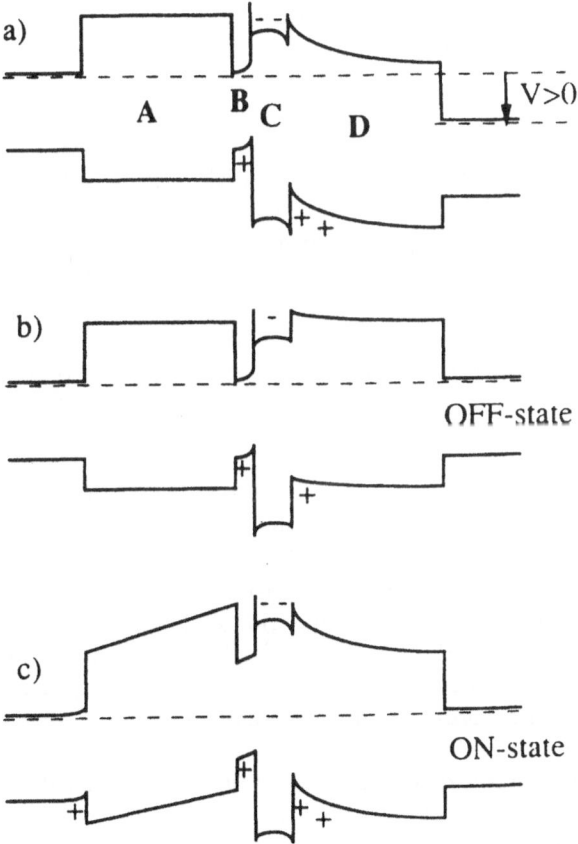

Fig. 3. a) Band profile in an open-circuited diode. Estimated charge densities are 7×10^{11} cm^{-2} in the B-layer, 3.7×10^{12} cm^{-2} in the C-layer, and 3×10^{12} cm^{-2} in the D-layer. b) Band profile in a short-circuited diode in the OFF-state. Estimated charge densities are 7×10^{11} cm^{-2} in the B-layer, 1.1×10^{12} cm^{-2} in the C-layer, and 4×10^{11} cm^{-2} in the D-layer. c) Band profile in a short-circuited diode in the ON-state. Estimated charge densities are 2×10^{11} cm^{-2} in the buffer, 5×10^{11} cm^{-2} in the B-layer, 3.7×10^{12} cm^{-2} in the C-layer, and 3×10^{12} cm^{-2} in the D-layer

We now discuss a possible mechanism for switching between the two states and for bistability, which rely on the distribution of charges and electric fields in the heart of the structure. At low P_{ex}, when the diode is short-circuited, D-holes in excess are allowed to flow towards the cap layer, generating a positive photocurrent. As long as the electric field in the A-layer is nul, the equalization of the Fermi level on each electrode leads to a reduction of the attractive triangular potential at the C-D interface. In other words, as charges accumulate, the two potential drops at B-C and C-D interfaces are kept opposite (see Fig. 3b). Then the D-hole population remains weak and does not lead to a high recombination; it corresponds to the OFF state.

When P_{ex} increases, the augmentation of the charge densities in the type II well pushes the X PFL up to the Γ level in the GaAs layer, then the possibility of Γ-HH recombinations drastically reduces the lifetime of B-holes. It means that the B-C electric potential difference cannot exceed the initial Γ-X separation (about 60 meV under flat band conditions). That leads to the emergence of an electric field in the A-layer to hold equal potentials in the two electrodes, when the D-holes density increases. It is essential, at this stage, to consider that this electric field does raise the electron X-state energy with respect to the Γ-state, which further reduces the lifetime of B-holes, thus providing a positive feed-back. Because of the large thickness of the A-layer (100 nm), a much larger potential difference may be created between the buffer electrode and the C-layer with the same charge quantity. As a consequence, the limit (due to the equalization of the potential drop with the B-C dipole) to the accumulation of holes at the C-D interface disappears; it will reach an intrinsic limitation related to the geometry of the C- and D- layers (experimentally: a potential drop of about 300 mV): it characterizes the ON-state (Fig. 3c). This abrupt change in the charge distribution and potential profile of the structure is detected by two PL features: i) a steep increase of the low energy PL, which indicates that holes from the A-barrier are collected in the B-layer with a better efficiency, typically twice higher (but this does not imply an increase of the B-layer population, due to the correlative decrease of their radiative lifetime) ; ii) a huge increase of the $Ga_{0.65}Al_{0.35}As$ barrier PL, due to the high accumulation of holes at the C-D interface. This recombination occurs with electrons in the barrier, which are not so numerous in this region; thus the lifetime of the whole D-hole population should remain much longer than an ordinary direct recombination time. In the same way, positive charges in excess at the buffer-A interface have a longer lifetime than B-holes. Sufficient charge densities to maintain a positive feedback (and to hold the structure in the ON-state) can then remain even for lower P_{ex}: essentially, this is due to the longer lifetime of the whole charge population in the ON state than in the OFF state. This explains the hysteresis loop.

We are now able to understand qualitatively the effect of non-zero load resistor. In the OFF-state, the potential drop in the external resistor allows a larger D-hole population for a given P_{ex}. As a consequence, the C-electron population will also be larger; the transition will then occur at lower P_{ex}. Furthermore, the contrast between OFF-state and ON-state will be blurred out. That is what we experimentally observe. If the resistor is so large that the D-holes population in the OFF-state reaches the saturation value of the ON-state for a lower P_{ex} than the threshold, we have no more bistability. It occurs for a load resistor of a few 10 kOhms.

To conclude, we have observed an all-optical bistability between two different charge distributions in the structure. In the OFF-state, we think that most charge accumulation occurs in the heart of the structure, while in the ON-state holes accumulate at the extremities of the structure. This charge transfer bistability is, in this actual structure, rather slow (we measured 10 ns for OFF/ON commutation and about 1 μs for ON/OFF commutation). But a key feature is that it needs no external polarization and then no electrical addressing. Moreover, threshold values are at least two orders lower than in current devices based on more classical optical non- linearities [10], and an important point is that the structure parameters and experimental conditions are not critical.

References

1. Finkman, E., Sturge, M. D., Tamargo, M. C.: Appl. Phys. Lett. **49** (1986) 1299.
2. Danan, G., Etienne, B., Mollot, F., Planel, R., Jean-Louis, A. M., Alexandre, F., Jusserand, B., Le Roux, G., Marzin, J. Y., Savary, H., Sermage, B.: Phys. Rev. B **35** (1987) 6207.
3. Moore, K. J., Dawson, P., Foxon, C. T.: J. Phys. (Paris) **C5** (1987) 525; *ibid.* Phys. Rev. B **38** (1988) 3368.
4. Meynadier, M. H., Nahory, R. E., Worlock, J. M., Tamargo, M. C., de Miguel, J. L., Sturge, M. D.: Phys. Rev. Lett. **60** (1988) 1338.
5. Jezewski, M., Teissier, R., Mollot, F., Planel, R.: Superlattices and Microstructures **8** (1990) 329.
6. Zrenner, A., Worlock, J. M., Florez, L. T., Harbison, J. P.: Appl. Phys. Lett. **56** (1990) 1763.
7. Bastard, G.: Phys. Rev. B **24** (1981) 5693; Bastard, G.: In Wave mechanics applied to semiconductor heterostructures, Les éditions de physique, Les Ulis 91-France (1988).
8. Yuh, P.-F., Wang, K. L.: Phys. Rev. B **38** (1988) 13307.
9. Scalbert, D., Cernogora, J., Benoit à la Guillaume, C., Maaref, M., Charfi, F. F., Planel, R.: Solid State Commun. **70** (1989) 945.
10. Sfez, B. G., Oudar, J. L., Michel, J. C., Kuszelewicz, R., Azoulay, R.: Appl. Phys. Lett. **57** (1990) 324.

Surface Emitting Laser Diodes and Wavelength Break Selective Photodetectors

T. Wipiejewski, K. Panzlaff, and K. J. Ebeling

University of Ulm, Department of Optoelectronics,
Oberer Eselsberg, W-7900 Ulm, Germany

The structure and fabrication of a vertical cavity surface emitting laser diode is described. The lowest threshold currents are 16mA. The light emission is single longitudinal mode at 915 nm wavelength. Under pulsed excitation the maximum output power is 3.5 mW. A wavelength selective photodetector of related structure shows a quantum efficiency of 10% at resonance with 1.5nm spectral width at half maximum.

1 Introduction

Vertical cavity surface emitting laser diodes (VCSELDs) are attractive light sources for optical interconnects and optical signal processing [1,2]. The vertical structure allows two-dimensional array formation, on wafer chip testing and efficient light coupling into single mode optical fibers [3]. VCSELDs exhibit low threshold currents [4,5] and show single longitudinal mode emission under high speed current modulation. Structures similar to VCSELDs can be applied as highly resonant wavelength selective photodiodes [6,7].

2 Device Structure

We have fabricated VCSELDs with AlGaAs-AlAs Bragg reflectors and strained InGaAs MQW active layers on semi-insulating GaAs substrates. Growth is done by molecular beam epitaxy (MBE). Figure 1 shows a microscope photograph of a slanted lapped wafer after MBE growth. The $Al_{0.38}Ga_{0.62}As$ buffer layer on top of the substrate is 278 nm thick. The following bottom mirror consists of 29.5 pairs $Al_{0.12}Ga_{0.88}As/AlAs$ layers and is p-doped with $N_A = 5 \cdot 10^{18} cm^{-3}$, the top mirror consists of 20 pairs and is n-doped with $N_D = 5 \cdot 10^{18} cm^{-3}$. The etalon region between the dielectric reflectors contains a 140 nm thick $Al_{0.38}Ga_{0.62}As$ top spacer layer, a 1144 nm thick $Al_{0.38}Ga_{0.62}As$ bottom spacer layer and the active zone with four pairs of 10 nm thick $In_{0.15}Ga_{0.85}As$ quantum wells embedded in GaAs barriers.

The quantum wells of each pair are separated by 10 nm thick GaAs barriers, the distance between the pairs is 100 nm. The total thickness of the active region is 650 nm and thus equal to five half wavelengths of the lightwave in

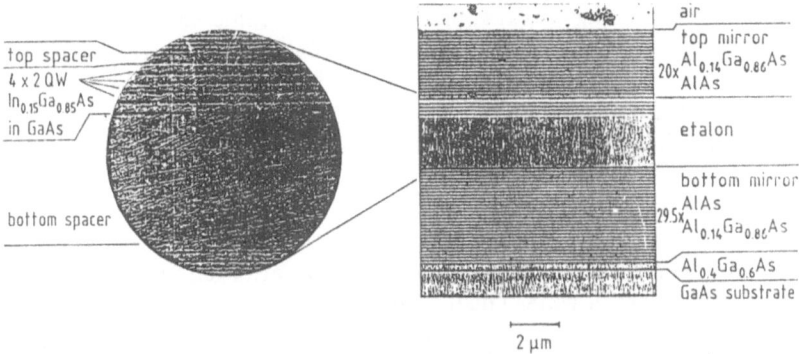

Fig. 1. Microscope photograph of a slanted lapped VCSEL wafer after MBE growth

the material. The quantum well pairs are placed in the inner four maxima of the electric field in the standing wave pattern. With this periodic arrangement of the active quantum wells it is possible to lower the threshold currents by a factor of two compared with an equidistant arrangement. The active region and the adjacent 70 nm of the spacer layers are nominally undoped, the outer parts of the spacer layers have the same doping level as the neighbouring dielectric reflectors.

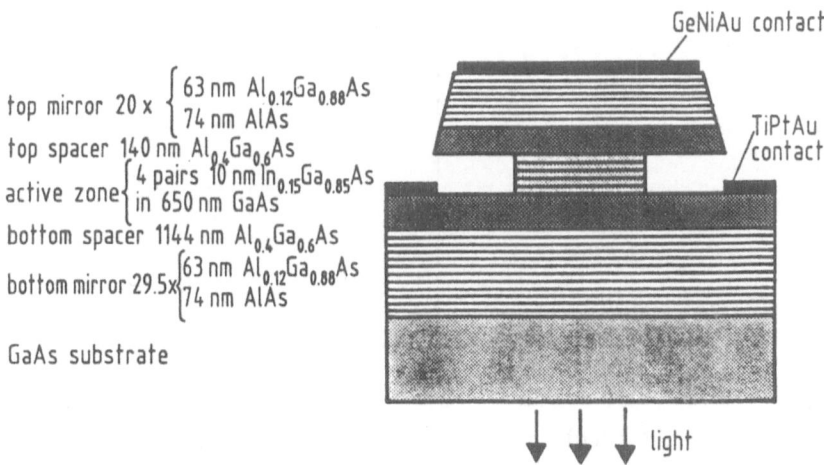

Fig. 2. Cross-sectional view of the VCSELD with etched mesa and metallic contacts

Microcavities are fabricated for current injection. Figure 2 shows a cross-sectional view of the VCSELD device. Mesa formation is done by a depth controlled wet chemical etching process [8]. The active volume containing the In-GaAs quantum wells and the GaAs barriers is defined by a material selective etching process using an $NH_4OH:H_2O_2:H_2O$ solution. Ohmic contacts of alloyed GeNiAu are deposited on top of the mesa and TiPtAu p-contacts are evaporated on the bottom spacer layer. Light output is through the GaAs substrate.

3 Results and Discussion

Above a threshold current of 16 mA lasing operation is achieved for a device with 12 μm diameter of the active zone. The light output versus current characteristics are shown in Fig. 3. Light emission is linearly polarized above threshold. The kink in the light output power at higher current levels is due to lasing of a higher transversal mode.

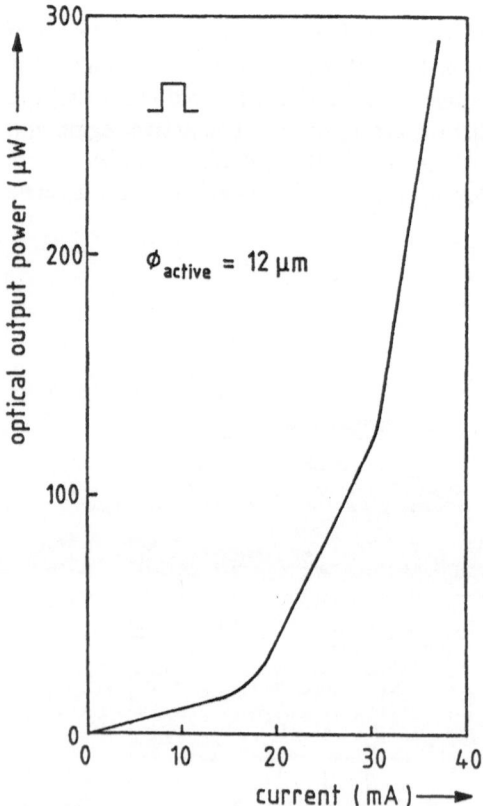

Fig. 3. Light output power as a function of injected current for a VCSEL diode with a diameter of the active volume of 12 μm

Above threshold lasing occurs in a single longitudinal mode at 915 nm wave-length as shown in the upper spectrum of Fig. 4. Below threshold the spectrum essentially contains two longitudinal modes at a wavelength separation of about 27 nm determined by the effective resonator length of ca. 5 μm.

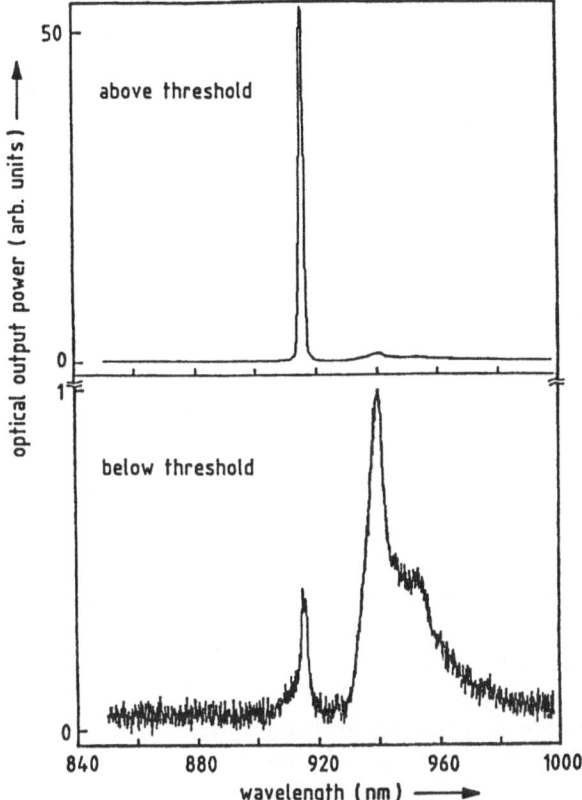

Fig. 4. Spectra of a VCSELdiode above and below lasing threshold

Figure 5 shows the light output versus current characteristics for VCSELDs of different diameters of the active area.

The threshold currents and the amount of spontanous emission increase with increasing mesa diameter. By taking the area of the active region into account it is seen that the threshold current densities increase with decreasing diameter of the active volume. This effect is mainly attributed to the relatively larger portion of surface recombination in devices with smaller active area.

The electrical resistances at lasing threshold range from 185 Ω for 24 μm diameter devices to 290 Ω for 12 μm devices. These fairly high resistances cause excessive heating of the device and finally prevent the laser diode from being operated continuously. Hole injection occurs through p-doped buffer layers but not across the multi-heterojunctions in the p-doped Bragg reflectors. Therefore the contribution of the transport of holes to the resistivity of the device is consid-

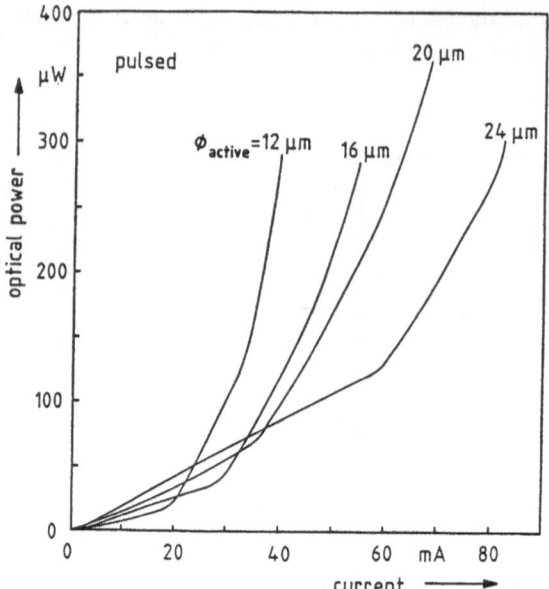

Fig. 5. Light output power versus current characteristic for VCSELDs with different diameters of the active area

ered to be significantly less than that of the contact resistances itself. Continuous device operation definitely requires improved current supply to the pn-junction.

We have fabricated some hundred VCSELDs on a sample of a few millimeters width. Threshold currents and output powers slightly vary for different devices which is attributed to imperfections in crystal growth and mesa formation. The emission wavelengths gradually shift across the wafer due to lateral inhomogenous growth conditions. The latter can be used to fabricate VCSELDs with different emission wavelengths on the same chip in a single production cycle. The maximum light output power observed so far is 3.5 mW for the largest devices under study. Some portion of the output light is absorbed in the GaAs substrate. The estimated absorption at lasing wavelength is ca. 40%. So the light power originally emitted by the inner laser diode is fairly higher than the measured values discussed so far. It is expected that this figure can be further improved using modulation doping techniques [9].

4 Wavelength Selective Photodetector

Wavelength selective photodetectors are extremely useful for wavelength division multiplexing systems for communications and switching. The structure and fabrication technology of the device are similar to VCSELDs. Ten absorptive 10 nm thick InGaAs strained quantum wells are embedded in a resonant cavity of a Q-factor of about 20. Both the top and the bottom reflector consist of 20 pairs

of AlGaAs/GaAs quarter wavelength stacks. The cavity has a free spetral range
of ca. 30 nm. The light reaches the absorption region of the detector through
the GaAs substrate.

Figure 6 shows the photocurrent I_{Ph} of the photodetector device as a function
of the wavelength λ of the incident light. The maximal quantum efficiency of the
detector is larger than 10% at the resonance wavelength of 916 nm. The spectral
width at half maximum is 1.5 nm. The quantum efficiency reaches values as low
as 0.1% for wavelengths out of resonance resulting in a maximal photocurrent
contrast of about 100.

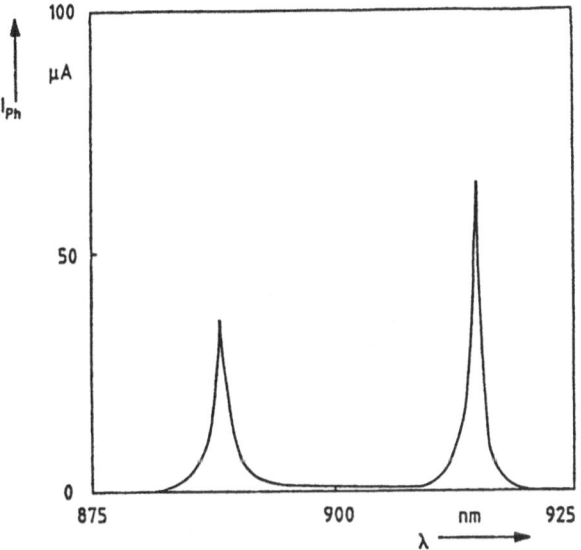

Fig. 6. Photocurrent I_{Ph} of a wavelength selective photodetector as a
function of the wavelength λ of the incident light

Theoretical analysis shows that the quantum efficiency of the highly resonant
photodetectors can be further increased by increasing the rear reflectivity R_o
above 99% and adjusting the front reflectivity R_i for vanishing total reflectivity
of the structure. Using this type of antireflection configuration simultaneously
maximizes the quantum efficiency of the photodetector.

Figure 7 shows the calculated quantum efficiency as a function of the normalized reciprocal wavelength $2n\pi L_{\text{eff}}/l$ with n as refractive index inside the cavity
with an effective length L_{eff} for $R_i = 0.9$ and various rear reflectivities R_o. In
all cases it is assumed that the product of the absorption coefficient α and the
thickness d of the absorptive region is adjusted for optimum quantum efficiency
at resonance.

Figure 8 shows the corresponding external quantum efficiency as a function
of the product. For the reflectivities considered maximal quantum efficiencies
are obtained for products $\alpha d \approx 0.1$. For instance, for $\alpha = 10^{-4}\text{cm}^{-1}$ the resulting

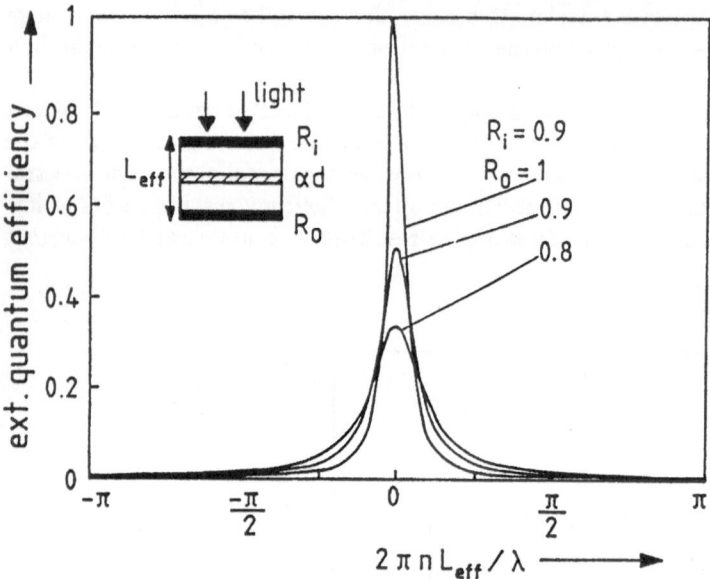

Fig. 7. Calculated external quantum efficiency as a function of the normalized reciprocal wavelength $2n\pi L_{eff}/\lambda$ for different values of the detector rear reflectivity R_o, the front reflectivity R_i is 0.9

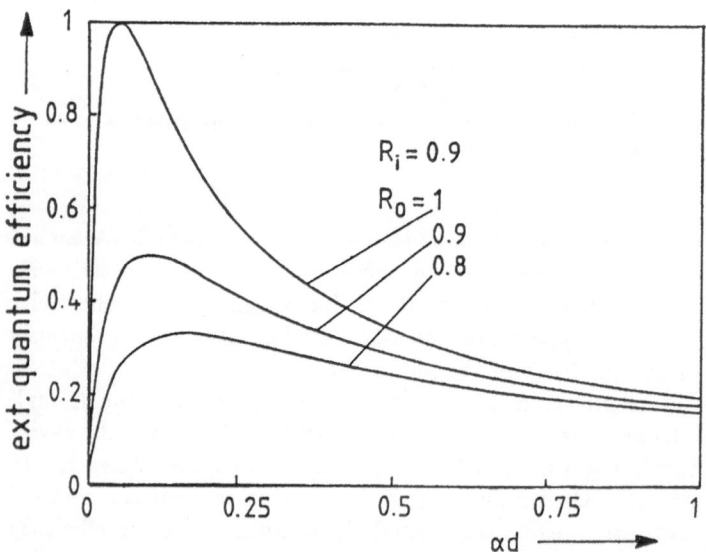

Fig. 8. Calculated external quantum efficiency as a function of absorption αd in the cavity

optimum thickness is $d = 100$ nm. For $R_o = 1$ the quantum efficiency reaches 100%. Larger front reflectivities further decrease the optimum thickness d and the halfwidth of the resonance curve. The calculations indicate that quantum well structures in resonant cavities can be well designed for efficient wavelength selective photodetectors.

5 Conclusions

We have fabricated InGaAs MQW VCSEL diodes by wet chemical etching. The lowest threshold currents are 16 mA for devices with 12 μm active area. The emission is longitudinally single mode at 915 nm wavelength. The maximum output power is up to 3.5 mW under pulsed excitation. A wavelength selective photodetector with 10 InGaAs quantum wells each 10nm thick shows a quantum efficiency of 10% at resonance with a spectral width at half maximum of 1.5 nm. For matched front reflectivity a quantum efficiency of more than 80% is expected although the total length of the absorptive region is just 100 nm.

Note added in proof: Meanwhile we have fabricated submilliamp tunable cw vertical cavity laser diodes with record low threshold currents of 650 μA.

References

1. Jewell, J. L., Harbison, J. P., Scherer, A., Lee, Y. H., Florez, L. T.: Vertical-Cavity Surface-Emitting Lasers: Design, Growth, Fabrication, Characterization. IEEE J. Quantum Electron. **27** (1991) 1332–1346.
2. Wipiejewski, T., Panzlaff, K., Zeeb, E., Ebeling, K. J.: Vertical cavity laser diodes with two-sided output and polarization control with external feedback. Post-Dead-line Paper, ESSDERC'92, Leuven, Belgium, (1992).
3. Tai, K., Hasnain, G., Wynn, J. D., Fischer, R. J., Wang, Y. H., Weir, B., Gamelin, J., Cho, A. Y.: 90% coupling of top surface emitting GaAs/AlGaAs quantum well laser output into 8μm diameter core silica fiber. Electr. Lett. **26** (1990) 1628–1629.
4. Geels, R. S., Coldren, L. A.: Submilliamp threshold vertical-cavity laser diodes. Appl. Phys. Lett. **57** (1990) 1605–1607.
5. Wipiejewski, K. Panzlaff, E. Zeeb and K.J. Ebeling: Low threshold inverted npn-InGaAs/GaAs vertical cavity laser diodes with lateral current supply. LEOS'92, Boston, USA, (1992).
6. Wipiejewski, T.: Diplomarbeit, Inst. f. Hochfrequenztechnik, Braunschweig (1990).
7. Kishino, K., Ünlü, M. S., Chyi, J.-I., Reed, J., Arsenault, L., Morkoç, H.: Resonant Cavity-Enhanced (RCE) Photodetectors. IEEE J. Quantum Electron. **27** (1991) 2178-2190.
8. Wipiejewski, T., Ebeling, K. J.: In Situ Control of Wet Chemical Etching of AlGaAs Multilayer Structures. German IEEE MTT-Chapter Workshop on Heterostructure Technology, Guenzburg, (1991).
9. Geels, R. S., Coldren, L. A.: Low threshold, high power, vertical-cavity surface-emitting lasers. Electr. Lett. **27** (1991) 1984-1985.

The Double Heterostructure Optical Thyristor in Optical Information Processing Applications

M. Kuijk,[1] *P. Heremans,*[2] *R. Vounckx,*[1] *and G. Borghs*[2]

[1] Vrije Universiteit Brussel, Departement TONA, Pleinlaan 2, B-1050 Brussel
[2] Interuniversity Micro Electronics Center vzw, Kapeldreef 75, B-3001 Leuven

The Double Heterostructure Optical Thyristor is a promising device for optical information processing. It has the important advantage of combining the receiving, transmitting, and memory functions in a single device, and simultaneously achieving a considerable optical gain. We show in this paper the dynamics of the switching of a single device and of several devices in a Winner-Takes-All network. We solve the problem of the dramatic surface recombination losses, without passivation or regrowth schemes, but by low doping of the outermost layers. This improvement led to arrays of sensitive devices (5×5, and 16×16 up to 32×64 elements) which are used in some interesting applications, such as locating the maximum light intensity (and optical dose) of optical input patterns, and transcribing parallel optical information from one array to another.

1 Introduction

Many interesting improvements in computing science evolve in the direction of conceiving parallel electronic computing systems. Huge improvements arise, but the connection-bottleneck seriously limits the increase of performance. Massive optical interconnects could solve this bottleneck. Even more promising can be the realization of certain functions, like memorization, shifting and logical AND/NAND/OR/NOR done in a parallel optical way, using arrays of smart pixels. One of the candidate pixels to achieve this, is the Double Heterojunction Optical Thyristor (DHOT), usually implemented as a **PnpN** structure in the GaAs/AlAs material system [1,2,3]. This element has an optical input, an optical output, and an electrical power supply in order to obtain optical gain. It is an element with an "off" state and a light emitting "on" state. A typical device structure, as we make it, is described in Sect. 2. Switching of this element is explained in Sect. 3 and the realization of improved optical sensitivity in Sect. 4. Finally (Sect. 5) we show how some interesting functions are achieved with arrays of these DHOT elements.

2 The Device

Our DHOTs consist of the following MBE-grown layers, starting from a p-type GaAs wafer: 1 μm p-doped (Be) buffer layer, 500 nm 10^{17} cm^{-3} p-doped (Be)

$Al_{0.2}Ga_{0.8}As$, 100 nm 10^{18} cm^{-3} n-doped (Si) GaAs, 100 nm 10^{18} cm^{-3} p-doped (Be) GaAs, 250 nm 10^{17} cm^{-3} n-doped (Si) $Al_{0.2}Ga_{0.8}As$, and finally a contact layer of n$^+$-doped (Si) GaAs.

The outer $Al_{0.2}Ga_{0.8}As$ layers are referred to as "emitters" and the middle GaAs layers the "bases", since they are the emitters and bases of the hetero junction bipolar transistors of the thyristor's equivalent model. The devices are defined by wet etching of mesas in a 3:1:50 solution of $H_3PO_4:H_2O_2:H_2O$. Nitride is used for electrical isolation, and a AuGe/Ni/Au contact is deposited on the cathode. A window left for light input and output is covered by the nitride film, which forms an anti-reflection coating. Figure 1 shows the cross-section of a processed DHOT. Arrays of these elements are obtained by connecting all top-contacts by means of an extra metallization stage in the processing cycle.

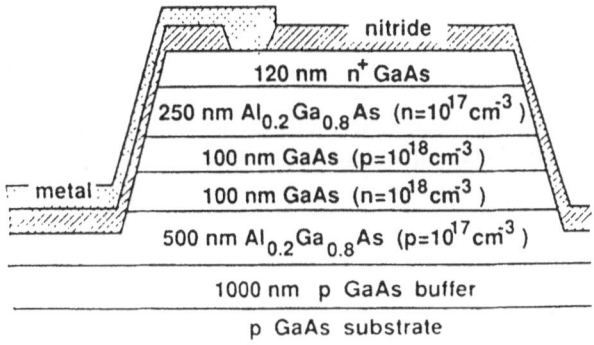

Fig. 1. Schematic cross-section of a processed DHOT. The nitride film serves the double purpose of electrical isolation and anti-reflection coating

3 Switching Principles

Several switching principles can be used, leading to a different optical input sensitivity, a different functionality, and a different switching behaviour.

3.1 Quasi Static Switching

In Fig. 2, a classical set-up is shown for an optically switched photo-thyristor. This is the manner in which the three-terminal Si-thyristor normally operates. The thyristor is connected in series with a load and a power supply. The voltage is applied carefully (such that dV/dt triggering is not possible) to V_s. This V_s is lower than the breakover voltage V_{Br}, to avoid triggering. The element is now in its 'off' state, ready for triggering. If a light pulse, containing sufficient energy, impinges on the device, the breakover voltage lowers, and the device switches on. In the 'on' state, the current I is several orders of magnitude larger than in the 'off'-state. If made in III-V, the device emits light in this 'on'-state. The main advantage of this switching principle is that the operation is asynchronous. The most significant drawback is that for optically *sensitive* devices, the maximum applicable dV/dt is very small, and hence, the time to get to V_s without firing (prior to switching), can easily take milliseconds.

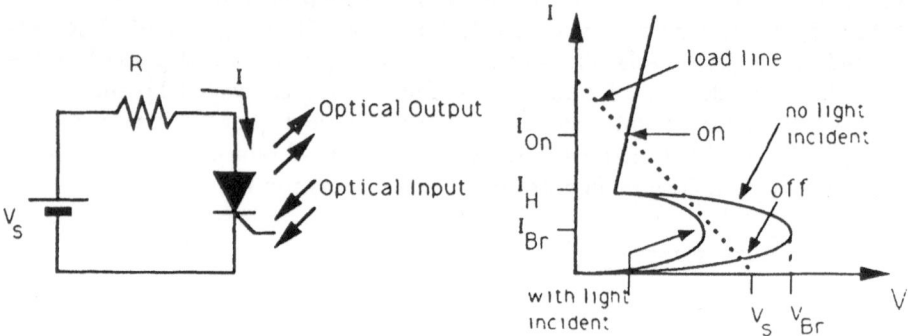

Fig. 2. The usual way of switching a photo-thyristor. The thyristor is connected in series with the load and the supply. Light input lowers the breakover voltage, and triggers the thyristor

3.2 *dV/dt* Switching of a Single DHOT

When V_s is applied for 10 ns to a $100 \times 50 \ \mu m^2$ device with a layer structure shown in Fig. 1 (over a $1 k\Omega$ resistance), the current and the voltage over the DHOT evolve as shown on Fig. 3.

The voltage is applied faster than the dV/dt limit, so that switching occurs at 2.9 V instead of at the breakover voltage of 4 V. The moment at which the voltage reaches its maximum depends on an eventual precharge in the base layers of the device. If, before applying V_s, an optical pulse has precharged the element, the voltage maximum is reached earlier and this maximum is lower.

3.3 Switching with Competition

To introduce competition, two or more DHOT's can be inserted in parallel, with a common external resistance (Fig. 4). dV/dt switching is used, but due to the competition, only one element fires. At first, the current rises in both elements. Then, DHOT1, which had a small precharge, drains a little more of the available current, and switches on. During the last stage of its 'on' switching, it removes the carriers of the losing element. This is represented on Fig. 4 as a maze-current which is positive for the winning, and negative for the losing element.

If there is no optical precharge, and if the two elements are very much alike, (e.g. when they are neighbouring elements on the same wafer) the winner is randomly chosen. Only a very small difference suffices to induce a correct switching

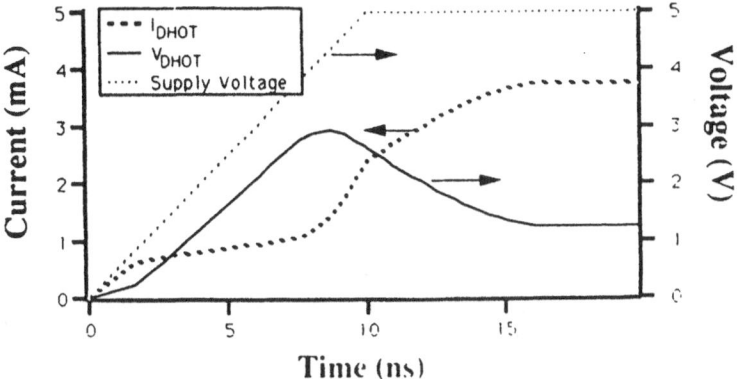

Fig. 3. V_s ramps from 0 to 5 V in 10 ns. After 8 ns, at a voltage of 2.9 V ($< V_{Br}$), the device triggers, and the current rises. This is a calculation with a layer structure as given in Fig. 1, for a $100 \times 50 \ \mu m^2$ mesa and an external resistance of 1 kΩ

behaviour. The parallel set-up of two elements is called a "Differential Pair". With more than two elements, one talks about a Winner-Takes-All network.

This way of switching is advantageous because it is several orders of magnitude more sensitive to optical inputs than the 'Quasi Static' way. Hara et al. [3] reported differential switching with a resolution of 0.1 nW. A second important advantage is that, due to the switching being induced electrically, there is no critical-slowing-down phenomenon [4]. Last, but not least, there is no critical biasing of the element(s) before switching. However, some type of electrical clocking is needed.

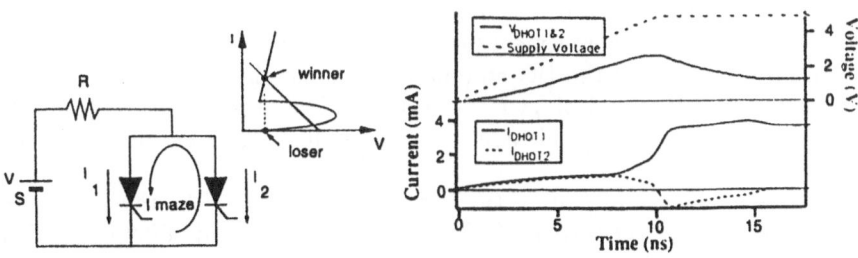

Fig. 4. On the left is shown the fundamental Winner-Takes-All network with two elements. On the right, the evolution of the currents and the voltage across the devices is shown. A small light pulse (not shown) has precharged DHOT1; as a result it wins. It drains (between 11 and 16 ns) even more than the available current. This current is delivered by the losing element (going negative), which itself is thereby inhibited from switching 'on'.

4 Improvement of the Optical Sensitivity by low Emitter Doping

The doping concentrations of the emitter layers and of the base layers have been chosen such as to improve the sensitivity of the DHOTs. For small anode currents, the base current of each bipolar transistor (of the two-transistor equivalent model of a thyristor) is dominantly determined by the surface recombination current [5] at the perimeter of the mesa within the space charge region between each emitter and base. Although the pinning of the Fermi level at the surface reduces the effective barrier height there, for small emitter-base voltages V_{EB} [6] this surface recombination current can still be expressed by the well-known formula [7]:

$$I_{surf,rec} \approx \frac{1}{2} q P S n_i \Delta W \exp\left(\frac{qV_{EB}}{2kT}\right) , \qquad (1)$$

where P is the mesa perimeter, n_i is the intrinsic concentration and ΔW is the width of that portion of the space charge region where the maximum recombination occurs – smaller than the complete space charge region [6]. Because of the pinning of the Fermi-level at the surface, S is an "effective" surface recombination velocity ($\approx 10^7$ cm/s [6]). Equation (1) assumes the recombination to occur over ΔW at maximum rate, i.e. at the rate that is reached when the electron and hole concentrations are equal. This rate is in fact only reached on one contour line (around the mesa) of the space charge region. Analogously to the reduction of the bulk recombination in the space charge region [8], it is crucial to position this contour line in the wide bandgap material, because n_i, and therefore the recombination current, is then much smaller.

In our devices, 90% of the emitter-base space charge region is located in the (wide bandgap) emitters, because the emitter doping is an order of magnitude smaller than the base doping. ΔW is therefore located in the wide bandgap layers, and the surface recombination currents are thus reduced.

To demonstrate the improvement for separate bipolar heterojunction transistors (BHT), we measured the DC amplification of the photo-generated current of a normal Npn BHT, and of an Npn BHT with low doped emitters (Fig. 5), for several optical inputs. An idententical β is reached at a 1000 times smaller current level.

Due to this increase in gain of the two transistors inside the thyristor, our DHOTs have a break-over current of 70 nA and holding current of 0.2 μA, which (see Fig. 6) are the lowest reported for thyristors with no mesa-passivation. The improved gain at low current levels thus obtained makes these DHOTs more sensitive to input light. Even the light of a similar pixel can lower the breakover voltage very strongly (Fig. 6), allowing fanout and cascadability. This method of dealing with surface recombination has the obvious advantage that the performance of the DHOTs does not degrade with time, which is a concern when mesa-passivation techniques [9,10] with dubious long-term stability are used.

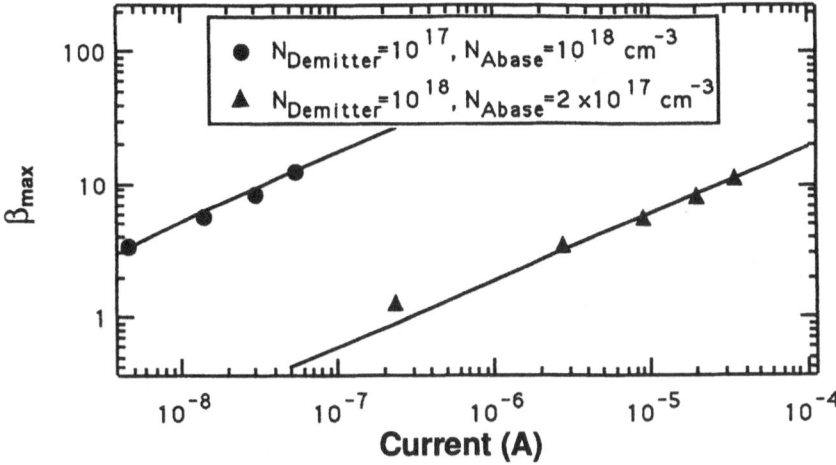

Fig. 5. The measured gain (= collector current / real photo-generated current) of the Npn transistor of DHOTs is plotted versus the collector currents. This for high emitter and low base doping and vice versa. To obtain a given gain, the current needed with the low emitter doped structure of Fig. 1 is three orders of magnitude lower

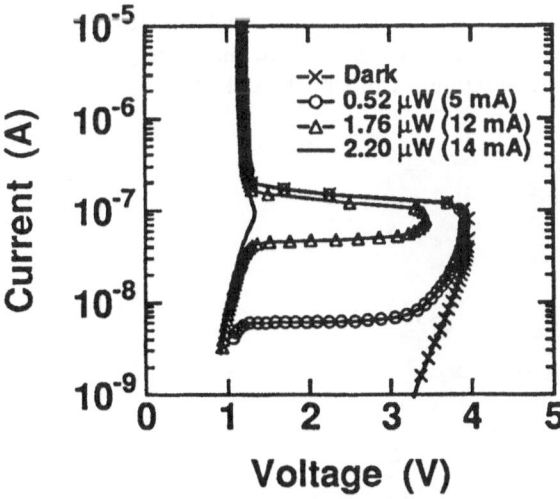

Fig. 6. Current-voltage characteristics of a DHOT (with 110×55 μm^2 mesa and 100×35 μm^2 window) for several optical inputs, generated by a neighboring identical DHOT, operated at the current levels mentioned between brackets. (Only 5% of the emitted light was focused on the measured DHOT)

5 Applications

5.1 Maximum Intensity Localisation

In several domains of optical parallel processing, such as optical neural nets and pattern recognition, there is a need to identify the location of the maximum light intensity in a 2D spatially distributed light pattern.

A monolithic array of 256 DHOTs (16×16) connected in parallel can achieve this [11], using the Winner-Takes-All principle of Sect. 3.1. In Fig. 7(a) photograph of a 16×16 DHOT array is shown. Figure 8 shows the array with a distributed light pattern, and the result after application of a 6 V step in 20 ns. The DHOT located at the place of the highest intensity wins, and emits tens of Watts of light.

The interval between the start of the voltage step application, and the end of the decision, is 150 ns.

Instead of a DC-distribution of light, an image can be applied before the voltage application. Figure 9 indicates the differential resolution of the system. Here we applied an optical pulse to pixel A, and simultaneously an optical pulse to pixel B (A & B randomly chosen) just before the voltage step application. When pixel A has a surplus of at least 80 pJ, pixel A wins, and vice versa. With smaller differences, the correct decision is not always made: sometimes, even both pixels switch on. This is because the elements also have their own (small) resistance, and the situation is therefore not completely ideal. Without light input a random pixel wins, showing that all pixels are identical. Interestingly, when each pixel is given its own (small) resistance, more than one pixel can be made to win, and more than one maximum can thus be detected.

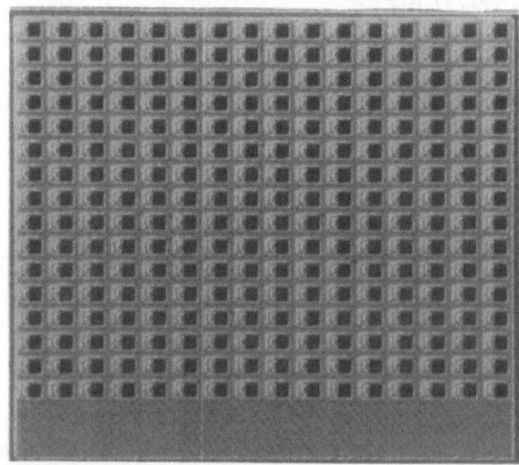

Fig. 7. A top view of the monolithic 16×16 DHOT array. The total area is 775×900 μm^2, inclusive of contact

(a)

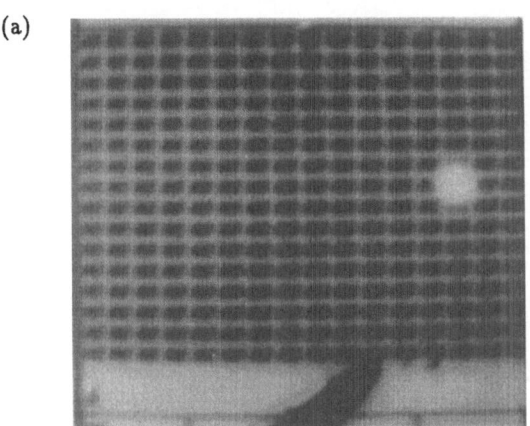

(b)

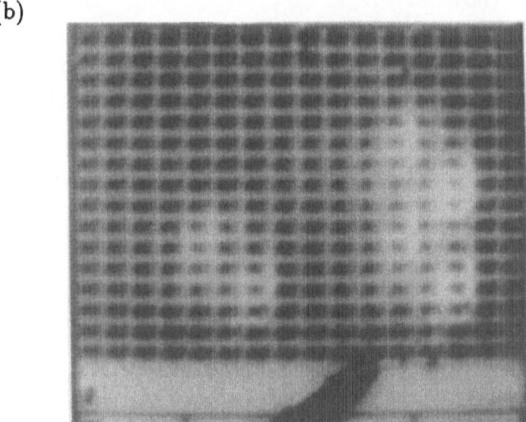

Fig. 8. (a) A distributed 2D light pattern is applied to the array. (b) Competition leads to a fast and correct decision of where the maximum intensity is located. The winning pixel can be found due to its light emitting 'on' state

5.2 Maximum Dose Localisation

In the previous paragraph, an optical image is exposed on an array of parallel connected DHOTs at 0V. Then a voltage step is applied, and a winner emerges. In order to demonstrate *optical dose* integration, similar experiments were made [11] with a variable time delay between optical inputs and the application of the voltage step. One would expect that the optically induced precharge would leak away during such a delay time, so that the required minimum optical energy with delay would be larger. Indeed, for delays longer then 1 ms, much larger optical pulses prove necessary (Fig. 10). However, for delay times smaller than 1 ms, the minimum optical input energy varies slowly (between 40 and 70 pJ).

This means that, in a first approximation all optical energy incident on the pixel within 1 ms before the application of the voltage step is integrated. This is

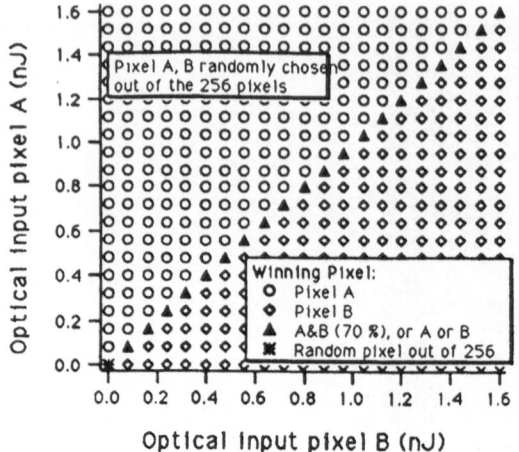

Fig. 9. Optical energy incident on a pixel A versus optical energy on pixel B. The pixel receiving the maximum input energy always switches on correctly if the difference between the input energies of the pixels is 80 pJ. For smaller differences, it is possible that both pixels switch on

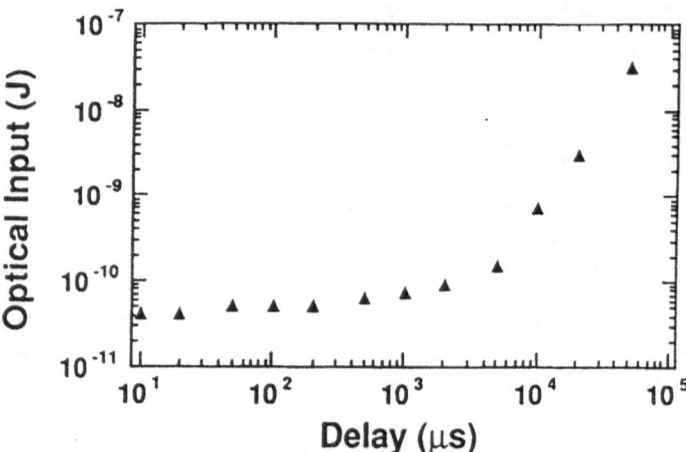

Fig. 10. The minimum optical energy necessary to reliably switch the illuminated pixel, as a function of the delay between this optical pulse and the application of the voltage step

true for all pixels of the array, so that the maximum optical dose can be localised. When more pixels win, the track of a maximum can be found.

5.3 Array to Array Transcription of Optical Information

To demonstrate that optical processing applications are feasible with DHOT arrays, we performed some preliminary experiments, in which optical information was transcribed from one array to another [2]. Each pixel of one array was imaged onto the corresponding pixel of another array on the same wafer (Fig. 11). We used monolithic arrays of 5 × 5 elements, 'loosely' connected in parallel (with every element having its own small resistance of about 7 Ω) with an external resistance of 100 Ω. By application of an upward voltage step (to 16 Volts in 20 ns), up to four pixels of an array could be switched 'on'. Subsequently the 'on'-state of these four DHOTs of the left array was transferred to the neighboring array by applying a voltage step to the receiving array (Fig. 12).

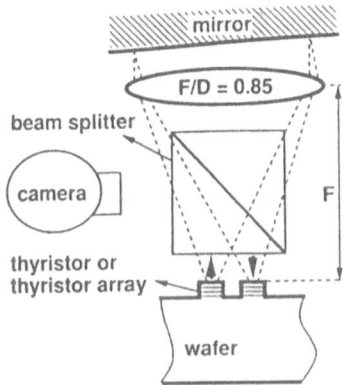

Fig. 11. Set-up used to obtain a mirrored image from one DHOT or DHOT array on another. The beam splitter is for obtaining an image on the CCD camera

Fig. 12. Photograph of two neighboring arrays; the right array has copied mirrored information (four points) from the left array with the dV/dt triggering scheme

6 Future Tracks

Winner-Takes-All arrays of 64×32 elements have now been processed and show the same performance as the above 16×16 element array. We are currently working on Optical Logic Planes based on differential pairs of DHOTs which can be conceived in such a way that an OR/AND/NOR/NAND functionality becomes feasible. Fast turn-off is also being studied, and results will soon be published.

7 Conclusions

We have demonstrated that DHOT elements are not only of academic interest. Arrays of specially conceived thyristors show promising functionalities. Not only memory functions were demonstrated, but also fanout, cascadability, maximum localisation of optical dose (and intensity), and transcription of optical information from array to array were proven to be practically feasible.

References

1. Pankove, J. I., et al.: A pnpn optical switch. In Optical Computing 88. Proc. SPIE **963** (1988) 191.
2. Heremans, P., Kuijk, M., Borghs, G.: Array to Array transcription of optical information by means of surface light emitting thyristors. IEDM Tech. Dig. **91** (1991) 433.
3. Hara, K., Kojima, K., Mitsunaga, K., Kyuma, K.: Differential optical switching at subnanowatt input power. IEEE Phot. Tech. Lett. **1** (1989) 370.
4. Hara, K., Kojima, K., Mitsunaga, K., Kyuma, K.: AlGaAs/GaAs pnpn differential optical switch operable with 400 fJ optical input energy. Appl. Phys. Lett. **57** (1990) 1075.
5. Kuijk, M., Vounckx, R., Pereira, R., Mertens R. P., Borghs, G.: Scaling down the double heterojunction NpnP optoelectronic switching device. Technical Digest Series **7** (Optical Society of America, Washington D.C., 1990) 18.
6. Dodd, P. E., Stellwag, T. B., Melloch, M. R., Lundstrom, M. S.: Surface and perimeter recombination in GaAs diodes: an experimental and theoretical investigation. IEEE Trans. Electron Devices **38** (1991) 1253.
7. Grove, A. S.: Physics and Technology of Semiconductor Devices, Chapters 5 and 10, Wiley, 1967.
8. Tiwari S., Frank, D. J.: Analysis of the operation of GaAlAs/GaAs HBT's. IEEE Trans. Electron Devices **36** (1989) 2105.
9. Sandroff, C. J., Nottenburg, R. N., Bischoff, J. C., Bhat, R.: Dramatic enhancement in gain of a GaAs/AlGaAs heterostructure bipolar transistor by surface chemical passivation. Appl. Phys. Lett. **51** (1987) 33.
10. Kuijk, M., Heremans, P., Borghs, G.: Highly sensitive NpnP optoelectronic switch by AlAs regrowth. Appl. Phys. Lett. **59** (1991) 497.
11. Kuijk, M., Heremans, P., Vounckx, R., Borghs, G.: Maximum Optical Dose Detection with a 16×16 Monolithic Thyristor Array. IEEE Phot. Techn. Lett. **4** (1992) 399.

An Architecture for a General Purpose Optical Computer Adapted to PNPN Devices

N. Langloh, M. Kuijk, J. Cornelis, and R. Vounckx

Applied Physics, Brussels University (VUB), Pleinlaan 2, B-1050 Brussels, Belgium

An architecture for a massively parallel computer is presented. The basic processing element is a pnpn device. The privileged application area is two dimensional vision: each element of the processing arrays performs Boolean operations on one single pixel and its nearest neighbors. The structure is SIMD (single instruction, multiple data), which means that all pixels undergo the same Boolean operation simultaneously. The logic architecture and the controlling structure needed to perform operations on two dimensional binary images, and the extension to gray value images, will be presented.

1 Introduction

There are several kinds of devices which can be used to build an optical computer (e.g. SEEDs [1], PNPNs [2]). An optical computer can also be used in different domains. However, there is one domain that is of particular interest for optical computing: image processing. There are two reasons for this. Firstly, most of the algorithms in imaging can easily be parallelized. Secondly, the information carrier of the images and in the optical computer is the same, namely light.

Image processing is a domain where very high amounts of calculations are required. The operations on images can be grouped in different classes: e.g. linear and non-linear algorithms, local and global algorithms, etc. Only local operations will be considered in this paper: each pixel of the image undergoes the same operation involving the pixel itself and other (neighboring) pixels. Mathematically we can write the operation as

$$g(i,j) = \sum_{k=0}^{m} \sum_{l=0}^{n} h(k-i, l-j, f(k,l))$$

where $f(i,j)$ represents the input image, $g(i,j)$ represents the output image $h(i,j,f)$ represents the operation on the image. For such operations the output image has the same dimension as the input image. If the function $h(k-i, l-j, f(k,l))$ is linear in $f(k,l)$, we have a *linear local shift invariant operation*. In this case, we can write $h(k-i, l-j, f(k,l))$ as $H(k-i, l-j)f(k,l)$. If the function $h(k-i, l-j, f(k,l))$ is not linear in $f(k,l)$, we have a *non-linear local operation*.

Local operations can be calculated on SIMD (single instruction, multiple data) machines. Each output pixel of the image can be calculated at the same time.

The structure of the paper is as follows:

- in Sect. 2, the choice of the optical processing components will be discussed (pnpn devices);
- Sect. 3 contains a description of the architecture;
- Sect. 4 is a worst case analysis of the speed-up achieved by the architecture in comparison to sequential computing;
- in Sect. 5, different possible extensions of the architecture are discussed.

2 Basic Processing Elements

The general motivations for using a pnpn device ([3,4,5,2]) as processing element are:

- It works with non coherent light. This has several advantages: the components are easier to design (there is no need for lasers), the lenses are less critical, etc.
- The devices have a high fan-out (an output signal can be redirected to about 2,000 other similar components).
- The switching speed is very high (less than 10 ns).
- In stead of projecting two signals simultaneously on a processing element, it is possible to project them one after the other. This property of pnpn devices is essential for the proposed architecture (see Sect. 3) and it offers several advantages in comparison to elements which require a simultaneous projection of the input signals and which perform a direct thresholding in order to produce a binary output: an increased noise immunity and the provision of a memory state besides the two logical output states. A disadvantage of this sequential operating mode is speed.

The operations on the data are Boolean operations. For the proposed architecture, the devices have the following advantages:

- the devices can be integrated on a wafer to make a processing plane consisting of multiple processing elements (currently 64);
- each processing element can perform the following operations: (n)and and (n)or (see Fig. 1). The switching between "and" and "or" can be done optically. The inversion is controlled electrically;
- the processing elements can be in three different states:
  - the "keep" state, where the information present in the element cannot change, and there is no emission of light;
  - the "on" state, where the element is emitting the information it contains;
  - the "receive" state, where the information present in the device can be changed optically;
- the input signals can be applied sequentially, signal after signal, and the logical operation is executed afterwards.

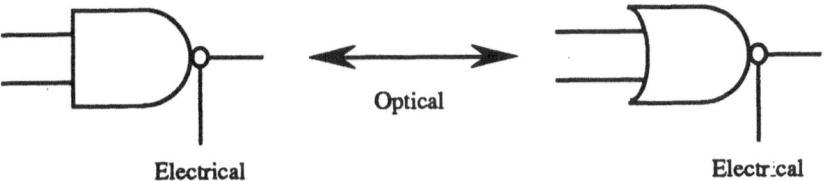

Fig. 1. The (n)and and (n)or operator

3 The Proposed Architecture

The following properties are essential to ensure its usefulness.

- We need a SIMD architecture capable of applying all possible local operations on images.
- We need an architecture that can deal with gray value images. The application area of binary image processing is very limited.

The pnpn's we decided to work with, have only two logic states (binary logic). In order to be able to evaluate all possible Boolean expressions, it is sufficient to do inversions, and-operations and or-operations. Every Boolean function can be written down in a canonical form containing only and-operations, or-operations and inversions.

For example the function

$$f(x_1, x_2, x_3) = \overline{\overline{x_1 x_2} x_3}$$

can be written in its canonical form as

$$f(x_1, x_2, x_3) = \overline{x}_1 x_2 x_3 + \overline{x}_1 \overline{x}_2 x_3 + x_1 \overline{x}_2 x_3$$

In Fig. 2(a), we give a first general diagram of the architecture. All possible

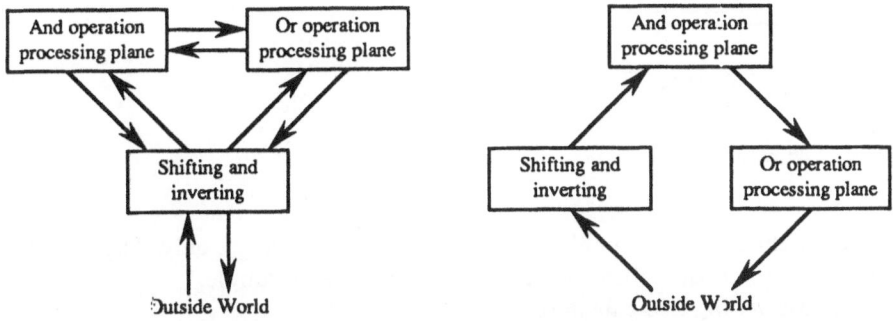

Fig. 2. Diagram of the logic structure of the architecture

Boolean functions can be executed if the data are then projected sequentially on the processing planes before doing the Boolean operations. The data flow is as follows: data are projected on the "shifting and inverting" plane where it is possible to shift and invert the data, data are projected on the "and operation plane" where the and-operations of the canonical form of the expression will be calculated. The results are projected on the "or processing plane", where the or-operations of the canonical form can be calculated.

It is clear that we do not need all the connections in the Fig. 2(a). It is sufficient to provide unidirectional connections as shown in Fig. 2(b).

The advantages of this processing scheme are that we need only one input/output channel and that the number of Boolean variables does not have to be fixed. A big disadvantage is the time loss due to the sequential operation. But with the massive parallel processing capacity at the pixel level, this will cause no problems.

The shift operations on the "shifting and inverting plane" are needed to do the local neighborhood operations. In the beginning we will use a shifter which only allows connections to all nine nearest neighbors of the pixel. This is not a serious limitation, because many local algorithms can be written so that these shift operations are sufficient. In a later stadium, we will try to perform higher order shifts.

A schematic representation of the architecture is shown in Fig. 3.

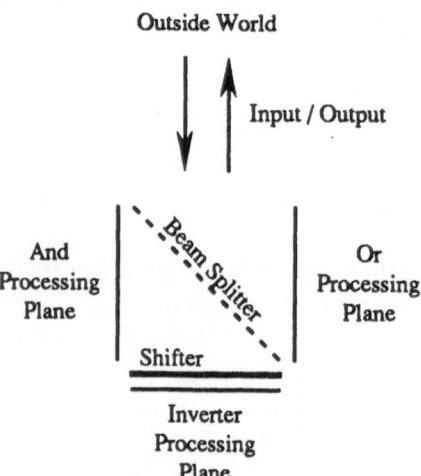

Fig. 3. The architecture

This architecture is capable of dealing with binary images only. The working principle of the architecture is demonstrated with an example.

Suppose we want to calculate

$$g_{i,j} = f_{i,j}\overline{f}_{i,j+1} + f_{i-1,j}f_{i,j-1}$$

The sequence of operations to obtain the result $g_{i,j}$ is:

- The image $f_{i,j}$ is projected on the "inverter processing plane" (see Fig. 4(a); the meaning of the planes is explained in Fig. 3).
- Then the image is projected on the "and processing plane" (see Fig. 4(b)).
- The inverted and down shifted image $(\overline{f}_{i,j+1})$ is projected on the "and processing plane" (see Fig. 4(c)).
- The and-operation is executed on the "and processing plane" (see Fig. 4(d))
- The result $(f_{i,j}\,\overline{f}_{i,j+1})$ on the "and processing plane" is projected on the "or processing plane" (see Fig. 4(e))
- The left shifted image $(f_{i-1,j})$ is projected on the "and processing plane" (see Fig. 4(f)).
- The up shifted image $(f_{i,j-1})$ is projected on the "and processing plane" (see Fig. 4(g)).
- The and-operation is executed on the "and processing plane" (see Fig. 4(h)).
- The result $(f_{i-1,j} f_{i,j-1})$ on the "and processing plane" is projected on the "or processing plane" (see Fig. 4(i)).
- The or-operation is executed on the "or processing plane" (see Fig. 4(j)).

The result $(g_{i,j})$ is then present on the "or processing plane". It can then be sent to the outside world (see Fig. 4(k)).

4 Performance Calculations

This Section contains the calculation of the number of operations needed to obtain a result and a comparison with the number of operations needed on a classical sequential architecture.

The largest canonical form of a Boolean expression containing n variables contains 2^n functions that are or'ed (this Boolean expression is always equal to 1 (true)). Each of these functions contains n variables or inverted variables that are and'ed. This means that with the process explained in Sect. 3 we need $n2^n$ projections on the "and processing plane", $(n-1)2^n$ and-operations on the "and processing plane", 2^n projections on the "or processing plane" and $2^n - 1$ or-operations on the "or processing plane". In the worst case, we need

$$O_{\text{par}} = (2n+1)2^n - 1$$

operations to obtain the result.

For a sequential architecture, we assume that the evaluation of the Boolean function needs $n - 1$ Boolean operations (best case). If the image contains k^2 pixels, we will copy the data nk^2 times from the memory to registers, we need $(n-1)k^2$ Boolean operations and we will copy the result to the memory k^2 times. In the best case, we need

$$O_{\text{seq}} = 2nk^2$$

operations to obtain the result.

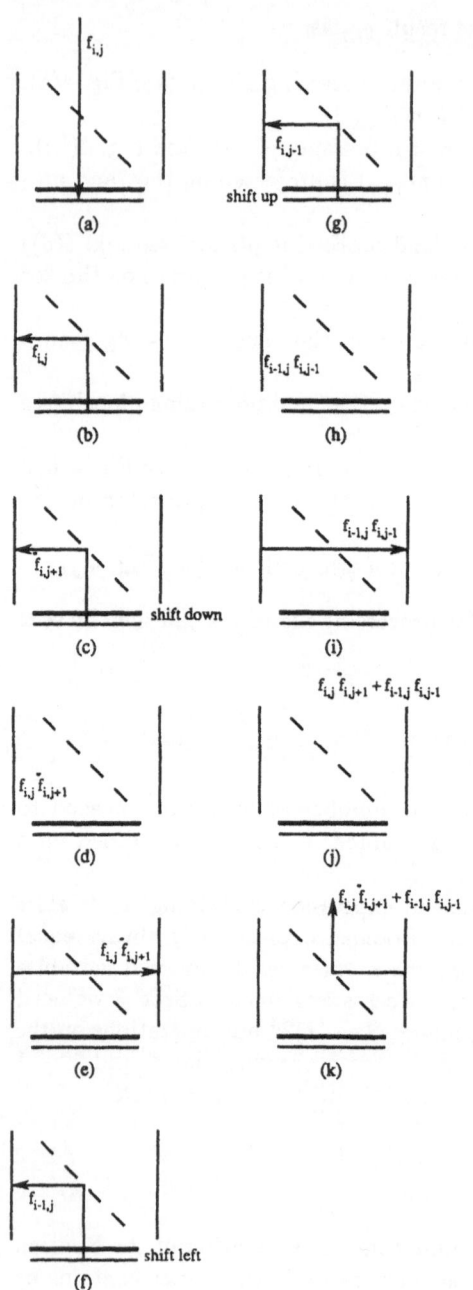

Fig. 4. Sequence to become $g_{i,j} = f_{i,j}\overline{f}_{i,j+1} + f_{i-1,j}f_{i,j-1}$

The gain, in the worst case, is thus

$$G = \frac{2nk^2}{(2n+1)2^n - 1}$$

The gain for operations on the central pixel and four neighbors ($n = 5$) and nine neighbors ($n = 10$) is given in the table.

k	$G_{n=5}$	$G_{n=10}$
1	0.0285	0.00093
2	0.11	0.00372
4	0.46	0.0149
8	1.82	0.0595
16	7.29	0.24
32	29.17	0.95
64	116.70	3.81
128	466.78	15.24
256	1,867.12	60.96
512	7,468.49	243.82
1024	29,873.96	975.28

5 Extensions of the Architecture

In this Section, we discuss possible extensions of the architecture, like working with gray value images and working with more than one image.

A first solution is to encode the gray values immediately on the processing planes. The advantage is the speed (additions can be done in one cycle), but the maximum number of gray values is determined by the hardware (no possibilities to extend the number of gray values).

A second solution is to represent every gray value bit on a different processing plane. The disadvantages are clear. There is a need for more processing planes (one for every gray value bit and one for the carry) and an addition needs more than one cycle. But the architecture is easily extensible (just add more processing planes), and the fabrication of a processing plane is less difficult. This second solution is used for the extension of the architecture to gray value images.

Suppose we want to calculate the sum $z = x + y$. If we write x and y as

$$x = \sum_{i=0}^{N-1} x_i 2^i, \quad y = \sum_{i=0}^{N-1} y_i 2^i$$

then we can calculate the sum with a carry ripple through addition

$$c_{-1} = 0$$

$$c_i = x_i y_i + x_i c_{i-1} + y_i c_{i-1}$$

$$z_i = x_i y_i c_{i-1} + \bar{x}_i \bar{y}_i c_{i-1} + x_i \bar{y}_i \bar{c}_{i-1} + \bar{x}_i y_i \bar{c}_{i-1}$$

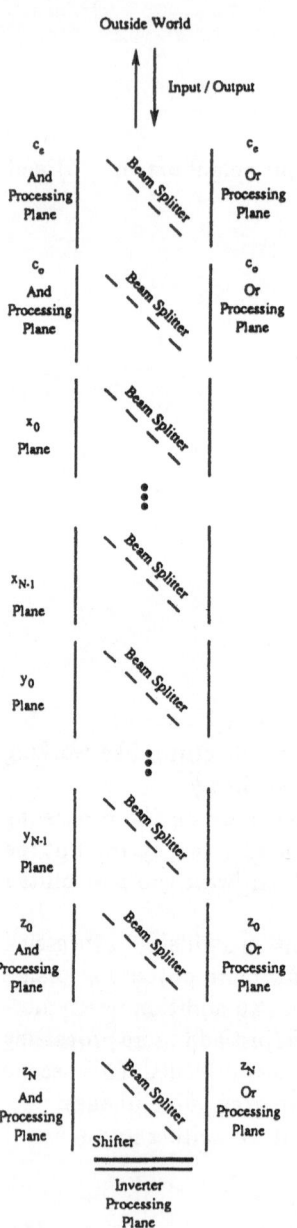

Fig. 5. An architecture capable to work with gray value images

$$z_N = c_{N-1}$$

The architecture capable of performing these operations is given in Fig. 5. We need two "carry operation planes" (c_e and c_o), because the carry c_i is a function of the carry c_{i-1}.

We will demonstrate the addition of two two-bit images.

- Project x_0 on the "inverter plane".
- Project x_0 on the "c_e and plane" and on the "z_0 and plane".
- Project y_0 on the "inverter plane".
- Project y_0 on the "c_e and plane" and on the "z_0 and plane".
- Do the and-operation on the "c_e and plane" and on the "z_0 and plane".
- Project the data from the "z_0 and plane" on the "z_0 or plane".
- Project $\overline{y}_0$ on the "z_0 and plane".
- Project x_0 on the "inverter plane".
- Project $\overline{x}_0$ on the "z_0 and plane".
- etc.

It will be clear that introducing more processing planes allows to execute more operations in parallel. It is then possible to do different and-operations and or-operations at the same time. The gain in speed will be determined by the compiler who writes the code for the controlling hardware of the architecture.

6 Conclusions

We have proposed an architecture for a general optical computer using pnpn devices for local operations in image processing. One of the main differences between the architecture proposed in this paper and "classical" optical architectures is the sequential projection of the data on the processing planes instead of the simultaneously projection.

The architecture is also easier to set up than most other architectures (e.g. the O-CLIP [1]). It requires less components and will hence reduce alignment problems. Since the data are not send in a loop, problems with loop-stability are avoided.

It is also shown that in a 3×3 neighborhood and for normal image sizes (at least 64×64 pixels) the number of operations is less than in sequential computers.

An extension of the architecture for gray value images is presented.

References

1. Wherrett, B. S.: O-CLIP - A demonstrator all-optical processor. In this volume.
2. Pankove, J. I., et al.: A pnpn optical switch. In Optical Computing, SPIE (1988) 963.
3. Heremans, P., Kuijk, M., Borghs, G.: Array to array transcription of optical information by means of surface light emitting thyristors. In IEDM Technical Digest Series, IEEE (1991) 91.
4. Kuijk, M., Heremans, P., Borghs, G.: Highly sensitive npnp optoelectronic switch by alas regrowth. Appl. Phys. Lett. **59** (1991) 59.
5. Kuijk, M., Vouncks, R., Mertens, R. P., Borghs, G.: Scaling down of the double heterostructure npnp optoelectronic switching device. In Conference on Lasers and Electro-Optics, Vol. 7, Optical Society of America, Washington, D.C. (1990).

Part VII

Architectural and Logic Structures

Towards Distributed Statistical Processing — Aquarium: A Query and Reflection Interaction Using Magic: Mathematical Algorithms Generating Interdependent Confidences

N. Langloh, R. Cottam, R. Vounckx, and J. Cornelis

Applied Physics, Brussels University (VUB), Pleinlaan 2, B-1050 Brussels, Belgium

The examination of inherent defects in classical computing structures leads to the proposition of an intuitive computational machine based on distributed statistical processing. The implications of distributed processing and inter-model statistics are considered, and the usual fundamental requirement for computational inversion is discounted. A possible form of primary relational database is proposed, and the possibilities of differential model-fit mapping and the auto-generation of model rules are suggested. The desirable decomposition of computation into interrelational and decision-making processes presupposes an intermediate structure capable of linking the two in a bi-directional communicative manner. We propose a query-reflection architecture to achieve this and describe its required characteristics. The pseudo-implementation of such a structure demands a statistical treatment of the combination of counter-propagating data and knowledge, which suggests a new approach to the design of fast optical computers.

1 Introduction

The initial intention is to formulate premises for the realisation of massively scaled processing of empirical data by relating pure data to its model representations. The proposition is of a set of *boundary conditions* within which such an empirical machine could possibly be constructed.

The most useful area of controlled electronic activity is in the intermediate region between unrestricted transport in metals and completely restricted transport in insulators. Similarly for optical devices the most interesting region is between the two extremes of perfect transmission and reflection. One of the primary techniques used in this work was to look for areas where there is the intuitive application of a polarised idea, for example the obvious choice of Analog or Digital, and to try and find other possibilities which lie between the two extremes. We require structures where formalised logic is subjugated to more data-based ideas, and this leads in the direction of *intuitive* computing. A fundamental question is whether useful processing can be carried out without the imposition on empirical data of externally defined models. Not to do this implies a requirement for the automatic generation of hypothetical models or rule structures, and we are now talking about some kind of *living* computer.

The characteristics of measuring instruments are defined statistically by their own measurements in comparison with those of other instruments. Descriptive models should be evaluated in a similar manner. The classical answer to *what is a good book* is a set of rules which, if followed, lead to *good* literature. Unfortunately, these rules only describe books which already exist, and they will change if other universally acclaimed *good* books appear whose styles violate or extend the set of rules. The usual definition of *living things* is similar: a set of rules is derived which describes a subset of all things, and these rules are tailored to fit as nearly as possible the set which we *already know are alive*! These are POST-imposed rules which do not necessarily apply to future circumstances.

A normal sequential computer program is written in advance of being used, but it is only accepted for use after testing. The testing takes the form of running the program within a set of known boundary conditions to see if it gives *the correct results*. This is analogous to checking if the set of rules for *good books* holds for an already available set of books. Such a program is therefore itself a set of post-imposed rules, and has no predefined validity in as-yet unmet situations. In the sense that computing system rule-bases are set up by reference to a set of predetermined conditions, the same argument holds here too, and if we include the interchangeability of soft and hardware structures, then the computer itself is subject to just the same restriction. The only way to avoid these problems is presumably to work in an environment where the only structures which exist are a function SOLELY of the data itself. This is, needless to say, not an easy task.

It is difficult to drastically improve the performance of a complex interlinked system by removing individual parts and replacing them with new ones, as the remaining externally imposed interrelations will define to a large degree the function of each new component (Fig. 1). This argument applies not only to hardware, but also to the ideas upon which a computer is founded. A possible solution to this dilemma is to remove simultaneously as many preconceptions about computer structure as possible. However, it then becomes extremely difficult to *navigate* towards a more effective structure by logical means. In place of using logical paths to arrive at a logically processing computer, an analogous choice is to follow intuitive routes to arrive at intuitive structures. A consequence of adopting this approach is that it is not initially possible to justify resultant conclusions on the basis of logical derivation from previously demonstrated bases.

2 Empirical Premises

It is possible to formulate empirically a number of useful guiding descriptives for intuitive computation. These include the idea of Activity Thresholds, the difference between Relationships and Decisions, the requirement for Reversible Processing, and qualification of the Prediction of Data.

In perception/computation situations there appears to be a level of perceived problem complexity, or *Thinking Threshold*, below which pre-learned or defined

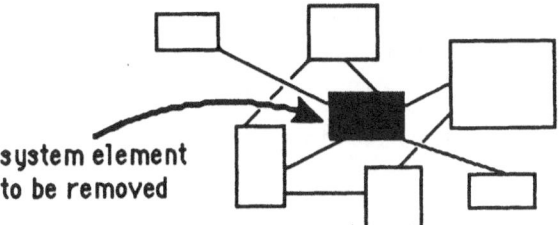

**system element
to be removed**

Fig. 1. Externally imposed effects on an element of a complex interlinked system

rules are automatically applied, and above which an attempt at a solution is made by more complex methods (Fig. 2a). Rules can only be applied if they are known: traditionally the level of rule-knowledge has been partially equated to intelligence, as in *intelligent people know how to do a lot of things.* There also appears to be a level of perceived problem complexity, or *Attempting Ceiling,* above which there is an inability to approach a solution on the basis of currently accessible tools (Fig. 2b). The level of this ceiling has also been partially equated to intelligence, as in *intelligent people can deal with complex problems.*

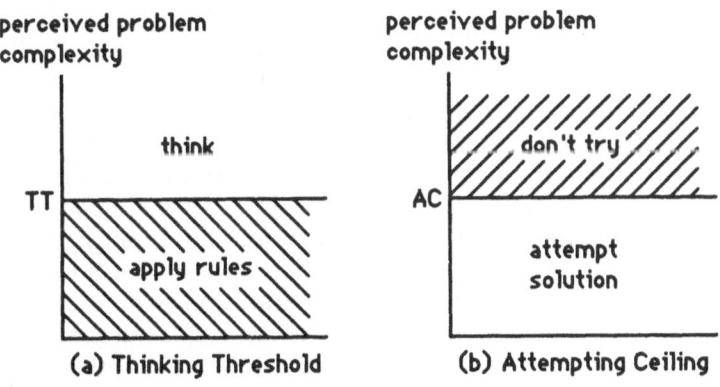

Fig. 2. Activity Thresholds in complex situations

Overlap of these two Activity Thresholds suggests that rules are applied whatever happens, and there is no capacity for relating the rules to the current context (Fig. 3a). This results in the application of rules in inappropriate situations, such as *the safest way to cross a road in England is to first look to the right, so this must be the case in France as well.*

A gap between the two Thresholds implies that there is a range of perceived complexity for which there is the ability to use rules where they are appropriate, to derive new rules where possible, or otherwise to continue by extended data manipulation (Fig. 3b). We should expect *extended capability* to be associated with the presence of a very big gap, as this implies that there is a wide range of

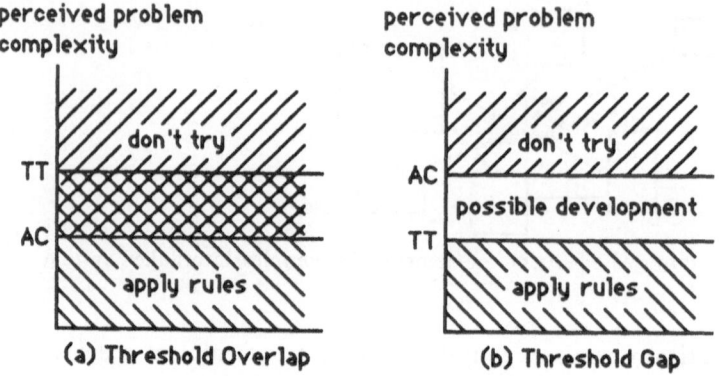

Fig. 3. Restrictive and non-restrictive Activity Threshold combinations

perceived complexity within which there is the capacity to attempt rule development. A gap of this kind is then a primary criterion for an *intelligent* computer,
but the advantages of the availability of a *rough and ready* set of simple easily
applied rules should not be neglected; it is faster to use an available rule than
to search for a more exact relation to an empirical situation. This equivalence
to the phenomenon of rapid reflexes in animals should be included if a machine
is to work in a real-time environment.

The result of environmental-reaction computation is usually some kind of
decision, in its simplest form a binary *yes or no*. This can only be arrived at from
data by a style of processing which reduces or effectively *destroys* the data on the
way: such a process is not reversible. Boolean logic itself is a case in point, and
as such is less than ideal for large-scale interrelational processing. In the Boolean
AND gate example of Fig. 4, if the output column Z is removed from the truth
table then it can be reconstructed correctly from the two input columns A and
B if the nature of gate (AND) is known; such is not the case if either of the input
columns A or B is removed; this is characteristic of an irreversible process. It
seems a good idea to partition a processing machine into two distinct areas, one
for *data-conservational* relating, and one for *data-destructive* decision-making.

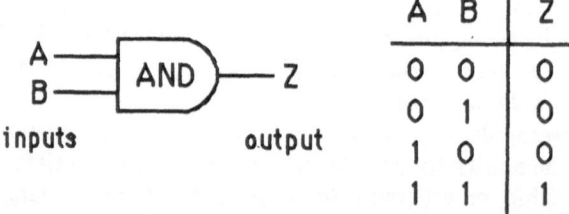

Fig. 4. The irreversible nature of the Boolean AND operator

The idea of modeling is to use current data to visualise the *real* nature of an effect, but this can never take account of whether a model will satisfy data in as-yet uncharted regions. The statistical relation between a dataset and its own corresponding model is therefore not sufficient for data prediction, and the relation should include an indication of the success of the generalised modeling process itself, which would require interaction between the statistics of all available dataset/model environments. This clearly implies the use of massively-scaled databases where the different datasets are in some way integrated.

The usual approach is to propose a model for a given situation and then to test it using currently available data. A more useful technique would be to allow the local data to interact with other datasets to autonomously generate a suitable model. A problem here is that the usual acceptance of a model in terms of *simplistic sufficiency* is left out, and so the model complexity must in some way be related to the requirements of the context.

3 Structural Criteria

The next step is to investigate the implications of our set of guiding descriptives in the structural or dimensional context of space and time. We need to look at the relationships between Experts and Programming, between Data, Control and Physical Structure, the Character of the Processing itself, and the usual Requirement for Inversion.

In conventional computation, the processing depends on a predetermined set of sequential instructions or a currently applicable set of rules. Successful programming or rule design is confirmed by more or less complete testing; the program is run in a controlled environment, and the relation between inputs and outputs is checked by an expert in the field (Fig. 5). If the relation is satisfactory the program is accepted, and if not then it is corrected and re-tested. Unfortunately, the same kind of expertise appears here twice, once in the programming and once in the acceptance procedure. This brings with it great accuracy in areas which are closely related to the testing procedures, but the imposed consecutive assumptions that (1) the physical structure of the computer is error-free, and that (2) the program controlling the data manipulation is error-free, are in some cases catastrophically destructive of accuracy. In such (commonly occurring) cases a controllable degree of *normal uncertainty* would be much more acceptable than *normally high accuracy* and occasional disaster. Small errors are more user-friendly than large ones!

The data-transport medium and the control-transport medium should at the very least be capable of bi-directional interaction. Classically, the application of control results in data-manipulation, which generates new data, which through conditional jumps can influence the control. If the data- and control- transport media are identical, then in the absence of a physically structured environment constraining their interactions the data and control are indistinguishable. Ideally, in a data-based structure, control and static architecture should be integrated into a purely dynamic form which disappears in the absence of data.

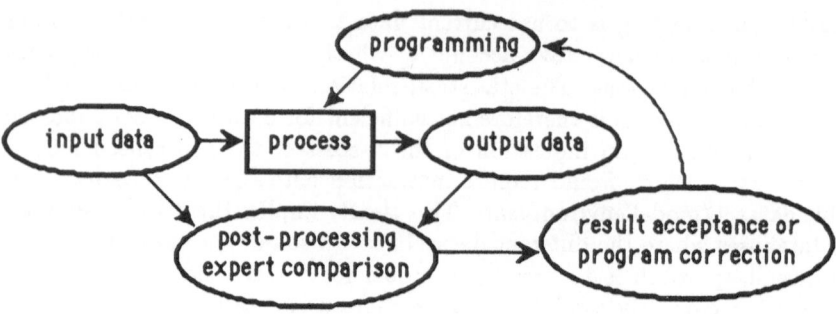

Fig. 5. Program creation and testing

The requirement in relational operations for conservation of data redefines this processing as a rearrangement of existing data. In the simple example of a planar pixeled binary image which passes through a processing plane to give a new planar pixeled binary image, the total information present in each of the two image planes must be the same. Any required pixel in the output plane must then be derived from a possibly spatially unrelated pixel in the input. This requires non-local connection between all of the input/output pairs in the processing plane, or totally distributed processing. This is illustrated for a simple one-dimensional case in Fig. 6, where the input and output *images* have each 20 pixels. The input plane has 60 units of information equally distributed, and the output is the result of some arbitrary processing function; the total information present in the output is the same as that in the input. In such a scheme individual inputs and outputs are no longer uniquely linked, and the complete image must be processed in a unified manner. It would not appear possible to carry out such an operation using discrete pixelated processing elements or devices.

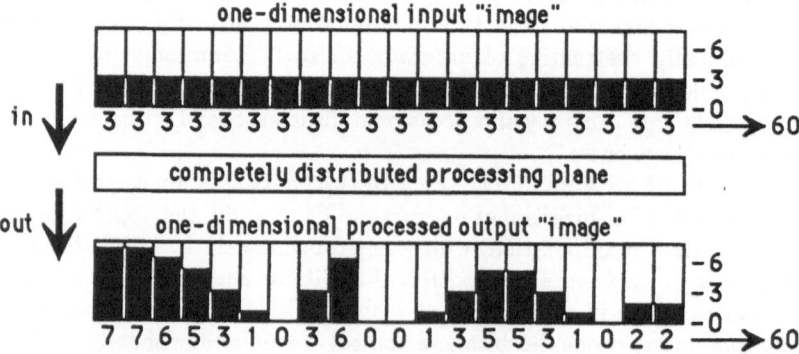

Fig. 6. A simple distributed processing example

In such a distributed processing plane the correctness of the resulting pixel array depends on the number of input/output pixel pairs. The probability of being able to generate a totally correct output array for any arbitrary processing function increases rapidly with the number of pixels involved. With only one pixel in each of the input and output there is only one possible processing result; some functions will require forward transmission of the input pixel, which is always available, but others will require inversion, which is impossible. Even with large pixel arrays, there is the possibility that output information densities will be required which differ from the input densities, but such cases appear always to be associated with decision-making processes and not with interrelationships.

The combination of large-scale totally distributed processing and data-conservation removes the usual computational requirement for discrete inverting devices. Inversion is only required if lateral processing access is limited, as is the case for conventional Boolean gates and discrete-neuron neural networks.

A simple if very restricted example of discrete element distributed processing is provided by a winner-takes-all circuit, as shown in Fig. 7. The input summation must rise above the circuit's threshold value, the output summation is fixed at unity, and the *processing-plane* must be completely laterally connected for the circuit to operate correctly. The *winning* output is equivalent to a (possibly scaled) forward transfer of the input, and all the other outputs give an inversion.

4 Resulting Shape

The desirable decomposition of computation into interrelational and decision-making processes presupposes an intermediate structure capable of linking the two in a bi-directional manner (Fig. 8). For the moment we will leave aside consideration of this intermediate communicating structure and concentrate on the characteristics of the *central* database region.

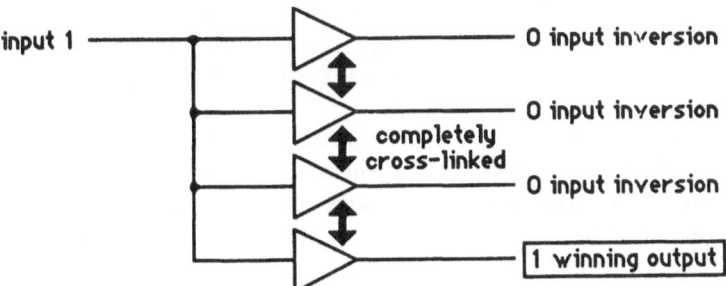

Fig. 7. Simple distributed processing in a "winner takes all" circuit

The primary requirement for the central region is that it should have the character of a purely empirical database, with totally distributed data-conserving interactions, no decision-making processes and no thresholding, which appears

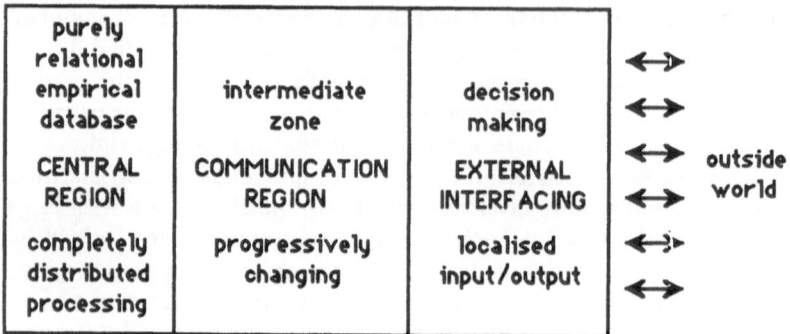

Fig. 8. Communication between the central region and external interfacing

to imply a completely chaotic structure. This conclusion, however, presupposes the absence of defined boundary conditions.

The presence of chaos or of explicate order (structure) [1] in a given environment will depend on the boundary conditions themselves, where these are defined in the most general terms and not solely spatially. The imposition of suitable conditions on such a data-assembly should of itself result in the development of structures in the internal interactions which may be interpreted as models by the more decision-based environment of the communicating intermediate zone.

An obvious boundary condition is the relation between the information content of the total dataset and the information capacity available. Data must in some way be *squeezed* into the central region to avoid dispersion and loss of interaction intensity. This also implies that the database region must be capable of growing as the total dataset increases in size.

Instead of mapping raw data into the central region, it is interesting to consider the possibility of filtering all datasets through their corresponding models and storing only the model-misfits in a distributed processing database. This would enable the generation of modeling of as-yet undescribed global phenomena, which can then be used to correct existing data descriptions and reclaim data in terms of a unification of modeling success statistics instead of using the specific dataset/model relations on their own.

Returning now to the intermediate communicating structure, we propose a query-reflection architecture capable of linking the central region to the necessary external interfacing. This structure also has wider implications, in that it is applicable to future development of current expert systems.

The intermediate zone will resemble a pyramid of rule-levels linking an extensive database (the pyramid base) to a single outside point (the pyramid summit), as shown in Fig. 9. Viewed simplistically, each level consists of a set of rules describing a model, which is itself derived from a more extensive rule-set one level farther down, and which enables the derivation of a smaller rule-set or simpler model one level higher up. The closer to the database the more individual rules there are, and the closer the rules are themselves to simple data descriptions where fewer approximating assumptions are made. This means that the models become simpler and their constituent rules are representative of more data on moving up the pyramid. In this context the extreme form of a model is just a decision, so the structure corresponds to the requirement of linking data at one end to decision-making at the other.

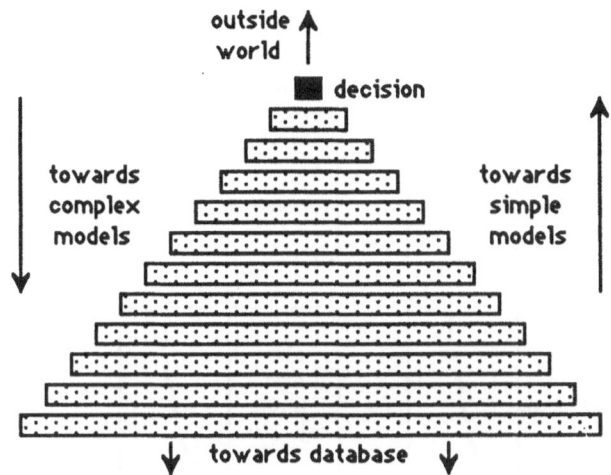

Fig. 9. Model complexities in a rule-base pyramid

A simple example of such a multi-level arrangement is that of the sequence of models which are used to describe the v-i characteristics of a p-n junction diode, labelled from 1 to 6 in Fig. 10. The simplest model (1) is a binary *decision*, with perfect forward conduction and perfect reverse insulation. Each model level (2, 3, ...) adds parameters and complexity while approaching more closely an exact equivalence between the model and the complete available dataset. An even more complex model further down the sequence would also include the statistics of the differences between the empirical and deterministic model values. Ideally, the model values corresponding to the most complex form would exhibit the same statistical properties as the empirical data, which correspond to a model-misfit mapping description of the storage of a *big zero*.

5 Real-time response

It is useful to formulate all incoming response requirements as queries to a central reaction-system. It is evident that a query must receive a suitable response in the time permitted by its context. At one extreme of the range of possibilities is the requirement for IMMEDIATE reaction in the case of imminent system damage. This clearly implies the availability of simple but rapidly-accessible models. Analogously, on finding that a car is about to run you down, it is not very useful to examine all the details of its shape before getting out of the way!

At the other extreme of the range is the requirement for highest possible reaction accuracy, for which models close to the database itself are suitable if plenty of time is available to access them.

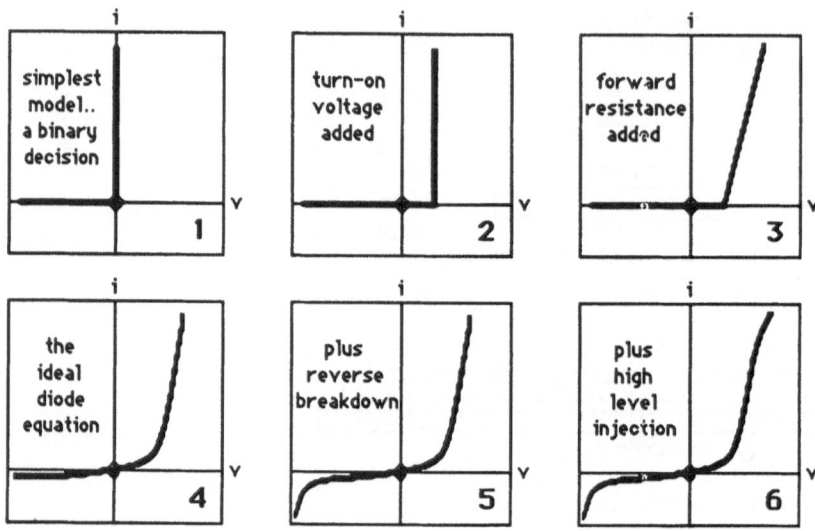

Fig. 10. The p-n junction diode model sequence

The multiple query-reflection structure we propose takes account of these possibilities (Fig. 11). An incoming query is *reflected* by each encountered model-layer. The first reflection is from a high level simple model involving probably gross approximations. It has the shortest path-length, which enables a rapid response. If sufficient time is available it is possible to wait for the second reflection to be returned, where the more gross approximations will have been split into their constituent parts in order to determine the applicability of these in the current context. An even longer wait-time will result in the appearance of reflection number three, and so on down towards the data level. Prior definition of the necessary penetration depth is not required; it is sufficient to wait as long as is permissible in the current context, at which point the most accurate response in the time available will have been received.

It is questionable whether newly arriving data should be dumped into the central database concurrently with operation of the intermediate zone in a query-reflection mode. An apparently more reasonable solution would be to hold new data temporarily and include it later during a *sleep-time*, as illustrated in Fig. 12.

This unfortunately leads to a spatially complex untidy structure, and a pseudo-one-dimensional form would clearly be more attractive. Closer exami-

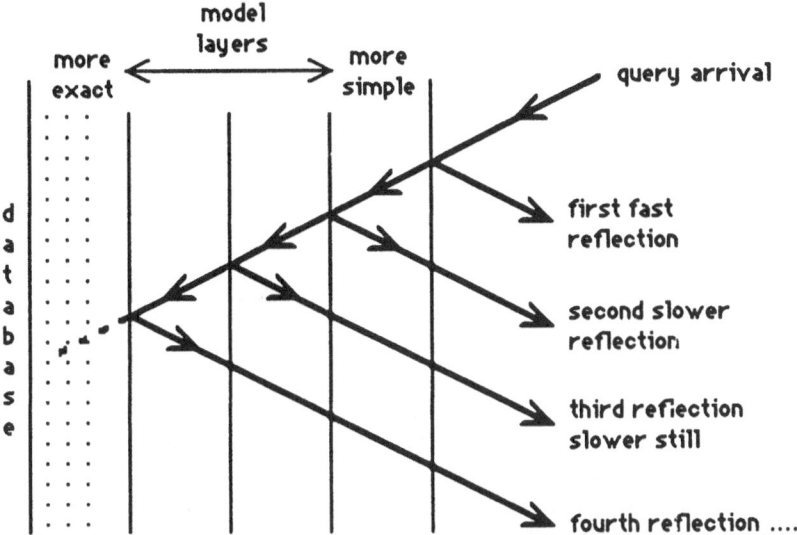

Fig. 11. Basic characteristics of the query-reflection structure

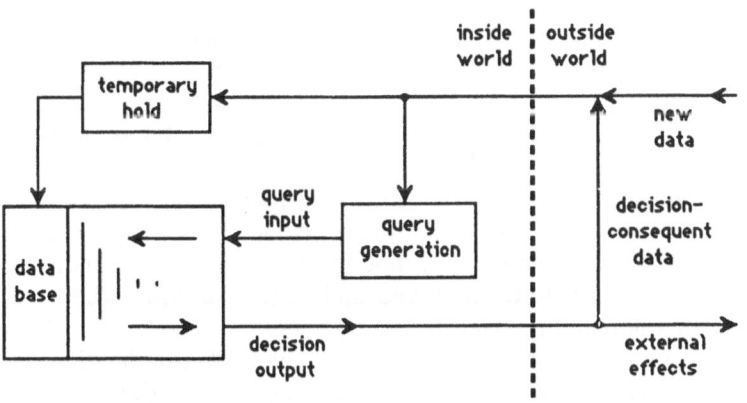

Fig. 12. Sleep-time inclusion of new data to avoid query-reflection interference

nation of the nature of the queries themselves leads to an elegant solution. If the multiple model-layer arrangement is extended outwards into the input region itself, we can describe the situation as an inward propagation of data-like queries and a simultaneous outward propagation of data-like models (Fig. 13). Longitudinally, this resembles two cross-coupled counter-propagating feed-forward neural networks, where each provides the weighting required for operation of the other, and it is pseudo-one-dimensional as desired. Transverse dimensional effects will be considered later.

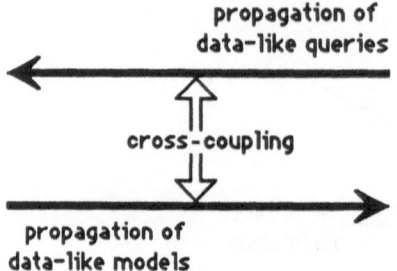

Fig. 13. Longitudinal counter-propagation of queries and models

6 Aquarium

If all successive reflections in the multiple query-reflection structure are exactly the same, then the simplest model is a perfect representation of the entire available dataset. If successively appearing reflections R1, R2, ... differ, then the nature and magnitudes of the differences DR12, DR23, ... describe discrepancies between their respective model-layers. This is illustrated in Fig. 14, where the relation between query and reflections is described in terms of time (vertically) and the model dimension (horizontally), which is related to distance through the structure. If the model layers are such that reflection is possible in both directions, then the multiple reflections themselves will be reflected back inwards, giving rise to information summations which correspond to summed inter-planar model differences. This permits correction of the model-layers themselves: as these are built up from the data level by progressive rule combination the resulting inter-planar *oscillation* of information can be used to modulate the layer reflectivities. Adjacent models will then also be correlated.

There are two major problems in modeling the interactions. Firstly, that of propagating data through relational knowledge, which means that data is mapped onto other data by relations and not by functions, and secondly, of propagating the data back again to check and update the database, which can be done by inverting the relational knowledge.

Two mathematical representations present themselves. We can describe the queries and reflections as the interactions of propagating waves with the model-layers. These interactions are partial reflections characterised by α and β in Fig. 15. It is also possible to describe a structure of this kind in terms of interaction centres, with information summation at the centres followed by re-emission (Fig. 16). In the limit of the difference between successive model-layers tending to zero the two formulations are equivalent, and may be described in a pseudo-one-dimensional equation by

$$\frac{\partial E}{\partial t} = a\left[1 - 2\alpha(E, M) - 2E(M, t)\frac{\partial \alpha}{\partial E}\right]\frac{\partial E}{\partial M} - 2E(M, t)a\frac{\partial \alpha}{\partial M} \qquad (1)$$

where

$- t = $ time,

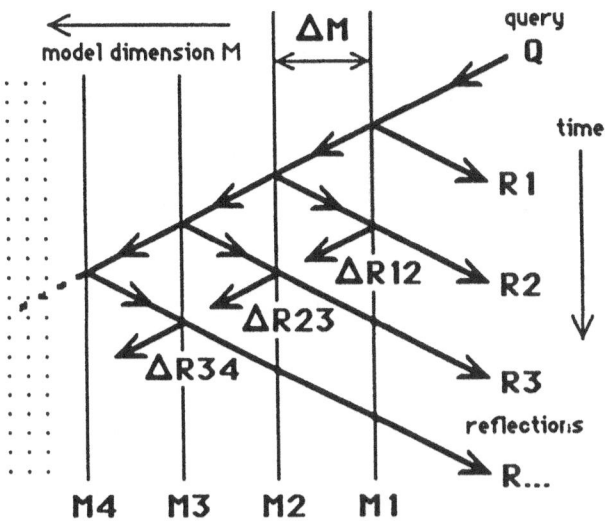

Fig. 14. Aquarium

- M = model,
- information $E = f(M, t)$,
- local reflectivity $a = f(E, M)$,
- information phase velocity a = constant.

The situation is obviously much more complex if the lateral dimensions are taken into account. Clearly, the degree or range of lateral interaction must increase progressively from the outside to the inside of the structure, to give a corresponding increase in the degree of distribution of the processing character. This brings with it a progressive reduction of the longitudinal information group velocity to zero as the central region is approached (Fig. 17). This means that the so-called central database is completely inaccessible!

A more realistic description of the combination of database and query-reflection region is that the one progressively changes into the other through the model-layer sequence, and that the data is distributed into the query-reflection structure itself. Progression towards the *data base end* is equivalent to filtration of incoming data through the successive model-layers, and it is only the model-misfits which will be eventually left over as a *central* non-deterministic chaos.

Decision-making consists of choosing one of a set of relevant decisions on the basis of a dataset and previous experience (knowledge). Every possible decision receives a confidence factor (or *probability*) that it is the best solution to the problem. Working with classical probability is frustrating in this case, as the knowledge must be built up in such a way that every data element corresponds to only one decision. If (as usual) this is not the case, rules are required to distribute the data confidence factors over their associated decisions, but it is now impossible to find one-to-one correspondences and a lot of practical problems

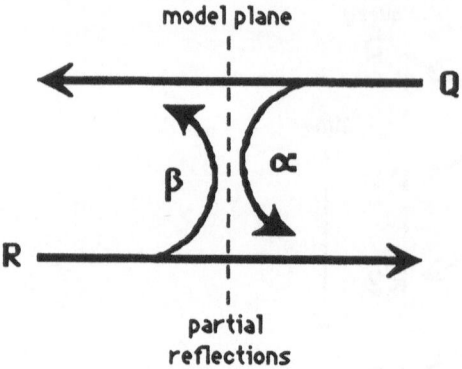

Fig. 15. Propagating wave representation

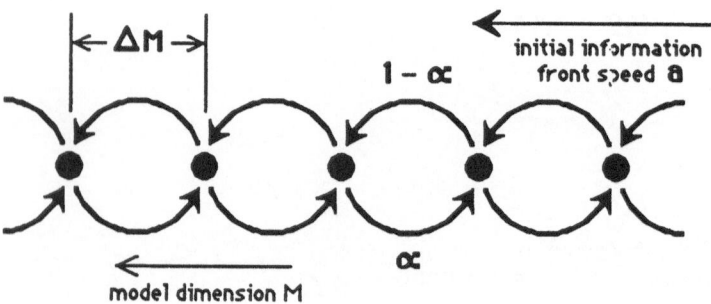

Fig. 16. Interaction centre representation

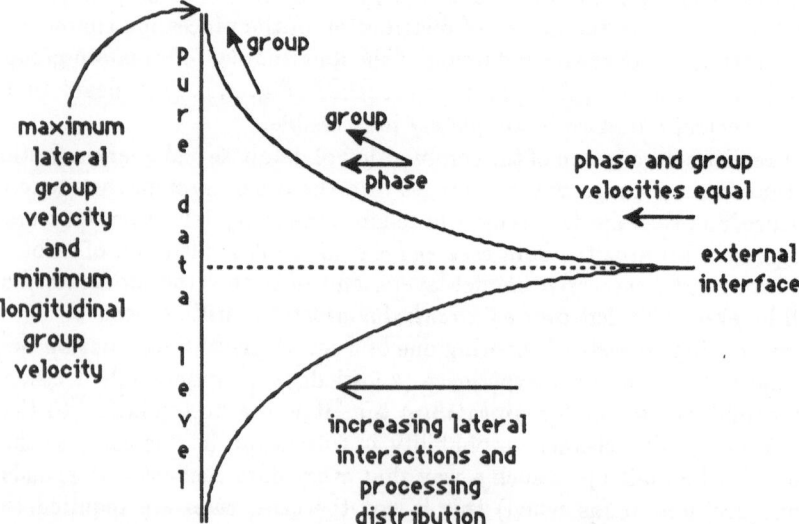

Fig. 17. Information group and phase velocities in Aquarium

result. As an example, consider a set containing 1,000,001 possible decisions of which 1,000,000 are associated with a single data element. A possible rule for distributing the element's confidence factor is to give each associated decision the same confidence. If the element's confidence factor is 99%, then each of the 1,000,000 decisions has a confidence of 9.9×10^{-5}%, but the single unassociated decision will have a confidence of 1%, making it the best decision!

There are different ways of tackling this problem. We can assign *weights* to subsets of a set instead of allocating confidence factors to each element [2,3]. This allows us to calculate the upper and lower probabilities of each element. In the example above, each of the 1,000,000 associated decisions would have a lower probability of zero and an upper probability of 99%, but now the unassociated decision would have a lower probability of zero and an upper probability of only 1%. The selection of *best decision* would be now very different, and more correct. There are still some problems remaining with this approach [4]. We have a strong belief that this is due to the fact that the theory of Dempster/Shafer is working with discrete elements and discrete mappings between them.

The architecture presented in this paper has by its nature a continuous data input set (which varies continuously over time) and a continuous representation of the knowledge (the contribution of the data to a decision can be changed continuously by a model). In 1 the reflection factor $a(M, t)$ is the confidence factor that the model M delivers the best decision.

Relational and decision-making processes may be represented by linear and sharp-threshold transfer characteristics, as shown at the two extremes of Fig. 18. The proposed structure, Aquarium, must progressively change in character between the two extremes, and this change may be represented by a *threshold-control* function which depends locally on the inter-model-layer difference ΔM and is obtainable through $a(M, t)$ from the local bi-directional reflection summations. A plot of the maximum transfer characteristic slope against the inverse model complexity is shown in Fig. 19. The remaining model-filtered data (the model misfits) are restricted to the central region in a manner which resembles the localisation of particles within a potential well (which are trapped). We believe that autonomous threshold-control can provide the boundary conditions necessary to force the generation from chaos of data-"structures" which may be recognised as models by more decision-based processing.

7 Conclusion

Aquarium is a possible implementation of a structure which is conceptually one level higher than a neural network, and where the interactions could be more realistically described as Continuous Distributed Processing (CDP) rather than Parallel Distributed Processing (PDP).

The design of extremely fast optical computers would be better approached by reference to a wave interaction architecture of this kind rather than by building clocked discrete element parallel processor arrays.

Recent publications in the area of Genetic Algorithms and Genetic Programming show that purely digital simulations of structures which are related to sim-

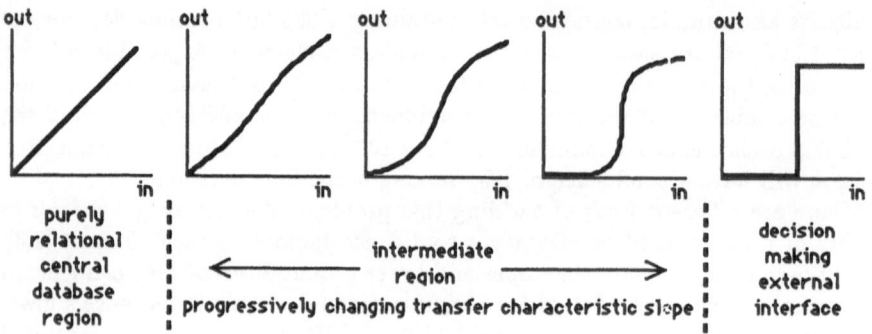

Fig. 18. Progressive change of the processing transfer characteristics through Aquarium

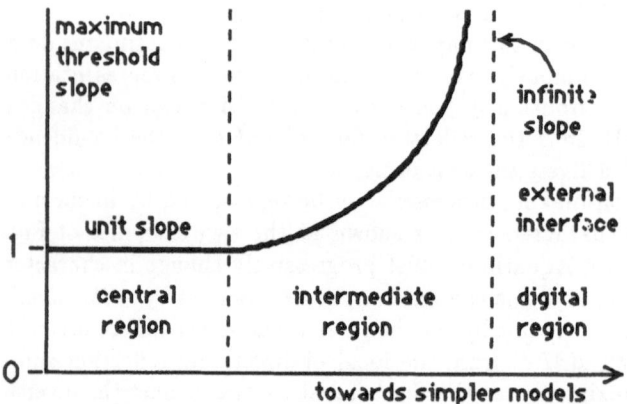

Fig. 19. Central region localisation of the model-filtered data

plified forms of Aquarium demonstrate not only that the inclusion of multiple-structured quality factors [5] improves the operation of algorithms, but also that in a large scale tree structure [6] the major remaining problems appear to be precisely those which are addressed by the non-logic-polarised continuously varying nature of Aquarium.

References

1. Bohm, D.: Wholeness and the implicate order. Routledge & Kegan Paul, London, 1980.
2. Dempster, A. P.: Upper and lower probabilities induced by a multivalued mapping. Annals of Mathematical Statistics. **38** (1967) 325–339.
3. Shafer, G.: A Mathematical Theory of Evidence. Princeton University Press, 1976.
4. Smets, P.: Varieties of ignorance and the need for wellfounded theories. Information Sciences **57–58** (1991) 135–144.

5. Lohman, R.: Structure evolution and incomplete induction. In Parallel Problem Solving from Nature, 2. Ed. Männer, R., Manderick, B. Elsevier, Amsterdam, 1992, 175–185.

6. O'Reilly, U.-M., Oppacher, F.: An experimental perspective on genetic programming. In Parallel Problem Solving from Nature, 2. Ed. Männer, R., Manderick, B. Elsevier, Amsterdam, 1992, 331–340.

Computer-Aided Design of Digital Opto-Electronic Systems with HADLOP

D. Fey

Physics Institute, Department of Applied Optics, Staudtstr. 7/B2, 8520 Erlangen, FRG

In order to design the very complex systems which occur in optical or opto-electronic interconnection and processing systems computer aided design tools are necessary. There are two main approaches to the design of such systems. One approach emphasizes a hybrid concept, known as smart pixels, in which communication is performed optically and processing is performed electronically. The other approach, known as symbolic substitution logic, tries to eliminate the electronics as far as possible. HADLOP (Hardware Description Logic for Optical Processing) is a software design tool for the modelling, simulation and evaluation of both approaches. In contrast to hardware description languages for pure electronic designs, with HADLOP it is possible to model the two-dimensional nature of optics. HADLOP works at the gate level because systems are described as a sequence of two-dimensional gate layers which are connected with optical connection modules. We present results for an opto-electronic broadcasting network, which has been evaluated in terms of the possible degree of parallelism, the energy requirements, and the speed of the system.

1 Motivation

The existing communications bottleneck in electronic computer systems leads to the transition to optical data transmission. An additional increase of efficiency is expected by systems with parallel connections between 2-dimensional electronic gate layers which exploit the high connectivity of optical connections. At the end of this development one hopes to eventually replace the electronic gates by faster optical logic gates. This demands system architectures which are adapted to the 2-dimensional nature of optics. As in electronics it is necessary to use design tools in order to manage the complexity of such architectures. The use of electronic design tools often fails because of the lack of sufficient support to model two-dimensional data processing. Therefore new design tools are necessary which have to fulfill the following requirements.

- consideration of the high parallelism in optics
- adaptation to technological progress
- interactive design
- functional verification

A design tool can help to compare different approaches for architectures with the same functional behaviour. Furthermore it allows the study of architectures which are not yet realizable but may be in the future. In this way a system designer can pass on his experiences to a device technician.

2 Modelling of digital optical systems with HADLOP

There are already different proposals for the data processing of data in digital optical systems, like smart pixels [1], optical programmable logic array (OPLA) [2] or symbolic substitution logic (SSL) [3]. All of them can be modelled, simulated and evaluated with the software system HADLOP. Therefore HADLOP considers both opto-electronic and all-optical approaches. HADLOP works at the gate level. This means that we lay emphasis on the digital behaviour of a system. In contrast to a model working on the physical layer, which produces analog values about voltage or light intensity, for example[1]. Furthermore at the gate-level we are not interested in a system description based on the exchange of data between registers (register-transfer level). Table 1 displays how, on the gate level, the logical operation of active components and the transfer operation of passive components are described.

Table 1.

	Smart pixels or OPLA	SSL
Logic (active components)	sum-of-product-forms $x = a * b + (\text{not } b) * c$	substitution rules
Data transfer (passive components)	explicit declaration of point-to-point connections	explicit declaration of point-to-point connections

Figure 1 displays the main modules of the software system. The design module allows interactive design of an architecture. It produces a formal description of the investigated architecture. With this information the simulation module can calculate the functional behavior. The evaluation module estimates the required time and energy for an architecture under investigation.

In the following we demonstrate the working of HADLOP for a broadcasting algorithm which can be used in a transputer or multiprocessor array. Interconnection networks based on the class of sorting networks are well-known in digital optics [5,6]. Therefore we selected for our demonstration a broadcasting technique which belongs to another class of interconnection topologies.

---

[1] SPICE, a well-known design tool for electronic circuits, is working at the physical level

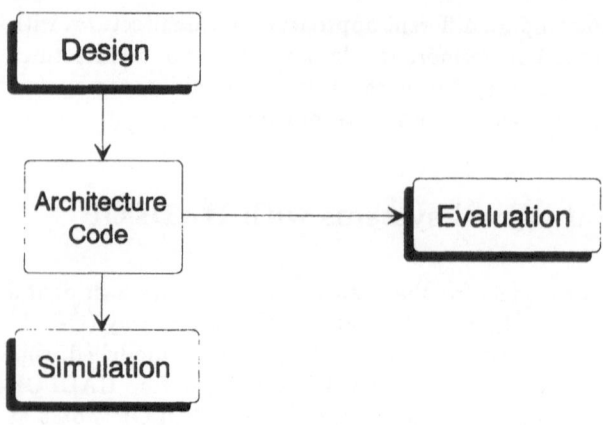

Fig. 1. Structure of the HADLOP software system

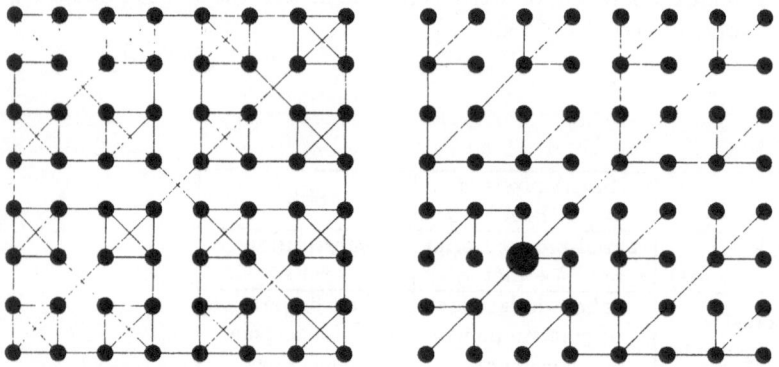

Fig. 2. Interconnection network (left) and spreading out of the broadcast information (right)

3 The broadcast algorithm

Figure 2 depicts the interconnection network and the spreading out of the broadcast information along this network. The interconnection network has a recursive set up. You start with n=4 nodes which are totally meshed. This new construction is called a virtual node. In the next stage you totally mesh $n = 4$ of these virtual nodes. This process is continued L times, where L is the maximum hierarchy level. Finally you get an array of n^L nodes in the whole array. The algorithm itself, which realizes a data transfer along the paths shown in Fig. 2, can be expressed by boolean equations. In total we need 16 AND gates with a fan-in of five and 6 OR gates with a fan-in of four for the example of Fig. 2

with 64 nodes. For this we propose the use of electronic logic within a smart pixel element. A more detailed description of the algorithm can be found in [4]. Let us now consider, how we can simulate this interconnection network using HADLOP.

4 The design module

With the design module you are able to generate a sequence of so-called architecture primitives. These architecture primitives can be, for example, a smart pixel element, a symbolic substitution rule, a permutation module, a control mask, a split/join construction or a feedback loop. Fig. 3 shows such a sequence for our broadcast algorithm in the HADLOP environment. At the left side you can see the start of a feedback element (FB_B), which could be realized with a Wollaston prism. The next architecture primitive is a permutation element (PERM). It represents an optical connection module between two 32×32 data layers. For every pixel position of the sender array the number of rows and columns the data is shifted, when it is mapped onto the receiver array, is clearly defined. The next two elements (STOP and DIAG) are so-called debugging primitives. The effect of the primitive (STOP) is to produce a breakpoint during the simulation process. Using the primitive (DIAG) it is possible to store the content of the data layer at a defined position within the architecture.

The next two elements (THRES) are logic modules, which represent our smart pixel logic. In order to reduce the complexity for one logic module we divided the whole execution process into two logic modules. Such a logic module models a two-dimensional gate layer. A gate layer contains several groups of logic gates, which all execute the same function (Single Instruction Multiple Data). The manner in which HADLOP models the complexity of such a group of gates, or a smart pixel in the case of an opto-electronic circuit, can be seen in Fig. 3. The dimension of the smart pixel is 2×2 (2 rows and 2 columns). This means one smart pixel element has four optical inputs/outputs, which are named 00, 01, 10, 11 (s. component DATA, Fig. 3). Furthermore the smart pixel element works as a finite state machine. Therefore it needs internal memory (s. component INTERNAL, Fig. 3). Finally the logic of the smart pixel element is expressed by sum-of-product-forms (s. component EQUATION, Fig. 3).

After these two logic modules our architecture in Fig. 3 shows an icon for a mask (CONT). This mask allows the simulation of the blocking (black square in Fig. 3) or non-blocking (white square in Fig. 3) of light. At the end of the architecture is placed a feedback element (FB_E), which can be practically implemented with a beam splitter.

The design module consists essentially of a little CAD-editor, which allows one to delete, to insert and to alter any of the architecture primitives shown in Fig. 3. After an architecture had been designed with the help of this CAD-editor, the simulation could be started.

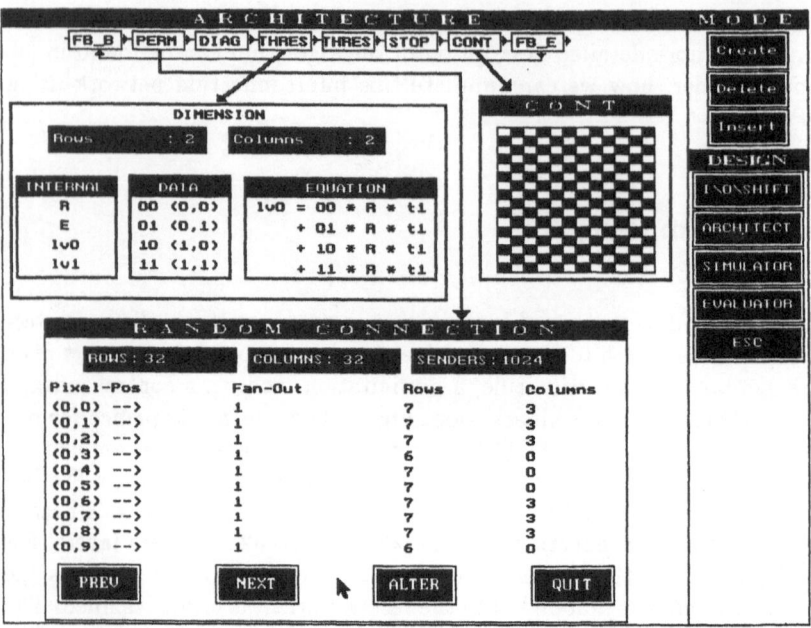

Fig. 3. Design of a broadcasting architecture with HADLOP

5 The simulation module

As mentioned in the previous section HADLOP has two debugging elements, ⟨STOP⟩ and ⟨DIAG⟩. ⟨STOP⟩ is primarily for use during the simulation. ⟨DIAG⟩ is for the study of the calculation steps after the simulation is terminated. Figure 4 displays some calculation steps of our broadcasting algorithm. We simulated the broadcasting on 256 nodes which are arranged in a 16×16 array. Every node is mapped onto a 2×2 pixel segment. From each pixel of this 2×2 segment an optical link leads to one of the four corresponding neighbouring nodes. The broadcast information starts at the lower left hand corner and spreads out to the upper right hand corner until every node has received the message. Finally we would like to know how much time this process needs. Our evaluation module can answer this question.

6 The evaluation module

To calculate approximate values for the necessary time and energy, each of the architecture primitives presented in Sect. 4 is assigned a certain amount of time and energy. In addition a distance between the modules is specified in order to calculate the time for the data transfer. Figure 5 displays the result for our broadcast algorithm in the case of 1ns switching time for the smart pixels and

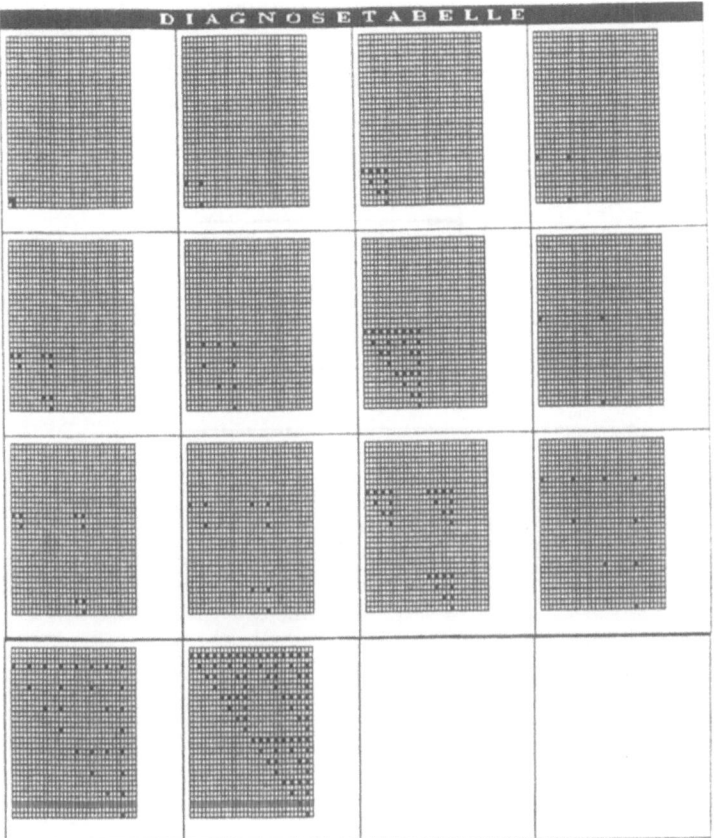

Fig. 4. Simulation of the broadcast algorithm

a distance of 10 cm between them. The whole process would then need 2.55 μs and 92.2 mW power.

In order to get values for other configurations with greater number of nodes we do not stop the data transfer when the information has reached the upper right hand corner. Instead the data are run through the loop many times. In this case no new simulation is necessary. It is sufficient to change the number of cycles, which is possible by simply entering the right values (s. component

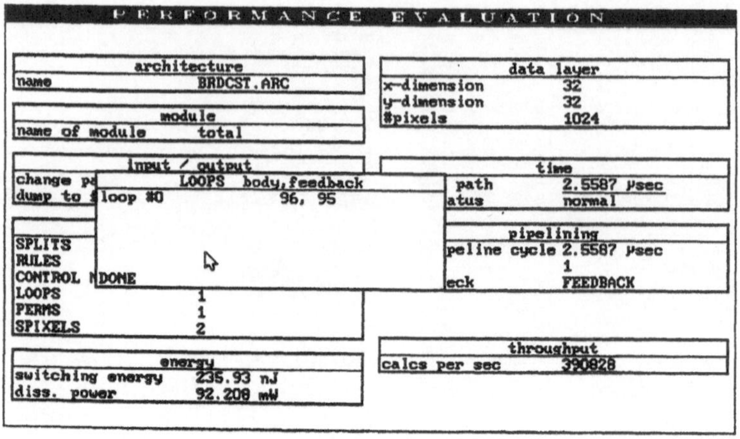

Fig. 5. Results of evaluation module

LOOPS, Fig. 5). In this way we can easily perform a time analysis. The following
results were produced.

# processors	#optical links	#cycles	#speed
16	64	15	425.33 ns
64	256	47	1.28 μs
256	1 024	95	2.55 μs
1 024	4 096	255	6.83 μs
4 096	16 384	512	13.61 μs
16 384	65 536	1 024	27.30 μs
65 536	262 144	2 048	54.61 μs
252 144	1 048 576	5 120	136.53 μs

The alteration of the parameter for the loop cycle is not the only variable
that can be changed without starting a new simulation. It is also possible to alter
the switching time for the logic modules. In Fig. 6 the results of the broadcast
algorithm latency versus the number of processors for different switching rates
are presented. The graph for switching rates of 1 GHz and 100 MHz shows the
logarithmic increase of the latency.

logarithmic increase of the latency.

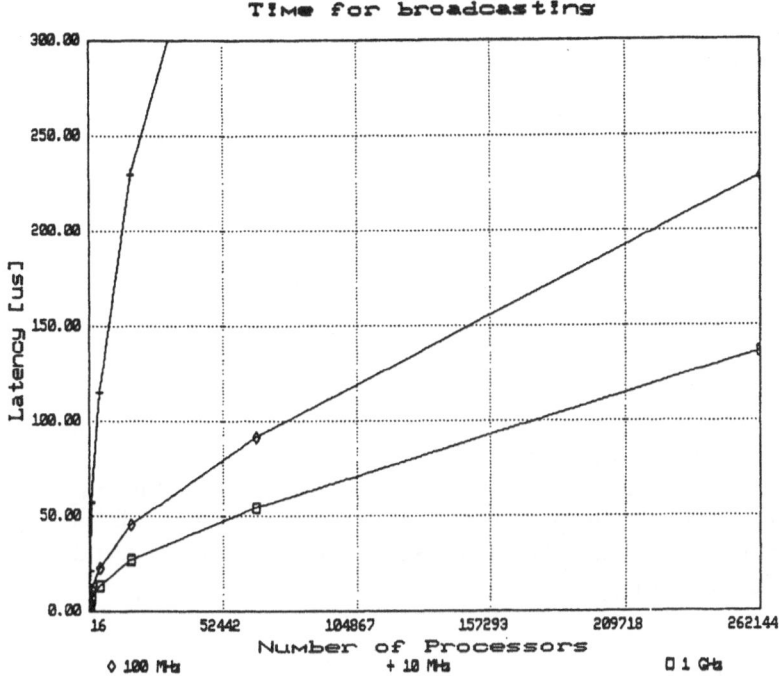

Fig. 6. Latency for broadcast algorithms versus number of processors at different frequencies

References

1. Midwinter, J.E.: A novel approach to the design of optically activated wideband switching matrices. Proc. IEE **134** Part J (1987).
2. Murdocca, M.J.: A digital design methodology for optical computing. MIT Press, 1991
3. Brenner K.H., Huang A., Streibl N.: Digital optical computing with symbolic substitution. Appl. Opt. **25** (1986) 3054.
4. Stirk C.W., Athale R.A.: Sorting with optical compare-and-exchange modules. Appl. Opt. **27** (1988) 1721.
5. Stucke G.: A complete 2D-shuffle-exchange-stage for large 1D data arrays. Optik **78** (1988) 84.
6. Vecchia D.G., Sanges C: An optimized broadcasting technique for WK-recursive topologies. Future Generation Computer Systems **5** (1989/90) 353–357.

Microwave Photonics

D. Jäger

Department of Optoelectronics, University of Duisburg, W-4100 Duisburg 1

In this paper, an overview is given on the fundamental concepts of a special kind of ultrafast photonic device based upon the interaction of propagating microwave signals with optical beams. The characteristic properties are due to coplanar transmission lines leading to cut-off frequencies in the THz range. In particular, high speed travelling-wave photodetectors are first discussed, followed by a presentation of novel modulators and SEED - self electrooptic effect device - elements with interesting properties.

1 Introduction

There has been continuing interest in the development of ultrafast ($>$ 100 GHz or $<$ 10 ps) photonic devices for optical information technology. Because optoelectronic components are at present among the most interesting structures, there is a need for the realization of elements based upon various microwave-optical interactions [1]. In order to avoid the limits of usual RC time constants, the electrical circuits have to be built in accordance with microwave techniques.

In this paper, a review is given of the increasing significance and the technical exploitation of this area of microwave photonics. As a first example, a travelling-wave photodetector [2] is discussed which can be utilized to generate microwave power with high efficiency. The results are briefly compared with those obtained by using an optoelectronic switch. The photodetector can also work as an optically controlled phase shifter, modulator or switch for microwaves [3]. A second example is a travelling-wave electrooptical modulator which can be employed for optical beam scanning, switching, etc. [4]. Thirdly, the vertical integration of a photodetector and a modulator leads to a travelling-wave SEED [5] with different interesting properties. In a final example, a high speed modulator is applied to laser beam testing or as a sensor for electrical fields [6].

2 Coplanar Waveguides

Figure 1 shows the basic coplanar structure on top of a semiconductor consisting of layers with different doping concentrations [1]. The ground metallizations (left and right) form the backside contact to the n^+-layer on top of the semi-insulating substrate. The center conductor on the mesa structure provides the second contact to the n-layers where the upper one determines the electrical and the lower the optical properties of the device (for examples see below). Consequently, the electrical field of the microwave propagating down the line is concentrated in the n region, particularly, in the depletion region when the center strip forms a rectifying Schottky-contact with the n-doped layer [7]. Then Fig. 1 reveals the cross section of a common planar Schottky diode [8]. Note also, that the cross section in Fig. 1 can be that of any pn-, pin-, pnp- or pnpn-diode, e.g. laser diode, pin-photodiode, phototransistor, switching element etc.

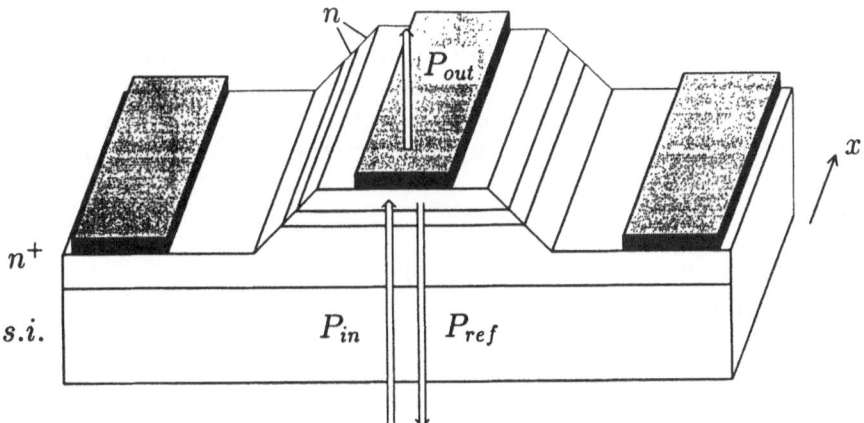

Fig. 1. Schematic cross section of the coplanar structure of a traveling-wave photonic device

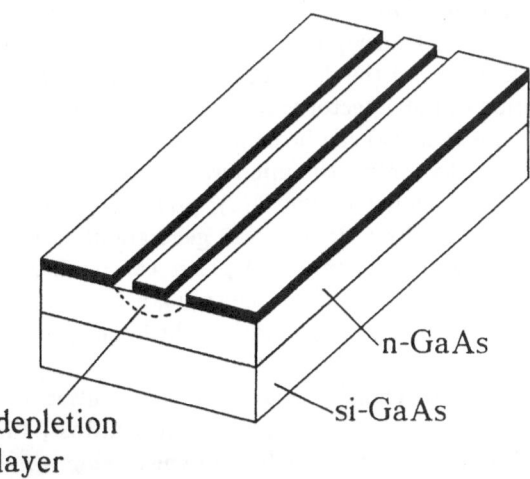

Fig. 2. Coplanar Schottky contact line

In this paper, the optical input is in vertical direction where different situations may be distinguished. In case of a photodetector the optical input power P_{in} is absorbed. In contrast, when the cross section of the device is that of an electrooptical modulator, the reflected (P_{ref}) or transmitted power (P_{out}) or the corresponding optical phases are controlled by the microwave signal. Note also that on the other hand such a modulator characteristic can be used to measure the electrical field of the microwave.

Up to now, the simple coplanar Schottky contact line of Fig. 2 has been investigated in detail [1-3,7,8]. The following key property is important for the behaviour of the device: When the depletion layer width is controlled by either electrical or optical signals the phase velocity of the propagating microwave is changed.

In the following, different photonic devices are described where Fig. 3 shows the optical beams with respect to the direction of microwave propagation. Fig. 4 consists of a schematic cross section with four layers together with a table of possible layer structures. Here DBR is a distributed Bragg reflector consisting of alternating layers of quarter-wavelength epitaxially grown AlAs and GaAs films. MQW means multiple quantum well structures of GaAs between AlGaAs barriers.

3 Travelling-wave Photodetector

Figure 3(a) shows the configuration of a very interesting travelling-wave photodetector (TW-PD) [9]. A possible realization is shown in Fig. 4 which is basically a planar InAlAs/InGaAs/InP Schottky-diode. In this case, two optical waves with frequencies ω_1 and ω_2 interfere in the detector at oblique incidence. As a result, the interference pattern moves at a phase velocity, given by the difference frequency and the wavelength of the envelope wave. Using heterodyne mixing, the generated photocurrent is a moving source for a microwave with frequency $\omega = \omega_2 - \omega_1$ propagating down the line. Under phase matching conditions, the efficiency is expected to be an optimum and the detector should exhibit a large bandwidth. Because the cross section of the detector is that of a diode, the propagation characteristics (e.g. the phase velocity) can be controlled externally [3,7] in order to achieve phase matching or to control the microwave phase.

When using only one optical input signal in Fig. 3(a), preferably a short pulse, then the transmission line, charged by a dc bias, can generate ultrashort electrical pulses via optoelectronic switching. The initial pulses can further be converted into desired microwave signals by special pulseforming networks [10]: Pulse compression, signal generation and microwave radiation which are optically controlled can be obtained.

As can be seen, by optoelectronic switching special microwave pulses and bursts or other modulated signals can preferably be formed. In contrast, the generation of cw microwaves can directly be realized by heterodyne mixing where the frequency and phase difference of the optical waves can easily be varied and converted into frequency and phase of the microwave.

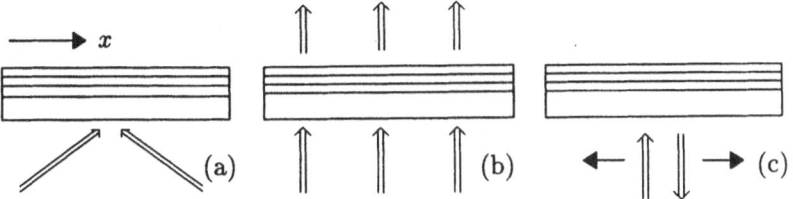

Fig. 3. Configurations for optical input and output. x is the direction of microwave propagation

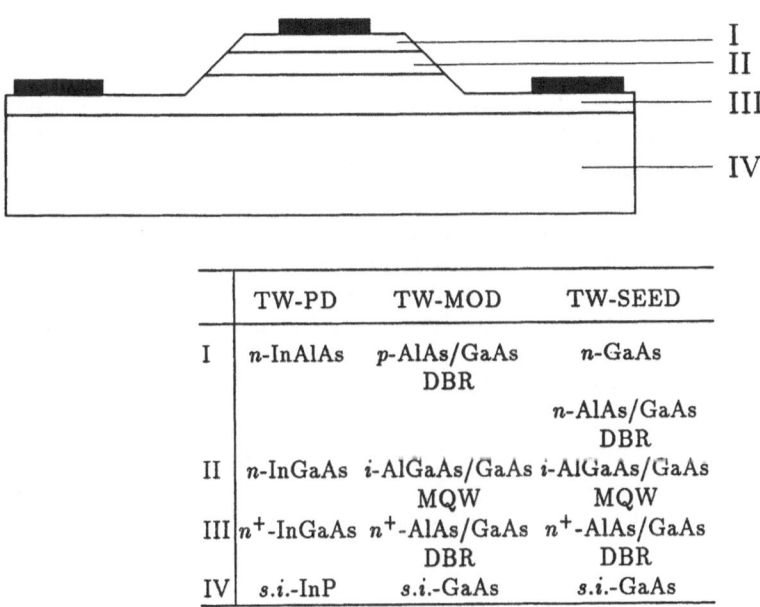

	TW-PD	TW-MOD	TW-SEED
I	n-InAlAs	p-AlAs/GaAs DBR	n-GaAs
			n-AlAs/GaAs DBR
II	n-InGaAs	i-AlGaAs/GaAs MQW	i-AlGaAs/GaAs MQW
III	n^+-InGaAs	n^+-AlAs/GaAs DBR	n^+-AlAs/GaAs DBR
IV	$s.i.$-InP	$s.i.$-GaAs	$s.i.$-GaAs

Fig. 4. Semiconductor layers of different devices

4 Travelling-wave Modulators and Coplanar SEED

The optical input and output in the concept of Fig. 3(b) are distributed along the line. Provided that the cross section of the device is that of an electrooptical phase modulator, a special travelling-wave modulator (TW-MOD, see Fig. 4 for realization of a vertical Fabry-Perot microresonator [11]) results which is capable of showing the following phenomena. As a result of the electrical signal propagating along the structure, adjacent regions of the modulator structure produce optical waves with different phases as determined by the microwave field. This behaviour can be useful for optical beam scanning, switching, frequency shifting, pulse synthesis, serial-parallel conversion, etc. [4]. A structure similar to that of Fig. 4 has been realized; it is described elsewhere in this book.

When the cross section of the device in Fig. 1 resembles that of a bistable element, such as a SEED, the propagating microwave creates a switching wave. In the present case, the SEED structure (see Fig. 4) results from the vertical integration of an electrooptical microresonator as the modulator and a Schottky diode as the photodetector [5,12]. The electrical signals are again propagating waves and due to the winner-takes-all principle, optical filaments are generated [5] together with the wellknown current filaments in electronics when S-type current voltage characteristics are present [13]. Under certain conditions a beam array is produced which can be moved as a wave along the line.

5 Microwave Field Sensor

In the configuration of Fig. 3(c), only a focussed optical input beam is used. Provided that the device exhibits electrooptical, i.e. modulator properties, the reflected optical beam can now be used to measure the electrical potential at that position [6]. When moving the optical beams, the method can be used to detect electrical signals in an integrated circuit at different nodes, the upper frequency limit being well above 100 GHz, beyond any electronic measuring capabilities [14]. It should finally be mentioned that the structure in Fig. 3(c) can be microminiaturized to give a tiny field sensor for many applications. Note also that the optical signals can be transmitted through glas fibres with only minor influence on the measured field distribution.

Obviously, such an electrooptical field sensor can quantitatively determine electrical potential distributions at ultrahigh frequencies and without any mechanical contact. It is therefore suitable for direct testing and characterization of MMICs and MOEMICS, i.e. monolithic-optoelectronic-microwave integrated circuits.

6 Conclusion

It is demonstrated that by using special coplanar metallization structures, ultrafast photonic devices are imaginable. High speed photodetectors, modulators and switching devices can be constructed producing extremely interesting phenomena for optical information technology. Work is in progress to experimentally prove the various predictions given in this paper in a quantitative manner.

References

1. Jäger, D., Block, M., Kaiser, D., Welters, M., von Wendorff, W.: J. Electron. Waves and Appl. 5 (1991) 337.
2. Soohoo, J., Yao, S.-K., Miller, F. E., Shurtz, R. R., Taur, Y., Gudmundsen R., A.: IEEE Trans. Microwave Theory Techn. MTT-29 (1981) 1174.
3. Kaiser, D., Block, M., Jäger, D.: Electron. Lett. 25 (1989) 1575.

4. Kobayashi, T., Hirasawa, S., Morimoto, A., Sueta, T.: Proc. Int. Top. Meeting Opt. Computing, Kobe (1990) 397.
5. D. Jäger.: Proc. SPIE **1230**, ICOESE '90, Bejing 1990, (1990) 787.
6. Kolner, B. H., Bloom, D. M.: IEEE J. Quantum Electron. QE-22 (1986) 79.
7. Jäger, D.: Int. J. Electron. **58** (1985) 649.
8. Dragoman, M., Block, M., Kremer, R., Buchali, F., Tegude, F. J., Jäger, D.: Proc. ESSDERC '92, Leuven (in press).
9. Block, M., Jäger, D.: Proc. 21th Europ. Microw. Conf., Stuttgart (1991) 22.
10. Paulus, P., Stoll, L., Jäger, D.: IEEE Trans. Microwave Theory Techn. **MTT-35** (1987) 1014; Paulus, P., Brinker, W., Jäger, D.: IEEE J. Quantum Electron. **QE-22** (1986) 108; Heidemann, R., Pfeiffer, Th., Jäger, D.: Electron. Lett. **19**, (1983) 316.
11. Humbach, O., Stöhr, A., Auer, U., Larkins, E. C., Ralston, J. D., Jäger, D.: IEEE Photon. Techn. Lett. (submitted).
12. Zumkley, S., Wingen, G., Scheffer, F., Prost, W., Jäger, D.: In Photonic Switching II. Ed. Tada, K., Hinton, H. S. Springer, Berlin, 1990, 185.
13. Symanczyk, R., Jäger, D., Schöll, E.: Appl. Phys. Lett. **59** (1991) 105.
14. von Wendorff, W., Stopka, M., Jäger, D: Microelectron. Eng. **16** (1992) 305.

Part VIII

Si-Based Devices

Avalanche Photodiodes for Optical Bistability

A. Koster

Institut d'Electronique Fondamentale - CNRS URA 22
Université Paris-Sud, bât. 220, F-91405 Orsay Cedex - France

1 Introduction

This has been one of the first topics considered by sub-group I "Silicon devices for optical logic" in ESPRIT WOIT. If the active region of an avalanche photodiode (APD) is a plane parallel silicon plate with good optical quality faces, a Fabry-Pérot resonator (FP) is realized by putting a reflection coating in place of the usual AR one on the input face giving the front mirror, the back mirror of the FP being the rear metallization of the silicon plate. Optical nonlinearities in silicon have two different origins - electronic and thermal - if silicon temperature is fixed : its refractive index decreases proportionally to the photo-carrier density. High optical excitation can lead to excess carrier density of the order of 10^{18} cm^{-3} and optical switching can be observed in pulsed regime where competition between electronic and thermal effects is present [1]. Under cw illumination, optical nonlinearities in a silicon APD are of thermal origin and greatly enhanced by Joule effect due the large reverse applied voltage necessary to get photo-carrier multiplication in the preavalanche regime. In these conditions, an optothermal gain very much larger than in references [2,3] is expected giving much lower power thresholds for optical bistability. Such experiments have been carried out at IEF with a slightly modified RCA C30817 supplied by EG&G Canada. The experimental results have been analyzed and the potentiality of pixelated thin film APDs operating as optical bistable devices in 2D array has been evaluated.

2 Experimental Study

The experimental set-up is described in Fig. 1: the attenuated gaussian beam of a cw Nd:Yag laser ($\lambda = 1.064$ μm) is focused with a 20 cm focal lens on the resonator giving a 50 μm waist incident beam. The supplied APD (doping profile p$^+\pi$pn$^+$ and reach through structure, silicon plate thickness $D = 120$ mm) is AR coated giving about 20% of the incident optical power absorbed at 1.06 μm, the estimated reflectivity of the back mirror is $R_B = 80\%$ and the APD is used without the glass window whose optical quality is inadequate. An external mirror is then necessary as FP front mirror: the selected mirror has a reflectivity $R_F = 60\%$ close to the optimal one given by $R_F = R_B \exp(-2\alpha D)$, considering a silicon absorption coefficient $\alpha = 10$ cm^{-1}; with this condition the power reflected by the FP at resonance is 0. This mirror and the APD are mounted in a supple and airtight stand full of helium in order to protect the APD of moisture. The FP input mirror is parallel to the input face of the silicon plate. The APD case and the FP stand temperatures are fixed at $T_0 = 22°$C. The APD is reverse biased by a voltage V, the resistance R limits the current in the APD and gives also a voltage proportional to photocurrent I_p. The two power-meters give output voltages proportional to the incident and reflected powers.

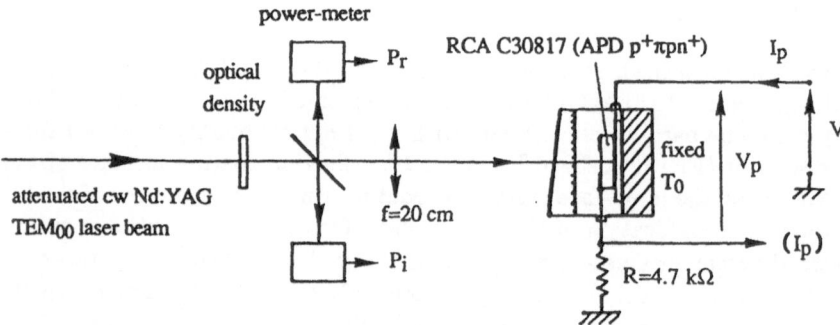

Fig. 1. Experimental set-up

2.1 Study in the linear regime

Characterization of the silicon thickness uniformity over the 800 μm diameter of the active region has been made at first. The FP resonator can be precisely centred relative to the incident beam at normal incidence using orthogonal translation stages giving analog voltages proportional to their positions. Figure 2 gives the evolution of photocurrent and reflected power versus the FP position along the vertical axis z: the six FP resonances almost regularly spaced correspond to a quasi-prismatic silicon plate with a thickness variation of 1 mm over 800 μm. Along the x axis, there are only two FP resonances. Owing to the distance $L \approx 2$ mm between the FP mirrors, the input beam cannot be more focused and the best measured finesse is only 3 instead of 6 under plane wave excitation and with an ideal FP.

2.2 Study in the nonlinear regime

The APD position being fixed in order to excite the resonator near the largest photocurrent resonance at low incident power, a bistable operation is observed with mW range incident optical power for APD reverse voltage V equal or greater than 200 V. Figure 3 gives the evolution of photocurrent and reflected power when the incident power varies between 50 μW and 2.4 mW in quasi-static regime for $V = 330$ V: the first switching down on the reflected power corresponds to a 580 mW threshold on the incident optical power. Decreasing the optical power gives a switching up for a threshold greater than the first one. Instabilities have been observed for $V = 300$ V and are related to this unusual transfer characteristic due to a smaller optothermal gain at higher temperature in the on state. These operating conditions of the APD being largely beyond normal photocurrent density and power dissipation given in the specifications, no measurements have been made for voltages nearer the 375 V avalanche voltage.

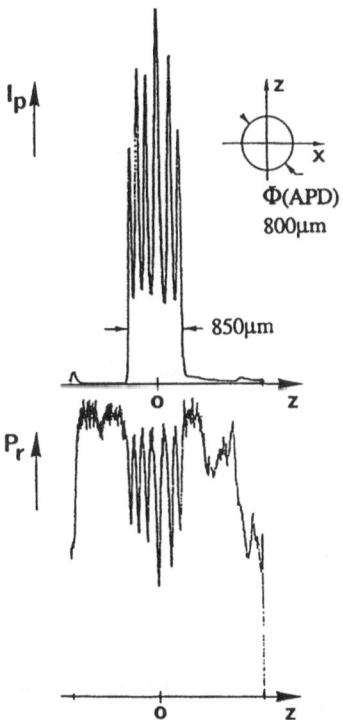

Fig. 2. Evolution of the photo-current and photo-current reflected power, along the z direction of the optical APD active region in the linear regime

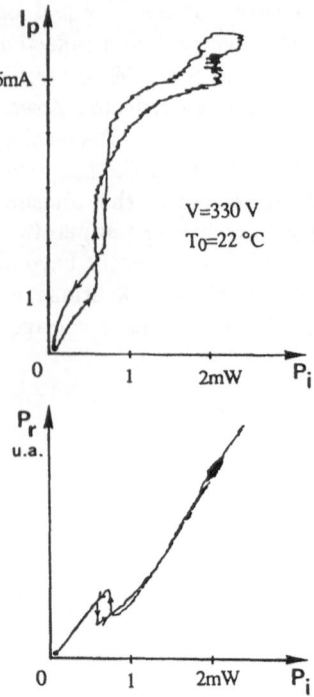

Fig. 3. Evolution of the photo-current and reflected power, versus the optical power in the nonlinear regime

3 Operation modelling

The optical thickness of the FP resonator versus the mean temperature T of the silicon plate along the optical beam is

$$\delta(T) = L + D(T)\,[n(T) - 1] \quad,$$

where $n \approx 3.5$ is the silicon refractive index. The change in optical thickness $\Delta\delta$ to get optical bistability is related to the FP finesse $\mathcal{F}$:

$$\Delta\delta = \frac{\lambda}{2\mathcal{F}} = D\left(\frac{dn}{dT} + (n-1)\frac{1}{D}\frac{dD}{dT}\right)\bar{\Delta T} \quad.$$

The thermo-optical coefficient dn/dT being approximately equal to 2.4×10^{-4} K^{-1} and the thermal expansion term to 6×10^{-6} K^{-1}, the latter will be neglected. An approximation of the mean temperature change along the optical beam in the silicon plate to get bistability is then obtained:

$$\bar{\Delta T}_{+} = \frac{1}{D}\int_{0}^{D}[T(u) - T_0]\,du = \frac{\lambda}{2\mathcal{F}}\left(D\frac{dn}{dT}\right)^{-1} \quad.$$

With $\mathcal{F} = 3$, $\bar{\Delta T}_+ = 8°C$, and for the experiment realized with $V = 330$ V, the Joule power P_J in the APD just after the first switching is about 1 W (3 mA $\times$ 315 V) taking into account the voltage $V_p = V - RI_p$ across the APD. A particular definition of the thermal resistance of the silicon plate (2×2 mm$^2 \times 120$ mm) useful with such opto-thermal effects can be given as: $R_T = \bar{\Delta T}/P_J$; a value of 8.5 KW^{-1} is found. The silicon plate is electrically insulated from the APD case, by a thin aluminum plate of unknown thickness, in order to reduce the parasitic capacitance, this plate acts also as thermal insulator and the heat power created in the enlightened region of the APD by Joule effect diffuses largely in the whole silicon plate, so the silicon temperature is fairly uniform. The following expression for the photocurrent I_P allows the determination of the optothermal gain:

$$I_p = q \frac{A(\delta) P_i}{\hbar \omega} M(V_p, T) \ .$$

$A(\delta) P_i$ is the part of the optical incident power absorbed in the FP resonator (A is the absorptance), $M(V_p, T)$ is the multiplication factor (more precisely, T is the temperature of the avalanche region). This expression is obtained assuming that every absorbed photon creates an electron and hole pair (as the photon energy is very close the band gap) and that all the photo-carriers are collected. The APD specifications give a multiplication factor $M(315 \text{ V}, 30°C) \approx 50$ just after the first switching. The heating power is

$$P_h = \left[1 + \frac{q}{\hbar \omega} M(V_p) V_p \right] A(\delta) P_i \ .$$

The first term corresponds to the absorbed optical power $A(\delta) P_i$, the second term is Joule power $V_p I_p$. The optothermal gain G is defined as the ratio of the total heating power to the absorbed optical power:

$$G = 1 + \frac{q}{\hbar \omega} M(V_p) V_p \ .$$

Just after the first switching, the optothermal gain is $G_+ = 13500$ and absorptance $A_+ = 0.12$. Just before the first switching, $I_p = 1.1$ mA and $V_p = 325$ V, with the previous value of the thermal resistance and with the corresponding heating power of 0.35 W, a temperature rise of 3°C is found, giving $M(325 \text{ V}, 25°C) \approx 70$. Just before the first switching, $G_- = 19500$ and $A_- = 0.03$ are then deduced. The serial resistance and the avalanche region heating largely decrease the opto-thermal gain after the first switching and these reasons explain why the second threshold power, observed for decreasing incident power is greater than the first one.

4 Potentiality of APDs in 2D Arrays

Conclusions of the previous analysis are: the avalanche region pn$^+$ has to be in close contact with the heat sink in order to keep the maximum optothermal gain, reduction of the heat power imposes laterally pixelated thin film APDs with transverse active region size comparable to the input beam diameter. Considering at first a single cylindrical pixel with a height equal to the thickness of the previous APD ($D = 120$ μm) and a diameter $F = 30$ mm, its thermal resistance is easily calculated considering the heating power density uniform along the pixel:

$$R_T = \frac{4D}{3\pi \Lambda_{Si} \Phi^2} = 377 \text{ KW}^{-1} \ ,$$

where $\Lambda_{Si} = 1.5$ Wcm^{-1}K^{-1} is the heat conductivity of silicon, this thermal resistance is 44 times larger than the one deduced from the experiment. The FP resonator is optimized with $R_B = 0.80$ and $R_F = 0.63$ then $A_+ = 0.6$, the finesse is $\mathcal{F} = 6$ and gives $\Delta T_+ = 4°$C which is obtained with $P_{h+} = 4°$C/377 = 10.5 mW. Considering an optothermal gain $G = 19000$ as in the experiment, an optical threshold $P_{i+} \approx 1$ mW is found and the 2D matrix size estimated with a maximum heat power of 10 W per cm^2 is 30×30 pixels per cm^2.

4.1 Matrix optimization

Heat power per pixel limits integration density; it can be reduced considering thin film APD. The expression of Ph+ is the following:

$$P_{h+} = \frac{\Delta T}{R_T} = \frac{3\pi \lambda \Lambda_{Si} \Phi^2}{8\mathcal{F}\frac{dn}{dt}D^2} = GA_+ P_{i+} \ .$$

If the ratio F/D of the beam diameter to the silicon plate thickness is fixed (as a geometrical factor of the FP related also to the incident beam), reduction of D is then only beneficial to integration density if the finesse is increased. The back mirror reflectivity being fixed by the metallization, the new optimum front mirror reflectivity is given by: $R_F = R_B \exp(-2\alpha D)$. Reduction of D decreases the cooling time τ which is the limit in speed of these optothermal devices. For the device geometry, the following expression of τ can be used:

$$\tau \approx \frac{4C_{Si}D^2}{\Lambda_{Si}\pi^2} \approx 60 \ \mu s \ \text{for} D = 120 \text{ mm} \ .$$

$C_{Si} = 1.6$ Jcm^{-3}K^{-1} is the volumic heat capacity of silicon. With a reduction by a factor of 10 of the pixel sizes, $D = 12$ mm and $F = 3$ mm, the previous value of the cooling time is reduced by a factor of 100 and $\tau = 0.6$ s. Considering a higher reflectivity metallization with $R_B = 0.90$, the FP is optimized with a finesse of 24 giving a maximum absorptance $A_+ = 0.23$. The heating power is $P_{h+} \approx 2$ mW, the optical threshold P_{i+} is close to 1 μW and has been calculated with a lower optothermal gain equal to 10^4 owing to the lower avalanche voltages in thin film APDs. The resulting 2D matrix size is 70×70 pixels per cm^2.

5 Conclusion

The integration density of the 2D matrix could be increased reducing the electrical power per pixel with thinner optimized FP resonator of higher finesse and using back metallization of higher reflectivity at 1.06 μm. The technological processes for realisation of matrix of thin film APDs for optical bistability require silicon plate thickness uniformity with specifications related to the high finesse FP resonators. Nevertheless, evaluations of the potentiality following the first experiments in optical bistability made with a commercial Si APD have shown: μ optical power thresholds, submicrosecond cooling time and pixel integration density close to 100×100 per cm^2, the high optical sensitivity and also the well known robustness of silicon devices are particularly interesting in optical information technology.

Acknowledgments

I thank Professor S. Cova from the Politechnico di Milano who in the WOIT context got me in touch with Dr. A. D. MacGregor and R. J. MacIntyre from EG&G Canada Ltd. Optoelectronics Division. I thank them for supplying the APDs and for their interest in this particular application of APDs. I thank Sylvain Meunier for his participation to the experiments, Bernard Abecassis for the realization of the FP set-up and my colleagues at IEF for numerous discussions on this topic.

References

1. Sauer, H., Paraire, N., Koster, A., Laval, S.: Optimization of a S.O.S. waveguide device for optical bistable operation. J. Opt. Soc. Am. B **5** (1988) 443.
2. Jäger, D., Forsmann, F.: Optical, optoelectronic and electrical bistability and multistability in a silicon Schottky SEED. Solid State Electr. **30** (1987) 6771.
3. Thienpont, H., Vanholder, S., Ranson, W., De Tandt, C., Vounckx, R., Veretennicoff, I.: In Thermo-optic SEED arrays for optical information processing. ECO3, The Haag, Proc. SPIE **1280** (1990).

On the Feasibility of Avalanche Devices for Optical Switching in Silicon

S. Cova,[1] A. Lacaita,[1] M. Ghioni,[1] and G. Ripamonti[2]

[1] Politecnico di Milano, Dipartimento di Elettronica e Centro di Elettronica
Quantistica e Strumentazione Elettronica - CNR
Piazza L. da Vinci 32, I-20133 Milano, Italy
[2] Università degli Studi di Milano, Dipartimento di Fisica,
Via Celoria 16, I-20133 Milano, Italy

We discuss some silicon Avalanche Photodiode (APD) structures where electronic optical nonlinearity could be exploited for optical bistability. The basic idea is to use the avalanching junction as a pilot of other devices where the carriers are effectively stored. We show that these structures can reach a switching power in the milliwatt range. However, thermal effects are always expected to overcome the electronic response if the sample temperature is not kept constant. We show the principle of operation of thermally-compensated structures with switching times in the microsecond range.

1 Introduction

Semiconductors operated near the band-edge are the most promising candidates as non-linear optical materials. Although compound semiconductors have better optical properties than silicon, the powerful silicon technology is still considered essential for monolithic integration of optical systems. Therefore, many efforts are devoted to develop all-silicon based optoelectronic devices.

From the physical point of view, it is well known that the only useful optical non-linearities in silicon can arise from interaction of optical radiation with the free carrier plasma [1], and thermal effects [2]. The use of free carrier absorption has been successfully demontrated in fast silicon optical modulators [3], while thermal effects have been exploited in Schottky self-electro- optic devices (SEEDs) [4]. Thermo-optical SEEDs show bistable characteristics with high contrast ratios and switching power in the milliwatt range. Some experiments have also shown that the switching times of silicon SEEDs are in the microsecond range [5] .

When an avalanche enhancement of thermal optical nonlinearities in silicon was reported [6] , scientists began to look for suitable Avalanche Photodiode (APD) geometries where electronic nonlinearity could be exploited. However, electronic effects in silicon are so poor that a free carrier density higher than 10^{18} cm^{-3} is needed to have a significant change of absorption and refractive index. Silicon modulators and bistable devices can exploit electronic responses only working under very intense optical excitation and high electrical power dissipation. In this paper we show that a simple APD etalon is not feasible. If *no* thermal effects are taken into account, the enormous current density of 10^6 Acm^{-2} with a power dissipation of 200 MWcm^{-2} would be needed for switching!

We have devised some structures to overcome the limits of the APD-based etalon. The basic idea is to use the avalanching junction as a pilot of other devices where the carriers are effectively stored. In these structures the APD is used as the optically sensitive device of the cavity, thus reducing the optical switching power in the milliwatt range. However, due to the poor optical properties of silicon, current densities as high as 10^4 Acm^{-2} are always needed. Therefore, in CW operation, thermal effects are expected to overcome the electronic response if the thermal dissipated power is not kept constant independently on the switch state. Here we show the principle of thermally compensated structures with switching times in the microsecond range.

2 Non-linear optical effects in silicon

In silicon, the refractive index changes, Δn, for near band-edge wavelengths, depends on the free carrier density, ΔN, and the sample temperature changes ΔT. The former contribution is given by [2]:

$$\Delta n = -k\Delta N \ , \tag{1}$$

where $k = 9 \times 10^{-22}$ cm^3; the latter is positive and given by:

$$\Delta n = h\Delta T \ , \tag{2}$$

where $h = 2 \times 10^{-4}$ K^{-1}.

Just to make an estimate, let us consider a simple Fabry-Perot structure given by a silicon sample, with a thickness $d = 500$ μm. The sample is excited by a CW Nd-YAG laser beam at 1.06 μm, with a spot size $w = 100$ μm, which induces a change, Δn, in the refractive index of the sample. This change is due to both free carrier photogeneration and temperature variation. The optical transmitted power shows an abrupt increase when the change of the optical cavity thickness, Δnd, is equal to $\lambda/2 = 0.5$ μm.

Let us consider first the effect due to carrier photogeneration. Based on (1) we estimate that a free carrier density, $\Delta N = 10^{18}$ cm^{-3} is needed to reach resonance. At this carrier density, Auger recombination is dominant and carrier lifetime, τ, is about 10^{-6} s. At steady state, the carrier recombination rate is equal to the photogeneration rate, that is:

$$\frac{\Delta N}{\tau} = \frac{\Phi\alpha}{h\nu} \ , \tag{3}$$

where Φ is the optical power density, $\alpha = 12$ cm^{-1} is the absorption coefficient at 1.06 μm, and $h\nu$ is the photon energy. From (3) a switching power density of 1.6×10^4 Wcm^{-2} is obtained, which corresponds to a laser power of about 3 W.

The absorbed power will rise the local temperature. Let us assume that the heat sink is on the edge of the sample. The heat flows from the area where the laser beam is focused through a thermal resistance, R_{th}, given by:

$$R_{th} = \frac{\rho}{2\pi d} \log\left(\frac{D}{w}\right) \ , \tag{4}$$

where, ρ is the material thermal resitivity, $D = 2.5$ cm is the outer sample diameter and $w = 100$ μm is the laser beam spot size. Taking, for silicon, $r = 1$ Kcm/W, R_{th} is 16 K/W. The overall absorbed power is about half the incident laser power, i.e. 1.5 W, the estimated average temperature increase is $\Delta T = 25$ K and the corresponding refractive index change is 5×10^{-3}. This means that the thermal effects will cause a change in the optical cavity thickness, $\Delta nd = 2.5$ μm, five times larger than the electronic contribution. Thermal resistances of a few tens of K/W are typical [4], therefore the above figures show that as long as the overall power dissipation is higher than 0.1 W per pixel, thermal effects overcome electronic nonlinearity in CW operation.

3 APD-based structures

Avalanche multiplication of photogenerated carriers was proposed for reducing the optical switching power. Let us take an APD with a 10 μm thick depletion layer and breakdown voltage of about 200 V. It should be pointed out that the photogenerated carriers are multiplied by impact ionization but they drift at saturated velocity, 10^7 cm/s, under the influence of the junction electric field. Therefore, taking an APD gain of 100, the free carrier density of $\Delta N = 10^{18}$ cm^{-3} is reached at a current density of 10^6 Acm^{-2}, with an absorbed optical power of 10 KWcm^{-2} and a corresponding power dissipation of 200 MWcm^{-2}. This result is due to both the high carrier velocity, which prevents carrier storage, and the high electric field, which is intrinsically needed to impact ionize.

As a following step we considered a structure where an APD is used as optical sensing element and the multiplied avalanche current is more effectively stored in a bipolar device. A schematic structure is shown in Fig. 1 where the avalanche current leads to a minority carrier storage in a p-n junction. Taking the above carrier lifetime of 10^{-6} s and a sample thickness of 500 μm, a current density of 8×10^3 Acm^{-2} is necessary to reach 10^{18} cm^{-3} stored carriers. This means about 1 A flowing through a 100 μm $\times$ 100 μm pixel. Taking into account that the whole current flows through the APD, it is clear that, the corresponding power dissipation is still very high, namely 200 W per pixel.

A significant reduction of both the optical switching power and the electrical power dissipated in the photodiode could be achieved providing an electronic amplification of the APD current. Taking a gain of 100 for both the APD and the following amplifier, the current of 1 A in the storage junction is obtained with an APD avalanche current of only 10 mA. Taking into account that APD collects carriers photogenerated in the neutral region within a diffusion leght from the depletion layer border we estimate that a photocurrent of 10 mA is obtained with

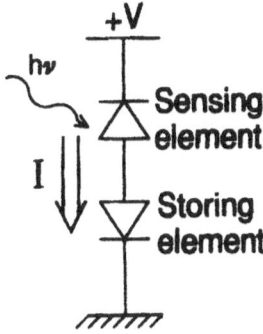

Fig. 1. Principle of operation of an APD-based etalon where carriers multiplied by impact ionization are stored in a separate p-n junction.

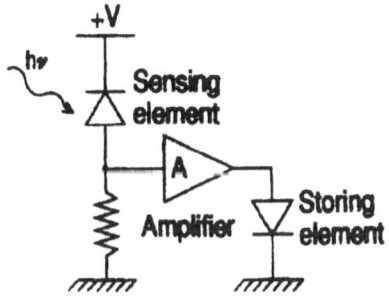

Fig. 2. APD-based structure. The APD current is only a fraction of the current flowing in the storing element.

an optical power of only 3 mW. The optical switching power for the structure in Fig. 2 is less than 10 mW per pixel, which is at least three orders of magnitude less than the switching power for the bulk silicon Fabry-Perot. However, at high current density, the voltage drop across any forward biased junction is at least 1 V. Therefore, also neglecting the power dissipated in the amplifying stages, a current of 1 A will cause at least a power consumption of 1 W per pixel. We conclude that, due to high power dissipation, any silicon-based bistable device is affected by thermal effects which are stronger than the electronic non-linearities.

In principle, thermal effects could be avoided by keeping constant the pixel temperature. Fig. 3 shows the basic idea of such devices. The current of an external source is alternatively switched between the arms of a differential stage. The photocurrent of an APD drives the transistor T_1 on, current flows through a p-n junction, thus storing the carriers needed for refractive index variation.

As the optical beam is switched off, the APD photocurrent becomes zero, T_1 is turned off and the external current flows through T_2 keeping constant the dissipated power. Note that the current flowing in T_2 does not cause storage of minority carriers, and the pixel is optically off. This solution could allow optical bistability due to electronic non linearities with switching times of the order of the minority carrier lifetime. However, a constant dissipated power of the order of 1 W per pixel is still needed.

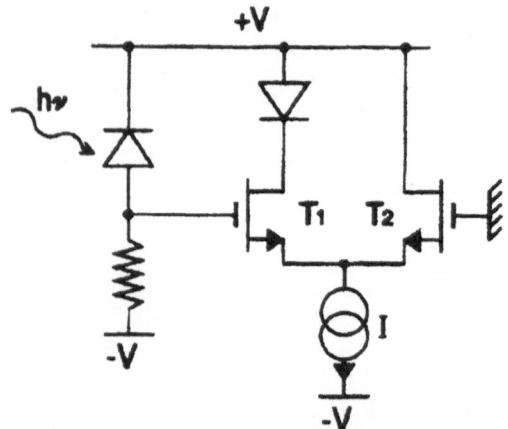

Fig. 3. Thermally compensated structure. The dissipated power is held constant by switching the current between T_1 and T_2

4 Conclusions

In conclusion, we have studied the feasibility of all-silicon optical bistable elements exploiting electronic nonlinearities. The use of simple devices based on avalanche multiplication is ruled out by the need of enormous optical power. By amplifying the electrical output of the APD, it is possible to achieve a switching power in the milliwatt range. However, electronic effects are submerged by thermal effects if the dissipated power is not held constant independently on the switch state. The dissipated power in this case is significant, but does not interfere with the optical properties of the material.

References

1. Soref, R. A., Bennet, B. R.: Electrooptical Effects in Silicon. IEEE J. Quantum. Electron. **QE-23** (1987) 123.
2. Eichler, H. J.: Optical Multistability in Silicon observed with a CW laser at 1.06 μm. Opt. Commun. **45** (1983) 62.

3. Treyz, G. V., May, P. G., Jean-Marc Halbout: Silicon Optical Modulators at 1.3 μm based on Free Carrier Absorption. IEEE Elec. Dev. Lett. **12** (1991) 276.

4. Forsmann, F., Jäger, D.: Thermo-optical SEED devices: External Control of Non-linearity, Bistability and Swtching Behaviour. Appl. Phys. B **45** (1988) 151.

5. Jäger, D., Forsmann, F.: Optical, Optoelectronic and Electrical Bistability and Multistability in a Silicon Schottky SEED. Solid State Electron. **30** (1987) 67.

6. Horbatuck, S. M., Prelewitz, D. F., Brown, T. G.: Avalanche enhancement of optical nonlinearities in semiconductor junctions. Appl. Phys. Lett. **56** (1990) 2387.

Sensitivity and Switching Contrast Optimization in an Optical Signal Processing Waveguide Structure

N. Paraire, P. Dansas, A. Koster, M. Rousseau, and S. Laval

Institut d'Electronique Fondamentale - Bât. 220, CNRS URA 22 - Université Paris Sud, F-91405 Orsay Cedex, France

In order to optimize an optical switching device, several quantities need to be considered; these include up and down switching times (connected with the nonlinearity mechanism), pixellation feasibility (which depends on device structure, heat dissipation and wafer uniformity), sensitivity and switching contrast (which is important for fan-out), technological complexity and cost.

Sensitivity and switching contrast optimization have been studied [1-3] for a nonlinear (NL) Fabry-Pérot working in the reflective mode. It has been shown that they depend on the reflectivity of the front mirror (R_F), of the back one ($R_B \approx 1$) and on the NL film thickness, and that linear resonance finesse and contrast must be simultaneously optimized.

Among the relevant variables for switching device optimization, we consider here the same parameters for a NL waveguide (WG) excited via a grating coupler and working, as a switch, either in the reflective mode (device transmission $T \approx 0$, reflectivity of the layers under the NL film $R_B \approx 1$) or in a mixed mode ($T \neq 0$, device reflectivity $R \neq 0$, $R_B \ll 1$), in order to define a figure of merit for both operating modes.

An optical signal processing device is all the more interesting, as a small input signal P_{in} (or signal variation δP_{in}) induces a large variation of the output signal(s) δP_{out}. Two quantities interfere here: sensitivity s, that we characterize in the following by the smallest input signal P_{in}^s inducing switching, and switching contrast c characterized by the relative output signal variation δS^s occurring during switching ($S = T, R$). Both quantities can be deduced from the transmissivity $T(n, \theta)$ and reflectivity $R(n, \theta)$ curves, where θ is the incidence angle and n the refractive index of the nonlinear film. The curves $T(n_0, \theta)$ and $R(n_0, \theta)$ – where n_0 is the linear refractive index – are determined experimentally; $T(n, \theta_0)$ and $R(n, \theta_0)$ are easily deduced from the former [4]. So, s and c can be optimized through the geometrical characteristics of the device.

Indeed, the absorption coefficient $A(n, \theta)$, defined [5] as $A(n, \theta) = 1 - R(n, \theta) - T(n, \theta)$ and deduced from the previous curves, is a classical resonance curve characterized by a maximum A_{max} and a full width at half maximum FWHM $= \Delta\theta_{|n_0}$ or $\Delta n_{|\theta_0}$. In the stationary regime, for an initial angular detuning from resonance $\delta\theta$, the device operating point [4] is on the curve $A(n, \delta\theta)$, with A and n verifying in other respects $n - n_0 = B(\lambda, \tau, e, \theta)A(n, \delta\theta)P_{in}$. The coefficient B depends on θ, on the wavelength λ, on the film thickness e and on the nonlinearity mechanism into action τ: $B = B_e < 0$ for electronic effects, $B = B_T > 0$ for thermal ones. For a given device, there is a critical detuning $\delta\theta_c$ for which the bistable cycle area is zero and the corresponding incident power P_{in}^s is minimum. The quantity $1/P_{in}^s$ is then proportional to the slope of the common tangent to the curves $A(n, \delta\theta)$ and $n - n_0 = f(A)$, which is well approximated by $A_{max}/\Delta n$ or $A_{max}/\Delta\theta$ which is used below to determine the device sensitivity.

If the device is used as a single switch, the switching contrast is measured by the change: $\delta S^s = S_{max}^s - S_{min}^s$ occurring during switching. On the contrary, if the device is used as a logic gate to control others or as an amplifier, the use of several input beams is necessary as the incident beam is always attenuated, $\delta P_{out} < P_{in}$. One holding beam defines the operating point and one (or several) other beam(s) act as control beam(s). Device 1 holding beam is device 1 output beam and device 2 control beam. S_{min}^s is then an interfering signal on device 2 which breaks in to define its operating point and limits its sensitivity. It must be as small as possible and it seems convenient to define the contrast as the ratio $\delta S^s / S_{min}^s$, a definition that we extend to all cases. In the following, we choose the bistable cycle associated with the detuning $\Delta\theta$ to define a mean contrast $c = \langle |\Delta S^s / S_{min}^s| \rangle_{\Delta\theta}$ on this cycle and a figure of merit for these devices as $F = (A_{max}/\Delta\theta) \langle |\Delta S^s / S_{min}^s| \rangle_{\Delta\theta}$ quantity which must be maximized.

We note that, for a given detuning $\delta\theta$, switching induces an index change δn (determined graphically on the $A(n, \delta\theta)$ curve) which, in turn, induces a change δS^s (on the $S(n, \delta\theta)$ curve which determines the switching contrast. This graphical method is represented in Fig. 1 for an air/silicon/sapphire device. The silicon film thickness is $e = 670$ nm (before grating etching), the diffraction grating at the interface air/silicon has a period $p = 308$ nm and a modulation depth $h = 140$ nm. It allows one to draw a transfer characteristic $P_{out} = f(P_{in})$, or $S = f(P_{in})$, with, here, $S = T$ and $\delta\theta = \Delta\theta$. As can be seen in Fig. 1, δS^s is not the largest S variation $\Delta S = S_{max} - S_{min}$ observable on the $S(n, \delta\theta)$ or the $S(n_0, \theta)$ curve: the latter is an apparent contrast which cannot be obtained during switching.

Moreover, $\langle |\Delta S^s / S_{min}^s| \rangle_{\delta\theta}$ is very sensitive to $\delta\theta$ and to the shape of the curve $S(n, \theta)$. For a device working in the reflective mode, $S = R = 1 - A$, so $|\Delta S^s| = \Delta A^s$: contrast and sensitivity are simultaneously optimized when $A_{max}/\Delta\theta$ is maximum.

On the contrary, for a device operating in a mixed mode (this case being very interesting as one can obtain simultaneously different logic operations on the various output beams) ΔS^s and ΔA^s are not related by a bijection and a trade-off between sensitivity and switching contrast must be found to optimize F. Moreover, S is not a symmetrical function of n (see Fig. 1) and so, for two opposite values $\pm\delta\theta$ of the initial angular detuning, two values of δS^s can be observed: indeed, the geometrical optimization of the device depends on the nonlinearity under consideration.

At IEF, theoretical analysis and a numerical model have been developed to describe the linear optical properties (R, T, A) of stratified structures including a deeply modulated sine-profiled diffraction grating, in order to optimize such devices. This has been performed in collaboration with F. Lederer and co-workers

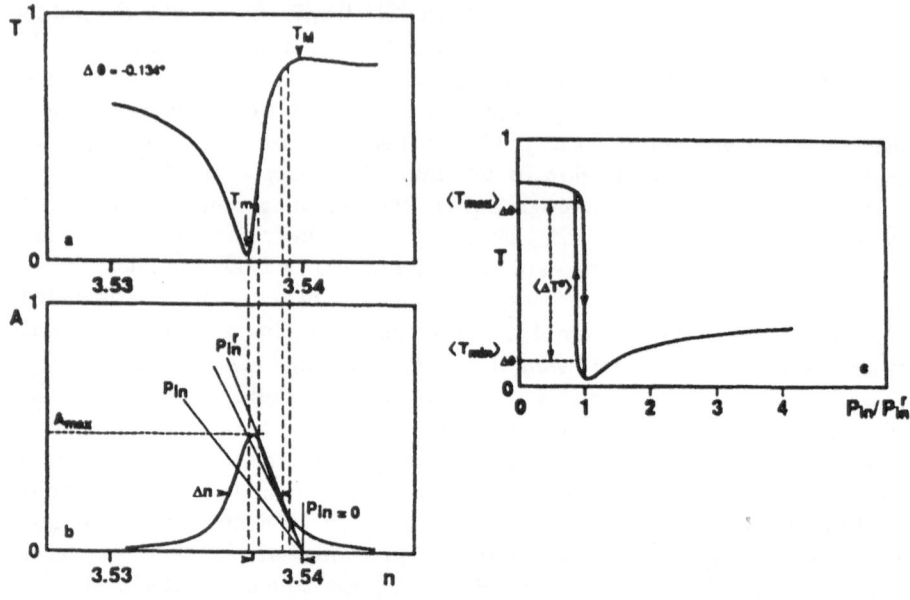

Fig. 1. a) Transmission $T(n)$ b) Absorption $A(n)$ c) Transfer characteristic $S(P_i)$ deduced from a) and b) for an air/silicon/sapphire device with $e = 670$ nm, $h = 140$ nm, $p = 308$ nm, $n_0 = 3.54$ and an initial angular detuning $\Delta\theta = -0.134°$.

from Jena University and confirmed by comparison with simulations from the more general and exact study of P. Vincent and co-workers at LOE in Marseille. For given characteristics of the incident beam (λ, divergence), three parameters can be optimized; the film thickness e, defined before grating etching, the grating modulation depth h and the reflectivity R_B of the layers under the nonlinear film ; these parameters are not independent.

It has been shown that a WG device working in the reflective mode exhibits, for any e value, at least one value $h_0(e)$ of h which minimizes R at resonance (R_m). For suitable values of R_B, R_m becomes nil. This can be seen in Fig. 2 where the reflection coefficient of a silver/silicon/sapphire WG with a grating at the silver/silicon interface has been reported as a function of θ for 3 TE propagation modes (white circles).

The parameter h_0 is very sensitive to the energy distribution in the WG section, especially near the modulated interface. Thus, it varies with the propagation mode under study and decreases as the mode order increases: in Fig. 2, $h_0(\mathrm{TE}_0) = 410$ A, $h_0(\mathrm{TE}_2) = 226$ Åand $h_0(\mathrm{TE}_3) = 146$ Å. However, the FWHM increases simultaneously and the best figure of merit is then obtained on the fundamental mode. h_0 also varies with R_B: in Fig. 2, curves $R(\theta)$ have also been plotted assuming no absorption in silver (black circles): this affects

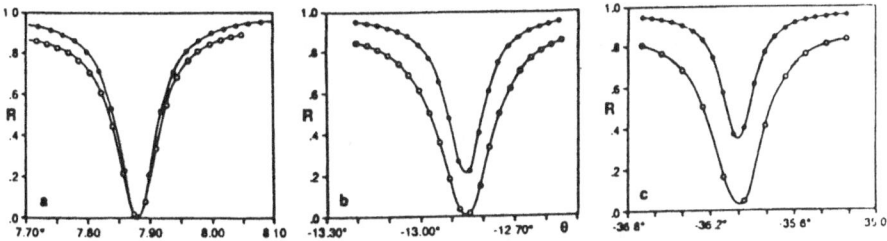

Fig. 2. Reflection curves R(q) for a sapphire/silicon/silver device with an absorbing (o) or non-absorbing (•) substrate (silver) and for 3 propagation modes a) TE_0: $h_0 = 410$ Å b) TE_2: $h_0 = 226$ Å c) TE_3: $h_0 = 146$ Å

particularly the highest order modes.

Furthermore, the larger the device reflectivity before etching, the larger is $h_0(e)$. This is shown in Fig. 3 where optimized $R(\theta, h_0)$ curves have been drawn for various e values and 2 propagation modes. Although the device reflectivity before etching varies from 0.62 to 0.90 for an optical thickness variation of $\lambda/4$, for optimized devices R_M varies only in the range 0.79–0.82 in the case of the TE_0 mode and 0.83–0.86 for the TE_1 mode. Moreover, coupling between TE_0 and TE_1 modes distorts some resonance curves (for example $a-1$) which do not seem any longer symmetric. Once h is optimized, the device figure of merit does not vary much with e. Such reflective mode devices are then very interesting when they operate in the fundamental mode and are easy to optimize as long as $h_0(e, R_B, TE_0)$ is well controlled.

For WG devices working in the mixed mode (i.e. with $R_B \ll 1$), usually curves $S(\theta$ or $n)$ are not symmetric: optimization then depends on the NL effect and output signal under consideration. Most of the numerical simulations have been performed assuming an air/silicon/sapphire device with a diffraction grating at the air/silicon interface.

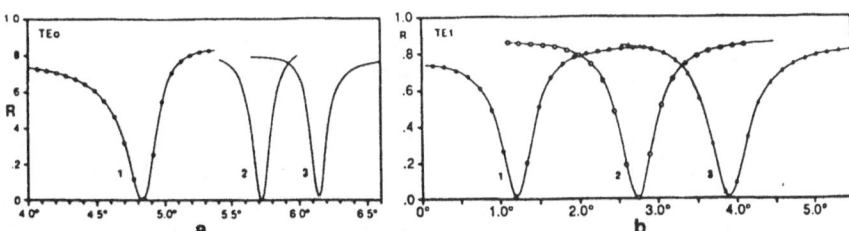

Fig. 3. Reflection curves $R(\theta)$ for optimized sapphire/silicon/silver devices, for 2 propagation modes and various initial thicknesses of the silicon layer. a) TE_0 mode, grating period $p = 325$ nm: 1) $e = 590$ nm, $h_0 = 1220$ Å; 2) $e = 620$ nm, $h_0 = 774$ Å; 3) $e = 650$ nm, $h_0 = 583$ Å b) TE_1 mode, grating period $p = 330$ nm: 1) $e = 580$ nm, $h_0 = 615$ Å; 2) $e = 610$ nm, $h_0 = 706$ Å; 3) $e = 640$ nm, $h_0 = 624$ Å

In Fig. 4, we have reported the absorption coefficient and the associated transmission coefficient for the TE_1 mode of devices where the mean geometrical thickness $e - h/2$ has been kept constant (600 nm) and the modulation depth h has been varied (from 100 to 220 nm), the grating period being 308 nm. In Fig. 5, the grating period is 330 nm, the modulation depth is kept constant ($h = 80$ nm) and the mean geometrical thickness is varied (from 575 to 675 nm), the TE_2 mode is observed. In both cases, it can be seen that A_{max} presents a maximum and so does the ratio $A_{max}/\Delta\theta$ when either e or h varies. The shape of the transmission curve changes with e and h and can be either symmetric ($e = 690$ nm, $h = 180$ nm in Fig. 4, $e = 635$ nm, $h = 80$ nm in Fig. 5), or asymmetric. In Fig. 4, the seeming contrast and the actual one $(\Delta T^s/T_{min})_{\Delta\theta}$ increase as sensitivity decreases but the figure of merit in transmission is optimum for the symmetric case ($F \approx 17$ for curve 1, $F \approx 33$ for curve 3, in arbitrary units, assuming electronic effects, i.e. an initial angular detuning < 0). In Fig. 5, contrast in transmission is maximum for the symmetric case (curve 4) and the figure of merit is maximum for various configurations (curves 2 and 4, for electronic effects). Indeed it is often interesting to choose the symmetric configuration, as it is more versatile (it can be used either for electronic or for thermal effects).

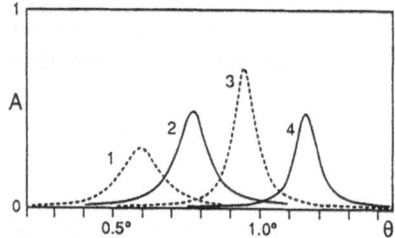 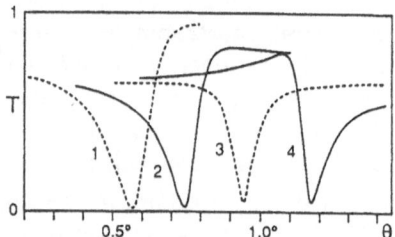

Fig. 4. Absorption $A(\theta)$ and transmission $T(\theta)$ curves for the TE_0 mode of air/silicon/sapphire devices: $e - h/2 = 600$ nm, grating period $p = 308$ nm – 1) $h = 100$ nm; 2) $h = 140$ nm; 3) $h = 180$ nm; 4) $h = 220$ nm

As expected, in this mixed mode, contrast and sensitivity maxima are not observed for the same geometrical parameters and a trade-off must be chosen, which is well described by the figure of merit.

Experimentally both devices (working in reflective and mixed modes) have been studied. A sapphire/silicon/silver one has been optimized for the TE_0 mode. For an initial film thickness $e = 627$ nm (known from ellipsometry measurements) and a grating period p = 330 nm, ho is equal to 80 nm and $R_{min} = 0$. Experimentally, $h_0 \approx 60$ nm and $R_{min} \approx 6\%$ (see Fig. 6a). For negative angular detuning, it allows the observation of very contrasted switchings [6] (Fig. 6b) on the reflected beam in the nanosecond time scale.

An air/silicon/sapphire device has also been constructed to work in the transmission mode on thermal effects. A quasi-symmetric transmission curve has been

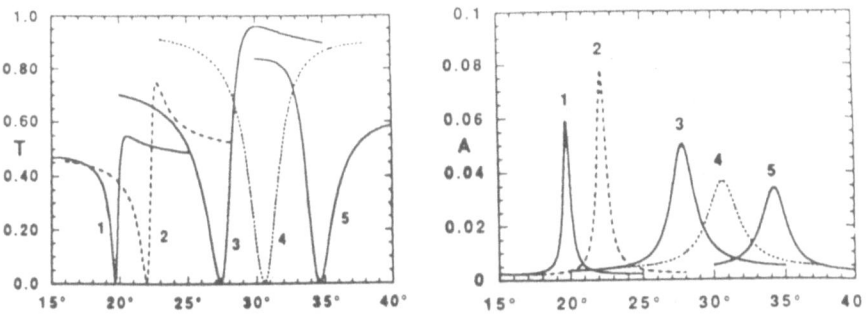

Fig. 5. Absorption $A(\theta)$ and transmission $T(\theta)$ curves for the TE$_2$ mode of air/silicon/sapphire devices: $h = 80$ nm, grating period $p = 330$ nm, the geometrical thickness $e - h/2$ is varied 1) $e - h/2 = 675$ nm; 2) $e - h/2 = 655$ nm; 3) $e - h/2 = 615$ nm; 4) $e - h/2 = 595$ nm ; 5) $e - h/2 = 573$ nm

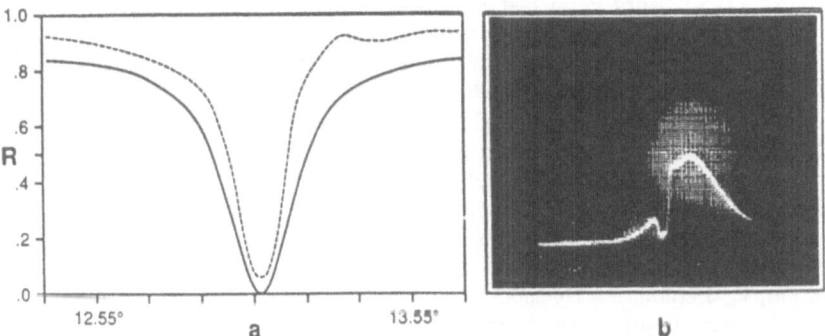

Fig. 6. Experimental sapphire/silicon/silver device a) Resonance reflection curve for the TE$_0$ mode (— theoretical optimized curve ($h_0 = 80$ nm); - - - experimental curve ($h_0 \approx 60$ nm)) b) Reflected power versus time, for an incident light pulse $\Delta t = 20$ ns and an initial angular detuning $\delta\theta = -0.20°$.

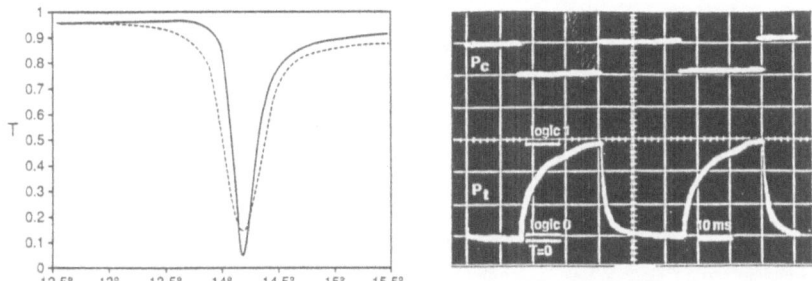

Fig. 7. Experimental air/silicon/sapphire device – a) Transmission curve for the TE$_0$ mode (- - - experimental curve; — theoretical curve ($h_0 = 80$ nm, $e = 635$ nm)) b) Operation of the device as a logic gate: the output beam is the reflected holding beam on the TE$_0$ mode, the control beam is on the TE$_1$ mode

chosen, which is obtained for a sample with $e = 635$ nm for $h \approx 80$ nm (Fig. 7a) and the corresponding device has been used as a logic NOR gate [7]. The holding beam is simultaneously the output beam. For a given operating point ($\delta\theta = 0.20$), a modulated signal introduced on the TE_1 mode, as a control beam, induces very contrasted switchings on the output beam (Fig. 7b).

In conclusion, it is possible to define a relevant figure of merit, taking into account the necessary trade-off between sensitivity and switching contrast, which is valid for NL waveguide switches working either in the reflective mode or in the mixed mode. The optimization depends on the operating mode under consideration, and the influence of the relevant parameters has been studied theoretically and experimentally in several configurations.

References

1. Wherrett, B.S.: Fabry-Perot bistable cavity optimization on reflection. IEEE J. Quantum Electron. **QE-20** (1984) 646.
2. Garmire, E.: Criteria for optical bistability in a lossy saturating Fabry-Perot. IEEE J. Quantum Electron. **QE-25** (1989) 289.
3. Sfez, B.G., Oudar, J.L., Michel, J.C., Kuszelewicz, R., Azoulay, R.: High contrast multiple quantum well optical bistable device with integrated Bragg reflectors. Appl. Phys. Lett. **57** (1990) 324–326.
4. Sauer, H.: Etude theorique du fonctionnement dynamique de dispositifs a guide d'onde non lineaire pour la commutation optique. These de Doctorat, Orsay (1990).
5. Vincent, P., Paraire, N., Neviere, M., Koster, A., Reinisch, R.: Gratings in nonlinear optics and optical bistability. J. Opt. Soc. Am. **2** (1985) 1106–1116.
6. Berard, D., Paraire, N., Chi, W.D., Koster, A.: Using silicon nonlinearities in waveguides for passive nanosecond optical pulse shaping at $\lambda = 1.064$ μm. Annales de Physique, Suppl. n°1 **16** (1991) 63–72.
7. Chi, W. D.: Etude et realisation d'un dispositif guide d'onde thermo-optique en silicium sur saphir pour la bistabilite et la logique optiques. Thèse de Doctorat, Orsay (1991).

A Passive Crystal Pixel Interchanger

C. De Tandt, W. Ranson, P. Schrey,* R. Vounckx, and R. Cottam

Applied Physics, Brussels University (VUB), Pleinlaan 2, B-1050 Brussels, Belgium

Preferentially etched silicon structures are proposed which would enable the partial interchange of nearest neighbour optical image pixels and a degree of binary optical computation. Simple image processing is simulated, and noise reduction and image matching are described.

1 Introduction

We propose an infrared-transparent silicon wafer structure which can be used to facilitate simple optical image processing of binary (*black or white*) pixeled images. The intention is to apply it to selected specific applications rather than to generalised two dimensional computing, with external programming and structural complications kept to a strict minimum. Currently envisaged processing is restricted to nearest neighbour interactions, although other more long range possibilities capable of global interactions are foreseen.

2 The Basic Pixel Exchanger

We have limited ourselves to using only the four very nearest neighbours in a conventional orthogonal image array, and have even left out of neighbour combinations the *original* pixel. This can be done by splitting each pixel into four parts and distributing them to the adjacent pixel sites (Fig. 1). The new four quarter-pixels can then be recombined to form new full-pixels, or in a more advanced scheme can be treated as separate entities. It is still possible to consider carrying out extensive image processing with this arrangement, although with a resolution lower than that attained by conventional techniques.

The pixels are split up into parts by projecting them onto regions of the wafer surface which are shaped so as to refract the individual partial-pixels in different directions inside the wafer. At the opposite wafer face the refraction is reversed, so that the partial-pixels regain their originally projected direction (Fig. 1). The result is that pixel beams can be transmitted straight through the structure, but the output beams will now consist of combinations of the input beam nearest

---

* Now with Alcatel-Bell, Antwerp, Belgium

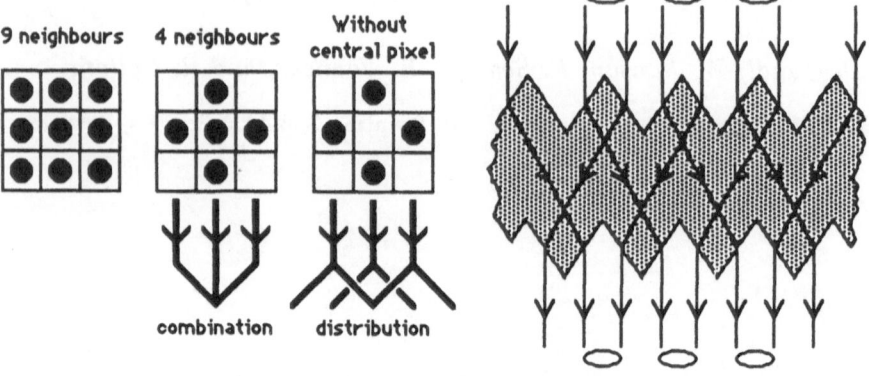

Fig. 1. Nearest neigbour distribution pattern and cross-section of the distributor structure

neighbours only. The three dimensional implementation of this idea consists of a parallel polished silicon wafer, with an array of pyramid pits on one side and an aligned array of pyramid hills on the other (Fig. 2).

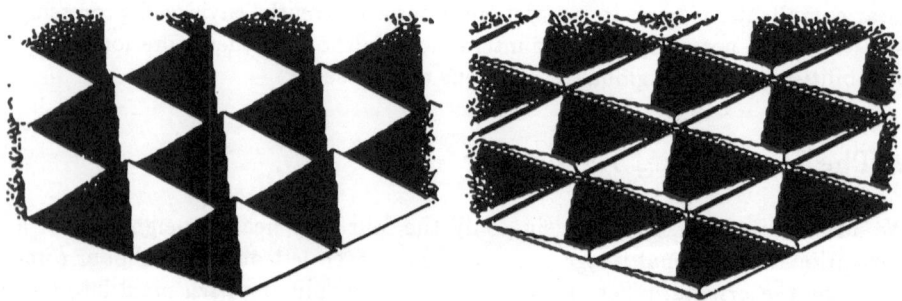

Fig. 2. Pyramid hill and pyramid pit arrays for the two wafer surfaces

There are two possible locations for incident pixel beams arriving perpendicularly to the first wafer surface (Fig. 3); they can be centred either on the tops of hills, giving *cross* (X) distribution (Fig. 4), or in the bottoms of pits, providing a different *spread* (S) distribution. Either cross or spread distribution results in the quarter-pixels being split out into their neighbour pixels, but their resultant locations depend on the kind of distribution. However, if now output full-pixels are constituted wherever there is a quarter-pixel inside the area corresponding to an input pixel, the final effect is the same. In Fig. 4 the final full-pixels appear where there is a single transferred quarter-pixel present. This is the lowest level version of the simplest kinds of processing which can be used to convert quarter-pixel combinations into new full-pixels, which are quarter-pixel combination thresholding and thresholding followed by inversion. Straightforward thresholding gives full-pixels for any 1 of 4 quarters (AND1), any 2 of 4 quarters (AND2), any 3 of 4 (AND3), or all 4 (AND4). Subsequent inversion yields four more *logic functions*, NAND1, NAND2, NAND3 and NAND4.

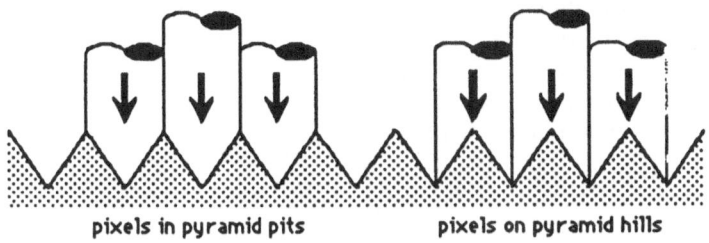

pixels in pyramid pits   pixels on pyramid hills

Fig. 3. The two possible incident pixel beam locations

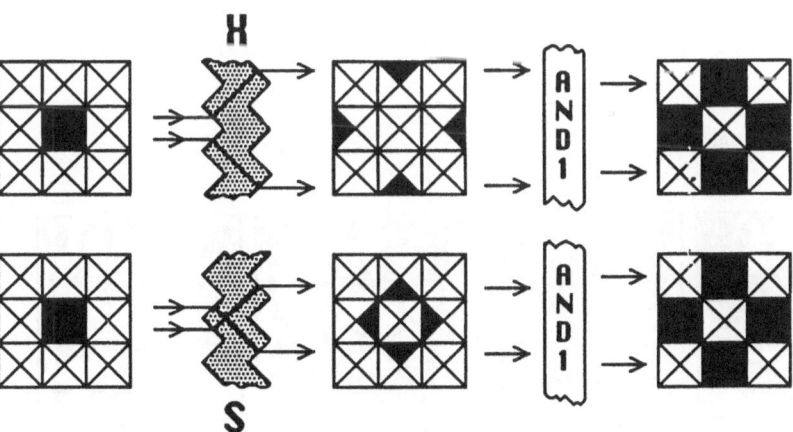

Fig. 4. Comparison of the final results of cross-distribution and spread-distribution, after full-pixel AND1 thresholding

3 Simple Image Processing

The possible combinations of the different distributions, the different logic functions, and the possibility of a re-cycling or looping operation which will pass the image repeatedly through the same set of silicon and logic planes result in a range of interesting non-programmed processes. We have carried out a number of simulations using different processing-component combinations to evaluate the usefulness of structures based on these ideas. The component arrangement is shown as on the left of Fig. 5, the initial image is put in once, and the subsequent operation sequence is indicated. For most of the following examples we have used the initial pixeled image shown in Fig. 5, or a part of it where it is useful to show more detail.

Cross or spread pixel distribution plus full-pixel thresholding gives a threshold-controllable degree of *noise* or detail removal depending on whether AND1, AND2, AND3 or AND4 is used (Fig. 6). Thresholding with AND2 gives a result similar to conventional single-pixel-noise removal, with some exceptions in detail but with both black and white pixels treated equally.

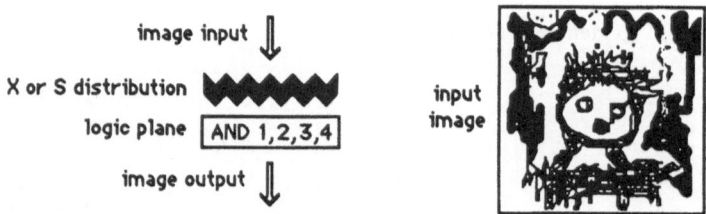

Fig. 5. Schematic representation of a processing sequence, and the initial image

Fig. 6. Detail removal: cross or spread distribution followed by AND1, AND2, AND3 or AND4

If the intention is to use structures of this type in a looping or circulatory processing scheme (as for example in an architecture related to the optical CLIP [1]) then a first requirement is for circulatory stability: going round the loop should not result in lateral dispersion of initially present detail. A logic-less loop which unfortunately exhibits such a dispersion is the combination of two successive X or S distributors, each followed by quarter-pixel inversion (structure A in Fig. 7). The problem is solved by changing one of the distributors, to give X-then-S or S-then-X with quarter-pixel inversions (structure B in Fig. 7). The first image shown in each of Figs. 8 and 9 is the first *intermediate* one, after just one X distribution and a quarter-pixel inversion. Structure A (Fig. 8) shows detail dispersion which increases with the number of circulatory passes. Although the first intermediate image of structure B (Fig. 9) is the same as that of structure A, the loop now gives a stable resulting image; the S distribution inverts the initial X distribution exactly. Adding in full-pixel NAND1 thresholding instead of quarter-pixel inversion removes the necessity for two kinds of distributor, and the sequence of X_NAND1_X_NAND1 in a loop provides a stable image after one stage of single-pixel noise reduction (Fig. 10), as does S_NAND1_S_NAND1.

Fig. 7. Loop processing stability: unstable (structure A) and stable (structure B) configurations

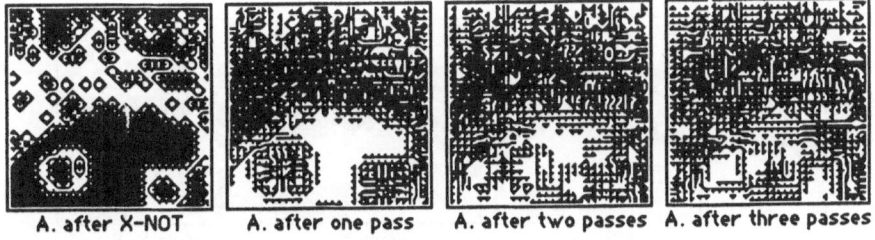

Fig. 8. Detail dispersion exhibited by structure A

The simplest stable or *do-nothing* loop consists of just a pair of opposite distributors, X-then-S (Fig. 11) or S-then-X. This provides the basis for a possible stable circulatory processing scheme, where the pixels are first distributed, operated on by a controllable logic plane inserted between the two distributors, and then *re-distributed* back to their own original pixel locations.

| B. after X-NOT | B. after one pass | B. after two passes | B. after three passes |

Fig. 9. Detail stability exhibited by structure B

input image   after every pass

Fig. 10. A stable loop configuration with two similar distributors

input image   after X   after X_S

Fig. 11. The simplest stable loop configuration

The four-fold split-up of the original image pixels means that not only individual quarter-pixels but quarter-pixel copies of the complete image are projected in the four directions of the image plane. After a number of passes, four full-pixel images the same as the single original one may be restored by reading the distributor output through an AND1 plane (Fig. 12). It should be noticed that the four images cross each other in the plane without mutual interference; each is confined to its own characteristic quarter-pixel at every stage.

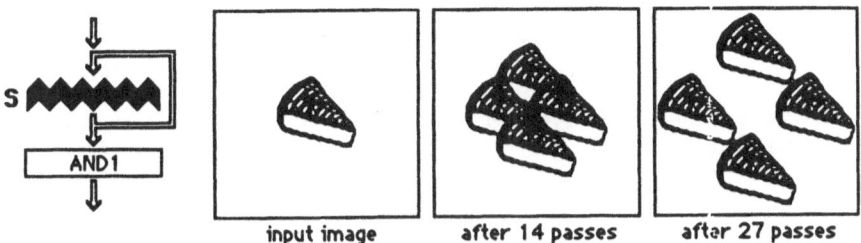

Fig. 12. Four-fold split-up of a single image to give four copies of the original

Conversely, a single image may be constructed out of four spatially separated different parts by projection of their quarter-pixel images into a central region and subjecting the resultant combination to an AND2 operation (Fig. 13).

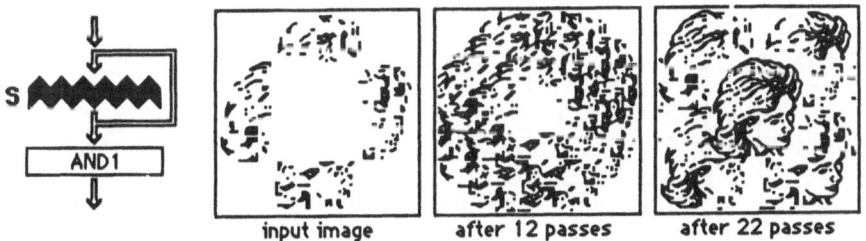

Fig. 13. Combination of four separate image-parts to give a single image

4 Image Matching

The most powerful possible application is in a matching scheme for feature recognition in complex images, where a large image is raster-scanned with a smaller reference feature to locate target features equivalent to the reference. We here make use of the complete independence of different quarter-pixel representations of the two images to enable them to be cross-scanned without interference.

Figure 14 shows a pseudo-landscape containing a target feature, which is first converted into *down* quarter-pixels and swept downwards, and a search feature (the small car) which is first converted into *up* quarter-pixels and then swept upwards.

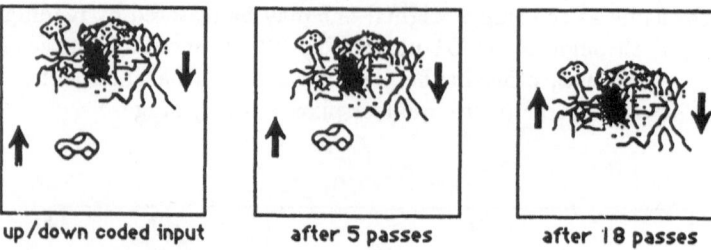

up/down coded input   after 5 passes   after 18 passes

Fig. 14. Cross-scanning of the pseudo-landscape and the reference feature to locate equivalent target features

The combined image is passed through an AND2 plane, and the total output intensity is observed. When the summed output from the AND2 plane equals the search feature pixel sum (in this case the car is constructed out of 63 pixels), an equivalent feature has been found in the pseudo-landscape, and it appears in the output plane in its correct location (Fig. 15).

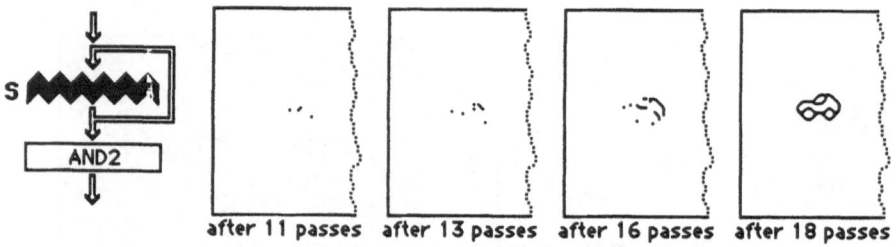

after 11 passes   after 13 passes   after 16 passes   after 18 passes

Fig. 15. Progressive build-up of the equivalent feature combination as the pseudo-landscape and the reference are cross-scanned

A basic problem with binary pixel image representation is evident if the progressive build-up of the output is observed (Fig. 15); ANY black-defined feature will fit to a solid black background; in this example a match for the back of the car is found over a wide range of combined images because of the completely black area in the original pseudo-landscape. The (possible) solution to this kind of problem lies in representing the image as grey-coded instead of binary pixels, but with conventional schemes this imposes a spatial incoherence between the image and its grey-coded representation. This is of fundamental importance in a processing arrangement where spatial relationships must remain the same throughout, as in the scheme described here; an applicable grey-coding representation must be spatially centrosymmetric (or at least four-fold symmetric).

5 Fabrication

For once, nature works in our favour in the fabrication of the necessary structures on the silicon wafer faces. First experiments indicate that a (100) surface wafer can be relatively easily preferentially etched to give the required (111) pyramids, and use of an infrared mask aligner makes it possible to align hill and pit masks on opposite faces in the usual manner. The incident angle of the incoming infrared pixel beams on the pyramid facets (54.736°) is determined by the cubic crystal structure of silicon. The refractive index (3.4929 for a wavelength of $1.395\,\mu m$ [2, page 641]) and the thickness T ($300\,\mu m$) of the parallel polished wafer define of themselves the lateral dimension W ($162\,\mu m$) of the pyramids if the correct straight-through relation between the input and output optical paths is to be maintained (Fig. 16). For pixels of this size we can neglect problems caused by diffraction at the facets, but to make a practically useable component anti-reflection coatings would be necessary to boost the total transmission through the structure.

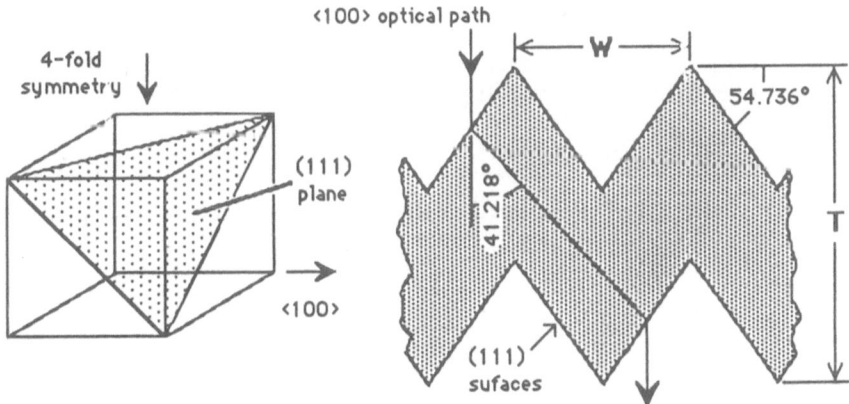

Fig. 16. Relationship between crystal directions and the (111) pyramid facets, and a cross-section of the required silicon wafer

The wafer surfaces are first covered with a 10 micron SiO_2 film, grown by wet oxidation at 1100 °C for 1 hour. This is masked and locally etched away with $HF/NH_4F/H_2O$ to define the regions where the silicon is to be later removed. The SiO_2 etching for the pit side requires a mask consisting of a pattern of crossed lines, which then make up the unetched regions in the resulting design, and the hill side mask is a square pattern of points, modified to take account of subsidiary directional etch distortions and under-etching. The silicon itself is etched with 0.78 mole KOH at 85 °C for 56 minutes [3], which corresponds to an etching rate of about 2 microns/minute.

Etching results correspond well with the originally proposed design, as can be seen from the SEM photograph of a pit-etched side in Fig. 17.

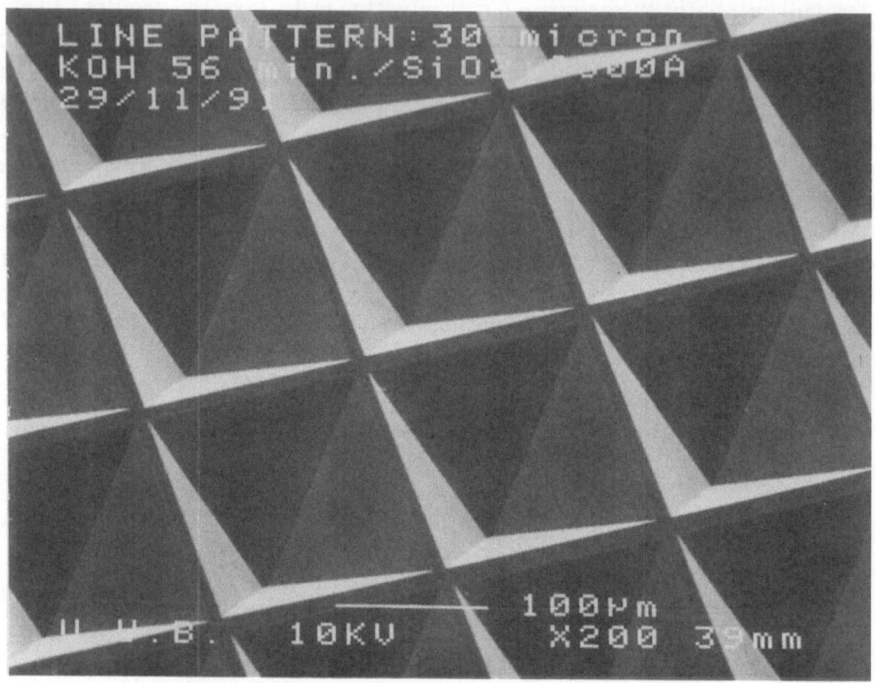

Fig. 17. SEM photograph of one side of the wafer, showing the regular array of pyramid pits

References

1. Wherrett, B. S.: O-CLIP - A demonstrator all-optical processor. In this volume.
2. American Institute of Physics handbook. McGraw-Hill, 1972.
3. McGuire, G. E., Ed.: Semiconductor materials and process technology handbook for VLSI and ULSI. Noyes Publications, 1988, p. 127.

Attendees

Dr. F. Mollot
Dr. R. Planel   Laboratoire de Microstructures et Microélectronique, CNRS, Bagneux, France

Dr. R. Kuszelewicz
Dr. J.-L. Oudar   Centre National d'Etudes des Télécommunications, Bagneux, France

Dr. R. Cottam
Dr. N. Langloh
W. Pfeiffer
Dr. H. Thienpont
Prof. I. Veretennicoff
Prof. R. Vounckx   Vrije Universiteit, Brussels, Belgium

Prof. D. Jäger
Dr. G. Wingen   Fachgebiet Optoelektronik, University of Duisburg, Germany

J. Bolger
Dr. G. Buller
Dr. J. Ehrlich
Dr. I. Galbraith
Dr. D. Goodwill
Dr. A. Kashko
Dr. A.D. Lloyd
G. MacKinnon
Dr. H.A. MacKenzie
N. McArdle
Dr. D. McKnight
J. Meredith
J.M. Miller
Ms. S. Molyneux
Dr. R. F. Neale
D. Neilson
Dr. I. Redmond
B. Robertson
N. Ross
G. Smith
Prof. S.D. Smith
Dr. J.F. Snowdon
Dr. M.R. Taghizadeh

Dr. F.A.P. Tooley
Dr. J. Turunen
Ms. S. Wakelin
Prof. A.C. Walker
Prof. B.S. Wherrett
Ms. R. Wilson   Dept. of Physics, Heriot-Watt University,
                       Edinburgh, UK

Dr. H. Bartelt   Siemens Corporate Research, Erlangen, Germany

Dr. K.-H. Brenner
Dr. W. Eckert
Dr. D. Fey
Dr. M. Kufner
Dr. T. Merklein
Dr. N. Streibl
Dr. K. Zurl   Physikalisches Institut, University of Erlangen,
                       Germany

Dr. S. Benner   Institut für Theoretische Physik, University of
                       Frankfurt, Germany

Dr. J. Grohs
Dr. M. Grün
Dr. K.-H. Schlaad
Dr. A. Uhrig
Prof. C. Klingshirn
Dr. U. Zimmermann Department of Physics, University of Kaiserslautern,
                       Germany

Dr. G. Borghs   Interuniversitair Micro Electronic Centrum, Leuven,
                       Belgium

Dr. T.J. Hall
A. Kirk
Dr. A.K. Powell   Dept. of Physics, King's College London, UK

Prof. S. Cova
A. Lacaita   Dipartimento di Elettronica, Politecnico di Milano,
                       Italy

B. Acklin   Institute of Micro-Technology, University of Neuchâtel,
                       Switzerland

Prof. J. Hvam   Fysisk Institut, Odense Universitet, Denmark

Dr. A. Koster
Prof. S. Laval
Dr. N. Paraire
Dr. M. Rousseau   Institut d'Electronique Fondamentale (CNRS), Paris,
                         France

Dr. S. Malik
Dr. N. de Beaucoudrey
Dr. I. Seyd Darwish   Institut d'Optique Théorique et Appliquée, Paris,
                         France

Dr. U. Olin   Institute of Optical Research, Stockholm, Sweden

Dr. J.B. Grün
Dr. R. Levy
Dr. J. Oberlé   Institut de Physique et Chimie des Matériaux de
                         Strasbourg, France

Dr. G. Lebreton   Groupe d'Etudes des Signaux et des Systèmes,
                         Université de Toulon et du Var, France

Dr. P. Churoux   Département d'Optiques, CERT, Toulouse, France

Dr. T. Wipiejewski   Dept. of Optoelectronics, University of Ulm, Germany

Dr. D. Rhein
Dr. A. Wachlowski   Alcatel, Vienna, Austria